T0309645

MATHEMATICAL FOUNDATIONS AND BIOMECHANICS OF THE DIGESTIVE SYSTEM

Mathematical modelling of physiological systems promises to advance our understanding of complex biological phenomena and pathophysiology of diseases. In this book, the authors adopt a mathematical approach to characterize and explain the functioning of the gastrointestinal system. Using the mathematical foundations of thin shell theory, the authors patiently and comprehensively guide the reader through the fundamental theoretical concepts, via step-by-step derivations and mathematical exercises, from basic theory to complex physiological models. Applications to nonlinear problems related to the biomechanics of abdominal viscera and the theoretical limitations are discussed. Special attention is given to questions of complex geometry of organs, effects of boundary conditions on pellet propulsion, as well as to clinical conditions, e.g. functional dyspepsia, intestinal dysrhythmias and the effect of drugs to treat motility disorders. With end-of-chapter problems, this book is ideal for bioengineers and applied mathematicians.

ROUSTEM N. MIFTAHOF is Professor and Head of the Department of Physiology at the Arabian Gulf University, Manama, Bahrain. He is a recognized leader in the areas of applied mathematics and gastrointestinal research, and has authored and co-authored two previous books in these fields. He has worked in both academia and industry across Europe, America and Asia.

HONG GIL NAM is Professor at Pohang University of Science and Technology, Pohang, South Korea, Director of the National Core Research Centre for Systems Bio-Dynamics and President of the Association of Asian Societies for Bioinformatics. Professor Nam has received numerous awards for his research contributions and he has made several media appearances in South Korea.

MATHEMATICAL FOUNDATIONS AND BIOMECHANICS OF THE DIGESTIVE SYSTEM

[illegible]

[illegible]

[illegible]

MATHEMATICAL FOUNDATIONS AND BIOMECHANICS OF THE DIGESTIVE SYSTEM

ROUSTEM N. MIFTAHOF

Arabian Gulf University

Pohang University of Science and Technology

HONG GIL NAM

Pohang University of Science and Technology

CAMBRIDGE
UNIVERSITY PRESS

University Printing House, Cambridge CB2 8BS, United Kingdom

One Liberty Plaza, 20th Floor, New York, NY 10006, USA

477 Williamstown Road, Port Melbourne, VIC 3207, Australia

314-321, 3rd Floor, Plot 3, Splendor Forum, Jasola District Centre, New Delhi - 110025, India

103 Penang Road, #05-06/07, Visioncrest Commercial, Singapore 238467

Cambridge University Press is part of the University of Cambridge.

It furthers the University's mission by disseminating knowledge in the pursuit of education, learning and research at the highest international levels of excellence.

www.cambridge.org
Information on this title: www.cambridge.org/9780521116626

First published 2010

A catalogue record for this publication is available from the British Library

Library of Congress Cataloging in Publication data
Miftahof, Roustem.
Mathematical foundations and biomechanics of the digestive system / Roustem Miftahof, Hong Gil Nam.
p. ; cm.
Includes bibliographical references and index.
ISBN 978-0-521-11662-6 (hardback)
1. Gastrointestinal system – Mathematical models. 2. Biomechanics. 3. Elastic plates and shells.
I. Nam, Hong Gil. II. Title.
[DNLM: 1. Gastrointestinal Tract – physiology. 2. Computer Simulation. 3. Gastrointestinal Motility – physiology. 4. Models, Biological. WI 102 M633ma 2010]
QP156.M58 v2010
612.3'2–dc22
2010004178

ISBN 978-0-521-11662-6 Hardback

To all those who gave so much, and were given so little

Contents

	Preface	*page* xi
	Notation	xv
	Introduction	1
	Exercises	4
1	The geometry of the surface	6
	1.1 Intrinsic geometry	6
	1.2 Extrinsic geometry	8
	1.3 The equations of Gauss and Codazzi	13
	1.4 General curvilinear coordinates	15
	1.5 Deformation of the surface	18
	1.6 Equations of compatibility	22
	Exercises	26
2	Parameterization of shells of complex geometry	28
	2.1 Fictitious deformations	28
	2.2 Parameterization of the equidistant surface	31
	2.3 A single-function variant of the method of fictitious deformation	33
	2.4 Parameterization of a complex surface in preferred coordinates	37
	2.5 Parameterization of complex surfaces on a plane	42
	Exercises	46
3	Nonlinear theory of thin shells	47
	3.1 Deformation of the shell	47
	3.2 Forces and moments	50
	3.3 Equations of equilibrium	55
	Exercises	60
4	The continuum model of the biological tissue	61
	4.1 Structure of the tissue	61
	4.2 Biocomposite as a mechanochemical continuum	62

4.3 The biological factor 71
Exercises 74
5 Boundary conditions 76
5.1 The geometry of the boundary 76
5.2 Stresses on the boundary 78
5.3 Static boundary conditions 81
5.4 Deformations of the edge 84
5.5 Gauss–Codazzi equations for the boundary 87
Exercises 88
6 Soft shells 89
6.1 Deformation of soft shell 89
6.2 Principal deformations 95
6.3 Membrane forces 97
6.4 Principal membrane forces 100
6.5 Corollaries of the fundamental assumptions 101
6.6 Nets 105
6.7 Equations of motion in general curvilinear coordinates 106
6.8 Governing equations in orthogonal Cartesian coordinates 109
6.9 Governing equations in cylindrical coordinates 111
Exercises 113
7 Biomechanics of the stomach 115
7.1 Anatomical and physiological background 115
7.2 Constitutive relations for the tissue 119
7.3 A one-dimensional model of gastric muscle 130
7.3.1 Myoelectrical activity 132
7.3.2 Decrease in external Ca^{2+} concentration 133
7.3.3 Effects of T- and L-type Ca^{2+}-channel antagonists 134
7.3.4 Acetylcholine-induced myoelectrical responses 135
7.3.5 Effect of chloride-channel antagonist 136
7.3.6 Effect of selective K^{+}-channel antagonist 136
7.4 The stomach as a soft biological shell 137
7.4.1 Inflation of the stomach 140
7.4.2 The electromechanical wave phenomenon 142
7.4.3 The chronaxiae of pacemaker discharges 145
7.4.4 Multiple pacemakers 147
7.4.5 Pharmacology of myoelectrical activity 155
Exercises 155
8 Biomechanics of the small intestine 157
8.1 Anatomical and physiological background 157
8.2 A one-dimensional model of intestinal muscle 158

8.2.1 Myoelectrical activity 159
8.2.2 Effects of non-selective Ca^{2+}-channel agonists 160
8.2.3 Effects of Ca^{2+}-activated K^+-channel agonist 160
8.2.4 Response to a selective K^+-channel agonist 161
8.2.5 Effect of selective K^+-channel antagonist 163
8.2.6 Conjoint effect of changes in Ca^{2+} dynamics and extracellular K^+ concentrations 164
8.3 The small intestine as a soft cylindrical shell 165
8.3.1 Pendular movements 166
8.3.2 Segmentation 168
8.3.3 Peristaltic movements 173
8.3.4 Self-sustained periodic activity 173
8.3.5 Effect of lidocaine 177
Exercises 180
9 Biomechanics of the large intestine 182
9.1 Anatomical and physiological background 182
9.2 The colon as a soft shell 184
9.2.1 Haustral churning 187
9.2.2 Contractions of the teniae coli 187
9.2.3 Peristalsis and propulsive movements 192
9.3 Pharmacology of colonic motility 192
9.3.1 Effect of Lotronex 192
9.3.2 Effect of Zelnorm 194
Exercises 194
10 Biological applications of mathematical modelling 196
10.1 Biomechanics of hollow abdominal viscera 196
10.2 Future developments and applications 201
Exercises 208
References 210
Index 217

Preface

Recent technological advances in various fields of applied science have radically transformed the strategies and vision of biomedical research. While only a few decades ago scientists were largely restricted to studying parts of biological systems in isolation, mathematical and computational modelling now enable the use of holistic approaches to analyse data spanning multiple biological levels and traditionally disconnected fields.

Mathematical modelling of organs and systems is a new frontier in the biosciences and promises to provide a comprehensive understanding of complex biological phenomena as more than the sum of their parts. Recognizing this opportunity, many academic centres worldwide have established new focuses on this rapidly expanding field that brings together scientists working in applied mathematics, mechanics, computer science, bioengineering, physics, biology and medicine. A common goal of this effort is to stimulate the study of challenging problems in medicine on the basis of abstraction, modelling and general physical principles.

This book is intended for bioengineers, applied mathematicians, biologists and doctors. It provides a brief and rigorous introduction to the mathematical foundations of thin-shell theory and its applications to nonlinear problems of the biomechanics of hollow abdominal viscera. It should be stressed that the text is not directed towards rigorous mathematical proofs of methods and solutions, but rather to a thorough comprehension, by means of mathematical exercises, of the essentials and the limitations of the theory and its role in the study of biomedical phenomena. The book can be used as a textbook for senior undergraduate and postgraduate students, although it may also be of interest to researchers and scientists who wish to obtain some general background knowledge of shell theory.

The approach is facilitated by the introduction of basic concepts of the theory of surfaces (Chapter 1), which are essential prerequisites for a subsequent understanding of the theory of shells. The mathematical techniques employed are reasonably elementary, and require knowledge of calculus and vector algebra. Throughout the text we deliberately use non-tensorial notation, since tensor calculus is not included in the normal curriculum for bioengineering and applied-mathematics students. There are only a few places in the book where the use of tensors was necessary.

In Chapter 2 the method of fictitious tangent deformation is introduced. Questions of parameterization of shells of complex geometry, which are common in practical applications, e.g. modelling of abdominal viscera, are considered.

The general theory and the governing system of equations for thin shells, without any restrictions on the magnitude of the displacements, rotations, or strains, are formulated in Chapter 3. The dimensional reduction is achieved by formal integration of the equations of equilibrium of a three-dimensional solid over the thickness of the shell and a successful application of the second Kirchhoff–Love hypothesis. The equations of equilibrium of the shell are derived in general curvilinear and orthogonal coordinates.

Part of the aim of the book is to gain insight into the biomechanics of soft biological tissues, namely the wall of the abdominal viscera, in particular by analysing their structure, morphology and electrical phenomena at multihierarchical levels of organization. Thus, in Chapter 4 the main consideration is given to a phenomenological continuum mechanics approach to derive constitutive relations for the biocomposite. It is treated as a three-phase mechanochemically active medium.

Chapter 5 is dedicated to boundary conditions. Organs of the digestive tract have a strong ligamentous support as well as multiple sphincters that are located at the site of organ junctions. Such arrangements, e.g. gastro-oesophageal, gastroduodenal and ileocaecal junctions, severely restrict the deformability of the organ in that region. Therefore, only small deformations on the boundary are considered.

Chapter 6 is devoted to the theory of soft shells, where fundamental hypotheses and general criteria for the co-existence of various stress–strained states are formulated and the governing system of equations is derived. The dynamic approach, which is oriented towards numerical solutions, is developed throughout.

The material about the nonlinear theory of thin shells presented in the book is well known. The reader is advised to consult texts by Galimov (1975), Ridel and Gulin (1990), Galimov *et al.* (1996), Ventsel and Krauthammer (2001), Taber (2004), and Libai and Simmonds (2005) for full information. The original contribution and the strength of this book, though, are in applications of the theory to model and to study the biomechanics of the gastrointestinal system, in particular the stomach, small intestine and colon. Attention is paid to the biological plausibility of the basic modelling assumptions which are crucial in selection of appropriate mathematical models. Electrical, chemoelectrical and mechanical phenomena have been integrated at various levels of models to simulate and analyse coupled processes, e.g. electromechanical waves of deformation and the propulsion of a solid bolus, that comprise the fundamental physiological principles of functionality of the gastrointestinal system.

Special attention is paid to simulations of clinically meaningful problems, namely the motility of the stomach and of the small and large intestines in normal and pathological conditions, e.g. functional dyspepsia, gastroparesis and intestinal dysrhythmia, and to the analysis of underlying physiological mechanisms from biomechanical perspectives. In Chapter 7 the problem of electromechanical wave activity in the stomach is analysed. The effects of localized and spatially distributed pacemakers, intraluminal pressure, mechanical characteristics of the tissue and pharmacologically active compounds on the dynamics of the stress–strain distribution in the organ are investigated.

Chapter 8 is dedicated to the dynamics of a cylindrical soft shell – a model of a functional unit of the small intestine. Patterns of myoelectrical activity in normal and pathological conditions are reproduced numerically. Particular attention is paid to the self-sustaining electromechanical wave phenomenon that mimics intestinal dysrhythmia and its pharmacological attenuation.

The biomechanics of propulsion of a solid bolus along the colon is studied in Chapter 9. The effects of various types of boundary conditions at the aboral end are investigated. The emphasis is placed on an analysis of the effects of promotility drugs on the dynamics of propulsion.

Biological and clinical applications of mathematical modelling are discussed in Chapter 10. Attention is paid to future developments and improvements of existing mathematical models of the abdominal viscera. Electrophysiological and neuropharmacological aspects of gastrointestinal motility, together with the key role of intrinsic neuronal pathways, multiple neurotransmission and extrinsic regulatory nervous control elements are emphasized.

Each of the chapters is followed by a set of problems, which can be used by the reader to check his or her understanding of the subject matter. Some problems bring out additional details that are not considered in the main body of the text.

This book was written in many places: Kazan State University (Tatarstan, Russia), the Arabian Gulf University (Kingdom of Bahrain), the National Core Research Centre for Systems Bio-Dynamics and Pohang University of Science and Technology (South Korea). We would like to thank our colleagues who have contributed both directly and indirectly to this book. We are particularly grateful to Dr W. Morrison for reviewing the manuscript and providing corrections and valuable comments. Finally, we wish to extend our thanks to the publisher, Cambridge University Press, and especially to Dr M. Carey, Publishing Editor, and Ms C. L. Poole, Assistant Editor, who supported the project from the very beginning and made its successful publication possible.

Professor Dr R. Miftahof
Professor H. G. Nam

Notation

Symbol	Description
$\overset{0}{S}, S, \overset{*}{S}$	cut, undeformed, and deformed (*) middle surface of a shell, respectively
S_z	surface co-planar to S such that $S_z \parallel S$
S^{r}	middle surface of a net
$\Sigma, \overset{*}{\Sigma}$	boundary faces of an undeformed shell and a deformed shell, respectively
$\mathrm{d}\Sigma, \mathrm{d}\overset{*}{\Sigma}$	differential line elements on Σ and $\overset{*}{\Sigma}$, respectively
h	thickness of a shell
x_1, x_2, x_3	rectangular coordinates
r, φ, z	cylindrical coordinates
$\{\bar{i}_1, \bar{i}_2, \bar{i}_3\}$	orthonormal base of $\{x_1, x_2, x_3\}$
$\{\bar{k}_1, \bar{k}_2, \bar{k}_3\}$	orthonormal base of $\{r, \varphi, z\}$
$(\alpha_1, \alpha_2), (\overset{*}{\alpha}_1, \overset{*}{\alpha}_2)$	curvilinear coordinates of the undeformed and deformed shell, respectively
$\bar{m}, \overset{*}{\bar{m}}$	vectors normal to S and $\overset{*}{S}$, respectively
$\bar{\tau}, \bar{\tau}^z, \overset{*}{\bar{\tau}}$	tangential vectors to lines on S, S_z, $\overset{*}{S}$ or their boundaries, respectively
$\bar{n}, \bar{n}^z, \overset{*}{\bar{n}}$	normal vectors to lines on the boundaries of S, S_z and $\overset{*}{S}$, respectively
$\bar{r}, \bar{\rho}, \overset{*}{\bar{\rho}}$	position vectors of points $M \in S$, $M_z \in S_z$ and $\overset{*}{M}_z \in \overset{*}{S}_z$, respectively
$\bar{r}_i, \bar{\rho}_i$	tangential vectors to coordinate lines on S and S_z, respectively
$\{\bar{r}_1, \bar{r}_2, \bar{m}\}, \{\bar{r}^1, \bar{r}^2, \bar{m}\}$	covariant and contravariant bases at a point $M \in S$, respectively
$\{\bar{\rho}_1, \bar{\rho}_2, \bar{\rho}_3\}$	covariant base at point $M_z \in S_z$

$\{\bar{n},\bar{\tau},\bar{m}\}$, $\{\overset{*}{\bar{n}},\overset{*}{\bar{\tau}},\overset{*}{\bar{m}}\}$	orthogonal bases on Σ and $\overset{*}{\Sigma}$, respectively
$\chi,\chi^z,\overset{*}{\chi}$	angles between coordinate lines defined on S, S_z, and $\overset{*}{S}$, respectively
γ	shear angle
$A_i,\overset{*}{A}_i,H_i$	Lamé coefficients on S, $\overset{*}{S}$ and S_z, respectively
a_{ik},g_{ik}	components of the metric tensor
$a,\overset{*}{a}$	determinants of the metric tensor
b_{ik}	components of the second fundamental form
$\bar{e}_i,\bar{e}_i^z,\overset{*}{\bar{e}}_i,\overset{*}{\bar{e}}{}_i^z$	unit base vectors on S, S_z and $\overset{*}{S}$, respectively
$ds, ds_z, d\overset{*}{s}$	lengths of line elements on S, S_z and $\overset{*}{S}$, respectively
ds_Δ	surface area of a differential element of S
$\Gamma_{ik,j},\Gamma^j_{ik}$	Christoffel symbols of the first and second kind, respectively
A^j_{ik}	deviator of the Christoffel symbols
$\bar{v}(\alpha_1,\alpha_2)$	displacement vector
u_1, u_2, ω	projections of the displacement vector on x_1, x_2 and x_3 axes, respectively
$\varepsilon_{ik},\varepsilon^z_{ik}$	components of the tensor of planar deformation through points $M\in S$ and $M_z\in S_z$, respectively
$\tilde{\varepsilon}_{ik},\overset{*}{\varepsilon}_{ik}$	physical components of the tensor deformation in undeformed and deformed configurations of a shell, respectively
$\varepsilon_1,\varepsilon_2$	principal physical components of the tensor of deformation, respectively
$\varsigma_{ij},\Delta_{ij}$	elastic and viscous parts of deformation, respectively
$\varepsilon_n,\varepsilon_\tau,\varepsilon_{n\tau}$	components of deformation of the boundary of a shell in $\bar{n},\bar{\tau}$ directions
$\underline{\varepsilon}_{ik}$, $\underline{æ}_{ik}$	components of the tensor of tangent and bending fictitious deformations, respectively
$\lambda_i,\lambda_{c,l}$	stretch ratios (subscripts c and l refer to the circular and longitudinal directions of a bioshell)
$\mathbf{\Lambda}_1,\mathbf{\Lambda}_2$	principal stretch ratios
$I_1^{(\mathbf{E})},I_2^{(\mathbf{E})}$	invariants of the tensor of deformation
$æ_n, æ_\tau, æ_{n\tau}$	components of bending deformation and twist of the boundary of a shell in $\bar{n}$, $\bar{\tau}$ directions
$e_{nn},e_{n\tau},e_{n\tau}$, $e_{\tau n},\omega_n,\omega_\tau$	rotation parameters
$\underline{e}_i^k,\underline{e}_{ki},\underline{\omega}_i$	rotation parameters in fictitious deformation
$k_{ii},\overset{*}{k}_{ii}$	normal curvatures of S and $\overset{*}{S}$, respectively

$k_{ik}, \overset{*}{k}_{ik}$	twist of S and $\overset{*}{S}$, respectively
$k_n, k_\tau, k_n^*, k_\tau^*$	normal curvatures in $\bar{n}, \bar{\tau}, \overset{*}{\bar{n}}$ and $\overset{*}{\bar{\tau}}$ directions, respectively
$k_{n\tau}, k_{n\tau}^*$	twist of the contours Σ and $\overset{*}{\Sigma}$, respectively
$1/R_{1,2}, 1/R_{1,2}^*$	principal curvatures of S and $\overset{*}{S}$, respectively
K, K_0	Gaussian curvatures of S and $\overset{*}{S}$, respectively
ΔK	increment of the Gaussian curvature
$e_{\alpha 1}, e_{\alpha 2}$	elongations in directions α_1 and α_2, respectively
$L'_j(\varepsilon_{ik}), L_i(T_{ik})$	differential operators
$\mathrm{d}\overset{*}{\Sigma}_z$	surface area of a differential element of $\overset{*}{\Sigma}$
$\mathrm{d}\sigma^z$	free surface area of a shell
$\mathrm{d}\Omega$	unit volume of an element of a shell
p	pressure
$\bar{p}_i$	stress vectors
F_c	contact force
F_d	force of dry friction
$\bar{R}_i$	resultant of force vectors
$\bar{M}_i$	resultant moment vector of internal forces
$\bar{p}_{(+)}, \bar{p}_{(-)}$	external forces applied over the free surface area of a shell
$\overset{*}{\bar{p}}{}_n^z$	normal stress vector on $\overset{*}{\Sigma}$
$\bar{F}$	vector of mass forces per unit volume of the deformed element of a shell
$\bar{X}$	resultant external force vector on $\overset{*}{S}$
$\bar{M}$	resultant external moment of external forces on $\overset{*}{S}$
$\overset{*}{T}_{ii}, \overset{*}{T}_{ik}, \overset{*}{N}_i$	normal, shear and lateral forces per unit length of a shell parallel to $\overset{*}{S}$, respectively
$T_{\mathrm{c,l}}$	total force per unit length
$T_{\mathrm{c,l}}^{\mathrm{p}}, T_{\mathrm{c,l}}^{\mathrm{a}}$	passive and active components of the total forces per unit length, respectively
T_1^r, T_2^r	forces per unit length of reinforced fibres
T_1, T_2	principal stresses
$I_1^{(\mathrm{T})}, I_2^{(\mathrm{T})}$	invariants of the stress tensor
$\overset{*}{M}_{ii}, \overset{*}{M}_{ik}$	bending and twisting moments per unit length of a shell perpendicular to directions α_1 and α_2 on $\overset{*}{S}$
$\overset{*}{M}_i$	projections of the moment vector on $\overset{*}{\bar{e}}_1, \overset{*}{\bar{e}}_2, \overset{*}{\bar{m}}$
$\overset{*}{X}_i$	projections of the external force vector on $\overset{*}{\bar{e}}_1, \overset{*}{\bar{e}}_2, \overset{*}{\bar{m}}$
$\bar{R}_{\overset{*}{n}}$	resultant force vector per unit length acting on $\mathrm{d}\overset{*}{\Sigma}_z$ in the $\overset{*}{\bar{n}}$ direction

$\overline{M}_{\overset{*}{n}}$	resultant moment vector per unit length acting on $d\overset{*}{\Sigma}_z$ in the $\bar{\overset{*}{n}}$ direction
$\overset{*}{G}, \overset{*}{H}$	bending and twisting moments in a boundary of a shell
$\overline{G}_i, \overline{M}_p, \overline{M}_q$	resultant moment vectors per unit length of a soft shell
σ_{ij}	stresses in a shell
σ_{ij}^{α}	stresses in phase α of a biomaterial
$c_1, \ldots, c_{14}$	material constants
d_{m}	diameter of smooth muscle fibre
L	length of bioshell/muscle fibre
$\tilde{S}_{\mathrm{c,l}}$	cross-sectional area of smooth muscle syncytia (SM)
$v_{x_1}, v_{x_2}, v_{x_3}$	components of the velocity vector
k_{v}	viscosity
η_{sp}	coefficient of viscous friction
$\rho, \overset{*}{\rho}$	densities of undeformed and deformed material of a shell, respectively
ρ_{ζ}^{α}	partial density of the ζth substrate in phase α of a biomaterial
m_{ζ}^{α}	mass of the ζth substrate in phase α of a biomaterial
$\upsilon, \upsilon^{\alpha}$	total and elementary volumes of a biomaterial, respectively
c_{ζ}^{α}	mass concentration of the ζth substrate in phase α of a biomaterial
η	porosity
$Q_{\zeta}^{\alpha}, Q_{\zeta}^{e}, Q_{\zeta}$	influxes of the ζth substrate into phase α, external sources and exchange flux between phases, respectively
$\nu_{\zeta j}$	stoichiometric coefficient in the jth chemical reaction
$U^{(\alpha)}$	free energy of phase α
$s^{(\alpha)}, S_{\zeta}^{1}$	entropy of phase α and partial entropy of the entire biomaterial, respectively
T	temperature
μ_{ζ}^{α}	chemical potential of the ζth substrate in phase α of a biomaterial
$\overline{q}$	heat-flux vector
$\boldsymbol{R}$	dissipative function
Λ_j	affinity constant of the jth chemical reaction
$\overline{J}_{\mathrm{i}}, \overline{J}_{\mathrm{o}}$	intracellular (i) and extracellular (o) ion currents
$I_{\mathrm{m1}}, I_{\mathrm{m2}}$	transmembrane ion currents (SM)
$I_{\mathrm{ext(i)}}$	external membrane current (ICC)
I_{ion}	total ion current (SM)

$\tilde{I}^{s}_{Ca}, \tilde{I}^{f}_{Ca}, \tilde{I}_{Ca-K}, \tilde{I}_{K}, \tilde{I}_{Cl}$	Ca^{2+}, Ca^{2+}-activated K^+, K^+ and Cl^- ion currents in smooth muscle, respectively
$I_{Ca}, I_{Ca-K}, I_{Na}, I_{K}, I_{Cl}$	Ca^{2+}, Ca^{2+}-activated K^+, Na^+, K^+ and Cl^- ion currents, respectively
Ψ_i, Ψ_o	electrical potentials
V_m	transmembrane potential
V_i	membrane potential of the interstitial cell of Cajal
$V_{c,l}, V^{s}_{c,l}$	membrane potentials in circular and longitudinal smooth muscle (SM), respectively
$\tilde{V}_{Ca}, \tilde{V}_{K}, \tilde{V}_{Cl}$	reversal potentials for Ca^{2+}, K^+ and Cl^- currents in smooth muscle, respectively
$V_{Ca}, V_{Ca-K}, V_{Na}, V_{K}, V_{Cl}$	reversal membrane potentials for Ca^{2+}, Ca^{2+}-activated K^+, Na^+, K^+ and Cl^- ion currents for interstitial cells of Cajal, respectively
C_m	membrane capacitance (SM)
C_s	membrane capacitance (ICC)
$R^{ms}_{i(0)}$	membrane resistance (SM)
R_{ICC}	input cellular resistance (ICC)
R_s	specific membrane resistance of a muscle fibre
$\hat{g}_{ij}, \hat{g}_{oj}$	intracellular (i) and extracellular (o) conductivities
$\hat{g}^{*}_{i}, \hat{g}^{*}_{o}$	maximal intracellular (i) and extracellular (o) conductivities
$\tilde{g}^{f}_{Ca}, \tilde{g}^{s}_{Ca}, \tilde{g}_{K}, \tilde{g}_{Ca-K}, \tilde{g}_{Cl}$	maximal conductances for ion currents of respective ion channels (SM)
$g_{Ca(i)}, g_{Ca-K(i)}, g_{Na(i)}, g_{K(i)}, g_{Cl(i)}$	maximal conductances of voltage-dependent Ca^{2+} (N-type), Ca^{2+}-activated K^+, Na^+, K^+ and Cl^- channels (ICC), respectively
$\tilde{m}, \tilde{h}, \tilde{n}, \tilde{x}_{Ca}$	dynamic variables of ion currents (SM)
$m_{Na}, h_{Na}, n_{K}, z_{Ca}, \rho_{\infty}$	dynamic variables of respective ion currents (ICC)
$\tilde{\alpha}_y, \ \beta_y$	activation and deactivation parameters of ion channels (SM), respectively
$\alpha_{y\infty}, \beta_{y\infty}$	activation and deactivation parameters of ion channels (ICC), respectively
$Z^{(*)}_{mn}$	'biofactor'
$[Ca^{2+}]$	intracellular concentration of Ca^{2+} ions (SM)
$[Ca^{2+}]_i$	intracellular concentration of free Ca^{2+} ions (ICC)
ϑ_{Ca}	parameter of calcium inhibition (ICC)
$\lambda, \hbar, \wp_{Ca}, \ \tau_{xCa}$	electrical numerical parameters and constants

R_{sp}	radius of a solid sphere
Z_c	position of the centre of a sphere
$\overset{0}{\breve{V}}, \breve{V}$	initial and current intraluminal volume, respectively
W	strain energy density function
SM	smooth muscle syncytium
ICC	interstitial cells of Cajal

Introduction

> But at some point it is necessary to go back again to the foundations and, this time, observe complete rigour.
>
> N. I. Lobachevsky

We define a thin shell as a body bounded by two closely spaced curved surfaces. Assume that every point of the shell is associated with the curvilinear coordinates α_1, α_2 and the unit normal vector $\bar{m}$, such that the distance along $\bar{m}$ is given by z $(-0.5h(\alpha_1, \alpha_2) \leq z \leq 0.5h(\alpha_1, \alpha_2))$ (Fig. 1). Then this body is called the shell of thickness h. Let the faces of the shell be smooth with no singularities. The shell is classified as thin or thick on the basis of the ratio h/R_i, where R_i are the radii of curvature of the middle surface S of the shell, i.e. the surface at $z=0$. Thus, the shell is considered to be thin if $\max(h/R_i) \leq 1/20$ and thick otherwise. However, it should be noted that the above estimate is very rough and in many practical applications other geometrical and mechanical characteristics should also be considered.

Most organs of the human body, including the eyeball, oesophagus, stomach, gallbladder, uterus, ureter and bladder, can be viewed as thin shells. Their high endurance and enormous functionality depend on biomechanical properties of the tissues they are made of and specific arrangements of constituents (proteins, fibrils, cells) within them. Biological tissues are regarded as anisotropic, heterogeneous, incompressible composites. They are inherently nonlinear in their mechanical response and undergo finite deformations. Additionally, most biological tissues, with the exception of bones, are soft. Thus, they (i) are thin, (ii) possess low stiffness in response to elongation, (iii) do not resist compression and bending, (iv) undergo large deformations, (v) generate lateral (shear) stresses that are small compared with tangent stresses, and (vi) may wrinkle during operation without loss of stability of the organ. The above properties define the high degree of variability of shapes that the organ can take on in the process of loading.

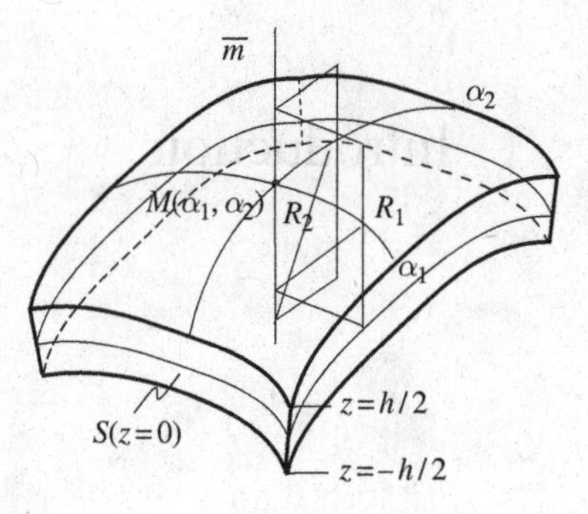

Fig. 1 A thin shell.

The distinctive anatomical appearance of organs is also correlated with their structural advantages. They contain the optimal space within and outside, exhibit high degrees of reserved strength and structural integrity combined with efficient biomechanical functionality, have optimal strength-to-weight ratios and are ideal to resist (support) the internal pressure and external loads. For example, the human stomach is the organ of the gastrointestinal tract located in the left upper quadrant of the abdomen. Its prime role is to accommodate and digest food. Even with small thickness of the gastric wall, which in normal subjects varies from 3 to 5 mm, and the characteristic radius of curvature of the middle surface within the range $10\,\text{cm} \leq R_i \leq 15\,\text{cm}$ it is capable of holding 2–5 l of mixed gastric content without increasing the intraluminal pressure.

The pregnant uterus is the organ of pear-like shape that occupies the lower and middle abdomen. Its prime functions are to accommodate and nurture the fetus (fetuses) during gestation, and to expel the baby during labour and delivery. The thickness of the uterine wall in different regions varies in the range 0.5–1.5 cm and the radii of curvature vary in the range 20–40 cm. Therefore, the pregnant uterus can also be approximated as a thin soft shell.

With the latest advances in mathematical modelling of biological systems, it has become possible to develop complex models of the abdominal organs and to gain insight into the hidden physiological mechanisms of their function (Miftahof *et al.*, 2009; Pullan *et al.*, 2004; Cheng *et al.*, 2007; Corrias and Buist, 2007; Pal *et al.*, 2004; Pal *et al.*, 2007). A first biomechanical model of the organ as a soft biological shell was developed by Miftakhov (1983c). Under general assumptions of curvilinear orthotropy and physical and geometrical nonlinearity, a mathematical formulation and a numerical investigation of the dynamics of stress–strain distribution in the organ under simple and complex loadings were performed. The dynamics of the development of uniaxial stress–strained states in the cardia and pylorus as a function of intraluminal pressure was demonstrated computationally and confirmed experimentally. The results provided a valuable insight into the mechanism of blunt abdominal trauma with rupture of the anterior wall of the stomach and gave a biomechanical explanation for the Mallory–Weiss syndrome. It was thought previously that atrophic changes in the gastric mucosa and submucous layer were responsible for longitudinal tears in the cardia-fundal region and life-threatening intragastric bleeding. The soft-shell-model studies demonstrated that the anatomical structure and configuration of the stomach *per se* make these regions more susceptible than the others to linear ruptures.

The biomechanics of the small intestine has been extensively studied experimentally and numerically. Miftahof was the first to construct a biophysically plausible model of the organ as a soft cylindrical biological shell. With the model it was possible to reproduce a variety of electromechanical wave phenomena, including the gradual reflex, pendular movements, segmentation and peristalsis. The model also contained intrinsic neuroregulatory mechanisms – the enteric nervous plexuses and multiple neurotransmitters. Thus, the model allowed one to study the effects of different classes of pharmacological compounds on the motility of the small intestine in normal and pathological conditions.

Mathematical models of visceral organs addressing various aspects of physiological functions have been proposed recently. However, all models, without exception, are based on a reductionist 'mechanistic' approach and thus have limited biological plausibility and implications for our understanding of the pathophysiology of various diseases. This is often due to indiscrete and erroneous applications of ideas and methods borrowed from the mechanics of solids to describe their mechanical behaviour. It should be emphasized that biomechanics is not just the transformation of general laws and principles of mechanics to the study of biological phenomena, but rather the adequate development and extension of these laws and principles to the modelling and analysis of living things. Therefore, accurate integrative models that incorporate various data and serve as the basis for multilevel analysis of interrelated biological processes are required. Such models will have enormous impact on unravelling hidden intricate mechanisms of diseases and assist in the design of their treatment.

Unfortunately, currently it is still common practice in the community of modellers to employ the system of Navier–Stokes equations when modelling the hollow abdominal viscera as a 'shell' structure. Using commercially available software and highly flexible graphical tools, they manage to fit results of numerical simulations to experimental data. The approach is utterly incorrect and the results cause confusion, rather than providing solutions to urgent clinical problems. Thus, it is erroneous to claim that antral contraction wave activity plays the dominant role in intragastric fluid motions, on the basis of results of computer simulations of a flow caused by prescribed indentations of the surface boundaries (Pal *et al.*, 2004; Pal *et al.*, 2007). The latter system is supposed to represent a two-dimensional model of the stomach. It is not surprising that the numerical method, which predetermines the desired patterns of flow, produces results that resemble those observed experimentally, in studies using magnetic resonance imaging. An adequate mathematical model of the above phenomenon should have comprised the combined system of the equations of motion of the bioshell – the stomach – and Navier–Stokes equations for the gastric content. Also, in view of the fundamental mechanical property of the tissue, its softness, it is unwise to argue for the dependence of stress–strain states on the radii of curvature of visceral organs (Liao *et al.*, 2004). It is the responsibility of an applied mathematician, a computer scientist and a mechanical engineer to suggest an adequate descriptor and to give a rigorous mathematical formulation of the model.

Although the models of the stomach and of the small and large intestine as biological shells described in this book have limited biomedical value, they are mathematically sound and are based on the accurate extension and application of general laws and hypotheses of the mechanics of thin soft shells. They incorporate electrophysiological and morphological data concerning the structure and function of human organs and reproduce quantitatively and qualitatively the dynamics of electromechanical wave activity and the stress–strain distribution in them. They can serve as a starting point for further expansions and biological improvements. We hope that with the publication of this book the approach to modelling of soft abdominal organs will be reconsidered.

Exercises

1. As noted once by Charles Darwin (1809–1882): 'I deeply regretted that I did not proceed far enough at least to understand something of the great leading principles of mathematics; for men thus endowed seem to have an extra sense'. Why is mathematical modelling in the life sciences so hard?
2. 'Reduction' versus 'integration' is the continual dilemma in mathematical modelling. Recall Albert Einstein (1879–1955): 'Models should be as simple as possible, but not more so'. How far should we go to achieve the balance?
3. Biology/medicine is an empirical science; nothing is ever proven. Explanations are given in terms of the concepts and prevailing perspectives of the time and available experimental facts regarding a particular phenomenon. What is the role of the mathematical sciences in medicine and biology?
4. The biological phenomena that an investigator seeks to understand and predict are very rich and diverse. They are not derived from a few simple principles. Should a mathematical modeller look for specific biological laws or learn to apply general laws of nature to study biological processes?
5. Biological systems are large and complex. Their dynamics can be hard to understand by intuitive approaches alone. Systems biology is a new paradigm that offers a holistic

approach to the investigation of interrelations and interactions at various structural levels of the system. What elements are essential in the study of the biomechanics of the digestive system, i.e. the stomach and the small and large intestine?

6. When does a mathematical model become satisfactory and useful? Formulate some criteria of a satisfactory mathematical model.
7. What is required in order for one to be a mathematical biologist?
8. Researchers rely on conventional solvers – commercially available packages such as MATLAB, BEM, MAPLE, etc. – in situations involving nonlinear systems of differential equations. However, one should be aware of imminent pitfalls that might not always be easy to recognize when dealing with mathematical formulations that are different from the classical ones. What problems should a modeller be aware of when using conventional software in solving new mathematical problems?

1

The geometry of the surface

1.1 Intrinsic geometry

Consider a smooth surface S in three-dimensional Euclidean space. It is referred to a right-handed global orthogonal Cartesian system x_1, x_2, x_3. Let S also be associated with a set of independent parameters α_1 and α_2 (Fig. 1.1) such that

$$x_1 = f_1(\alpha_1, \alpha_2), \qquad x_2 = f_2(\alpha_1, \alpha_2), \qquad x_3 = f_3(\alpha_1, \alpha_2), \tag{1.1}$$

where f_j ($j = 1, 2, 3$) are single-valued functions that possess derivatives up to any required order. By putting $\alpha_1 =$ constant and varying the parameter α_2 in $f_j(c, \alpha_2)$, we obtain a curve that lies entirely on S. By successively giving α_1 a series of constant values we obtain a family of curves along which only a parameter α_2 varies. These curves are called the α_2-coordinate lines. Similarly, on setting $\alpha_2 =$ constant we obtain the α_1-coordinate lines of S. We assume that only one curve of the family passes through a point of the given surface. Thus, any point M on S can be treated as a cross-intersection of the α_1 and α_2 curvilinear coordinate lines.

The position of a point M with respect to the origin O of the reference system is defined by the position vector $\bar{r}$,

$$\bar{r} = \bar{i}_1 x_1 + \bar{i}_2 x_2 + \bar{i}_3 x_3 = \sum_{i=1}^{3} \bar{i}_i x_i,$$

where $\{\bar{i}_1, \bar{i}_2, \bar{i}_3\}$ is the orthonormal triad of unit vectors associated with $\{x_1, x_2, x_3\}$. By virtue of Eqs. (1.1) it can be written in the form

$$\bar{r} = \bar{i}_1 f_1(\alpha_1, \alpha_2) + \bar{i}_2 f_2(\alpha_1, \alpha_2) + \bar{i}_3 f_3(\alpha_1, \alpha_2). \tag{1.2}$$

Equation (1.2) is the vector equation of a surface. On differentiating $\bar{r}$ with respect to α_i ($i = 1, 2$) vectors tangent to the α_1- and α_2-coordinate lines are found to be

$$\bar{r}_1 = \frac{\partial \bar{r}}{\partial \alpha_1}, \qquad \bar{r}_2 = \frac{\partial \bar{r}}{\partial \alpha_2}. \tag{1.3}$$

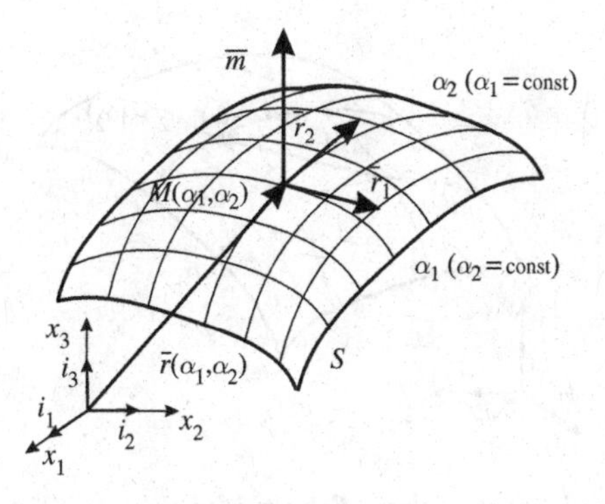

Fig. 1.1 Intrinsic parameterization of the surface.

Modules and the scalar product of $\bar{r}_i$ $(i = 1, 2)$ are defined by

$$|\bar{r}_1| = \bar{r}_1\bar{r}_1 = A_1 = \sqrt{a_{11}}, \qquad |\bar{r}_2| = \bar{r}_2\bar{r}_2 = A_2 = \sqrt{a_{22}},$$
$$\bar{r}_1\bar{r}_2 = A_1 A_2 \cos\chi = \sqrt{a_{12}}, \tag{1.4}$$

where χ is the angle between coordinate lines, A_i are the Lamé parameters and a_{ik} are the coefficients of the metric tensor **A** on S. Using Eqs. (1.4) we introduce the unit vectors $\bar{e}_i$ in the direction of $\bar{r}_i$ which are described by

$$\bar{e}_1 = \frac{\bar{r}_1}{|\bar{r}_1|} = \frac{\bar{r}_1}{A_1}, \qquad \bar{e}_2 = \frac{\bar{r}_2}{|\bar{r}_2|} = \frac{\bar{r}_2}{A_2}. \tag{1.5}$$

The vector $\bar{m}$ normal to $\bar{r}_1$ and $\bar{r}_2$ is found from

$$\bar{m} = \bar{r}_1 \times \bar{r}_2 \qquad \text{and} \qquad \bar{m}\bar{r}_1 = 0, \qquad \bar{m}\bar{r}_2 = 0, \tag{1.6}$$

where $\bar{r}_1 \times \bar{r}_2$ is the vector product. The vectors $\bar{r}_1, \bar{r}_2$ and $\bar{m}$ are linearly independent and comprise a covariant base $\{\bar{r}_1, \bar{r}_2, \bar{m}\}$ on S. The reciprocal base $\{\bar{r}^1, \bar{r}^2, \bar{m}\}$ is defined by

$$\bar{r}^1 = \frac{\bar{r}_2 \times \bar{m}}{\bar{r}_1(\bar{r}_2 \times \bar{m})}, \qquad \bar{r}^2 = \frac{\bar{m} \times \bar{r}_1}{\bar{r}_2(\bar{m} \times \bar{r}_1)}, \tag{1.7}$$

where $\bar{r}_i(\bar{m} \times \bar{r}_j)$ is the scalar triple product. Evidently, the vectors $\bar{r}^k$ and $\bar{r}_i$ are mutually orthogonal, i.e.

$$\bar{r}^k\bar{r}_i = \delta_i^k, \qquad \bar{r}^k\bar{m} = 0.$$

Here δ_i^k is the Kronecker delta such that $\delta_i^k = 1$ if $i = k$ and $\delta_i^k = 0$ if $i \neq k$.

Let $\bar{m}(\bar{r}_i \times \bar{r}_k) = c_{ik}$ and $\bar{m}(\bar{r}^i \times \bar{r}^k) = c^{ik}$. Hence,

$$c_{ik}\bar{m} = \bar{r}_i \times \bar{r}_k, \qquad c^{ik}\bar{m} = \bar{r}^i \times \bar{r}^k,$$
$$c_{ik}\bar{r}^k = \bar{m} \times \bar{r}_i, \qquad c^{ik}\bar{r}_i = \bar{m} \times \bar{r}^i. \tag{1.8}$$

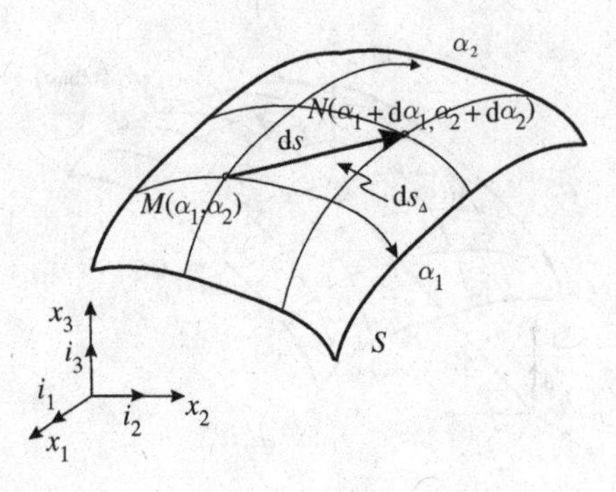

Fig. 1.2 The first fundamental form of the surface.

It follows that

$$\begin{aligned} c^{ii} &= 0, \qquad c^{12} = -c^{12} = 1/\sqrt{a}, \\ c_{ii} &= 0, \qquad c_{12} = -c_{12} = \sqrt{a}, \end{aligned} \tag{1.9}$$

where $a = (A_1 A_2 \sin\chi)^2$, and

$$c_{ik}c^{km} = \delta_i^m, \qquad c_{ik}c^{ki} = \delta_i^i = 2 \qquad (i, k = 1, 2). \tag{1.10}$$

The length of a line element between two infinitely close points $M(x_1, x_2, x_3)$ and $N(x_1 + dx_1,\ x_2 + dx_2,\ x_3 + dx_3)$ (Fig. 1.2) is given by

$$ds^2 = dx_1^2 + dx_2^2 + dx_3^2 = |d\bar{r}|^2 = |\bar{r}_i\, d\alpha_i|^2.$$

Using Eqs. (1.4) in the above, we have

$$ds^2 = A_1^2\, d\alpha_1^2 + 2A_1A_2 \cos\chi\, d\alpha_1\, d\alpha_2 + A_2^2\, d\alpha_2^2 = a_{ik}\, d\alpha_i\, d\alpha_k. \tag{1.11}$$

The quadratic form (Eq. (1.11)) is called the first fundamental form of the surface. It allows us to calculate the length of line elements, the angle between coordinate curves and the surface area

$$ds_\Delta = |\bar{r}_1 \times \bar{r}_2|\, d\alpha_1\, d\alpha_2 = \sqrt{a}\, d\alpha_1\, d\alpha_2, \tag{1.12}$$

and therefore it fully describes the intrinsic geometry of S.

1.2 Extrinsic geometry

Let Γ be a non-singular curve on S parameterized by arc length s (Fig. 1.3)

$$\bar{r} = \bar{r}(s) = \bar{r}(\alpha_1(s), \alpha_2(s)).$$

By differentiating $\bar{r}(s)$ with respect to s the unit vector $\bar{\tau}$ tangent to Γ is found to be

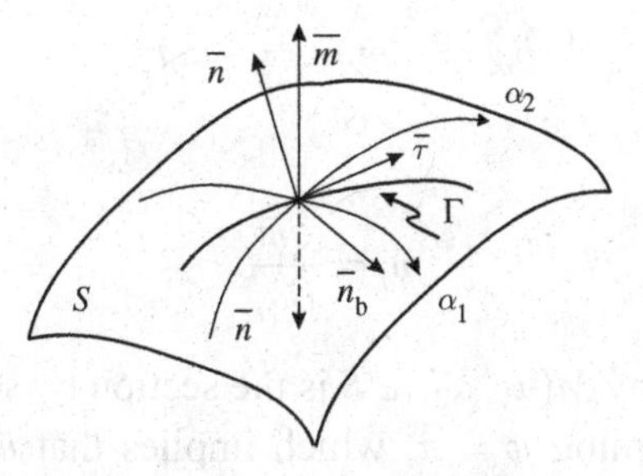

Fig. 1.3 The extrinsic geometry of the surface and a local base $\{\bar{n}, \bar{n}_b, \bar{\tau}\}$ associated with a curve Γ.

$$\bar{\tau} = \frac{d\bar{r}}{ds} = \bar{r}_1 \frac{d\alpha_1}{ds} + \bar{r}_2 \frac{d\alpha_2}{ds}. \tag{1.13}$$

By applying the Frenet–Serret formula for the derivative of $\bar{\tau}$ with respect to s we get

$$\frac{d\bar{\tau}}{ds} = \frac{\bar{n}}{R_c}, \tag{1.14}$$

where $1/R_c$ is the curvature and $\bar{n}$ is the vector normal to Γ. By substituting Eq. (1.7) into (1.8) we obtain

$$\bar{n} = \sum_{i=1}^{2} \sum_{k=1}^{2} \bar{r}_{ik} \frac{d\alpha_i}{ds} \frac{d\alpha_k}{ds} + \bar{r}_1 \frac{d\alpha_1^2}{ds} + \bar{r}_2 \frac{d\alpha_2^2}{ds}, \tag{1.15}$$

where

$$\bar{r}_{ik} = \frac{\partial^2 \bar{r}}{\partial \alpha_i \, \partial \alpha_k} = \frac{\partial^2 \bar{r}}{\partial \alpha_k \, \partial \alpha_i}, \qquad \bar{r}_{ik} = \bar{r}_{ki}.$$

Let φ be the angle between the vectors $\bar{m}$ and $\bar{n}$ such that $\bar{m}\bar{n} = \cos\varphi$. Then the scalar product of Eq. (1.15) with $\bar{m}$ yields

$$\frac{\cos\varphi}{R_c} = \frac{b_{11}\, d\alpha_1^2 + 2b_{12}\, d\alpha_1\, d\alpha_2 + b_{22}\, d\alpha_2^2}{ds^2}, \tag{1.16}$$

where

$$b_{11} = \bar{m}\bar{r}_{11}, \qquad b_{12} = \bar{m}\bar{r}_{12} = \bar{m}\bar{r}_{21}, \qquad b_{22} = \bar{m}\bar{r}_{22}. \tag{1.17}$$

The quadratic form

$$b_{11}\, d\alpha_1^2 + 2b_{12}\, d\alpha_1\, d\alpha_2 + b_{22}\, d\alpha_2^2$$

is called the second fundamental form of the surface. On differentiating Eq. (1.6) with respect to α_i we find

$$b_{ik} = -\bar{m}_i\bar{r}_k = -\bar{m}_i\bar{r}_i, \tag{1.18}$$

where

$$\bar{m}_i = \frac{\partial \bar{m}}{\partial \alpha_i}. \tag{1.19}$$

A normal section at any $M(\alpha_1, \alpha_2) \in S$ is the section by some plane that contains the vector $\bar{m} \perp S$. Assuming $\varphi = \pi$, which implies that $\bar{m}$ and $\bar{n}$ are oriented in opposite directions, from Eq. (1.10) for the curvature of the normal section $1/R_n$, we obtain

$$-\frac{1}{R_n} = \frac{b_{11}\,d\alpha_1^2 + 2b_{12}\,d\alpha_1\,d\alpha_2 + b_{22}\,d\alpha_2^2}{A_1^2\,d\alpha_1^2 + 2A_1A_2\,d\alpha_1\,d\alpha_2 + A_2^2\,d\alpha_2^2}. \tag{1.20}$$

Henceforth, we assume that the coordinate lines are arranged in such a way that $\bar{m}$ is positive when pointing from the concave to the convex side of the surface. On putting α_2 = constant and α_1 = constant in Eq. (1.20), for the curvatures k_{11} and k_{22} of the normal sections in the directions of α_1 and α_2 we find

$$\frac{1}{R_{\alpha 1}} := k_{11} = -\frac{b_{11}}{A_1^2}, \qquad \frac{1}{R_{\alpha 2}} := k_{22} = -\frac{b_{22}}{A_2^2}, \tag{1.21a}$$

and the twist k_{12} of the surface

$$\frac{1}{R_{\alpha 1 \alpha 2}} := k_{12} = -\frac{b_{12}}{A_1A_2}. \tag{1.21b}$$

It becomes evident from the above considerations that the second fundamental form describes the intrinsic geometry of the surface.

At any point $M(\alpha_1, \alpha_2) \in S$ there exist two normal sections where $1/R_n$ assumes extreme values. They are called principal sections. The two perpendicular directions at M belonging to the corresponding tangent plane are called the principal directions and the principal curvatures are $(1/R)_{max} = 1/R_1$ and $(1/R)_{min} = 1/R_2$ (Fig. 1.4). Thus, there is at least one set of principal directions at any point on S. A curve on the surface such that the tangent at any point to it is collinear with the principal direction is called the line of curvature. Thus, two lines of curvature intersect at right angles and pass through each point of S. We assume that the coordinate lines α_1 and α_2 are the lines of curvature ($\chi = \pi/2, b_{12} = 0$). Such coordinates have an advantage over other coordinate systems since the governing equations in them have a relatively simple form.

Let $\bar{r}_{ik}$ be second derivatives of the position vector with respect to $\alpha_{i(k)}$ $(i, k = 1, 2)$. On decomposing $\bar{r}_{ik}$ with respect to the covariant base $\{\bar{r}_1, \bar{r}_2, \bar{m}\}$ we get

$$\bar{r}_{ik} = \Gamma_{ik}^1\bar{r}_1 + \Gamma_{ik}^2\bar{r}_2 + \bar{m}b_{ik}. \tag{1.22}$$

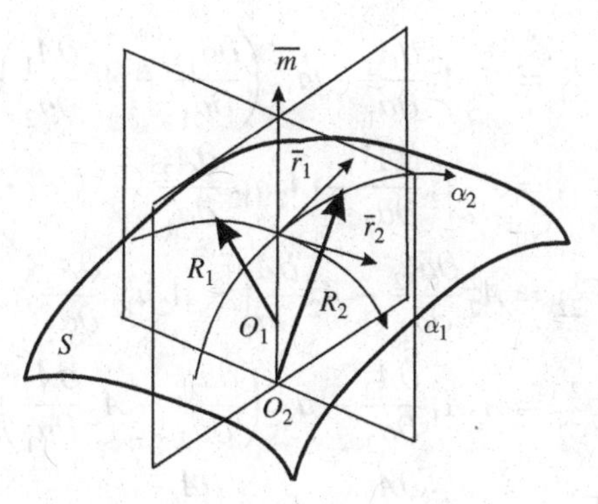

Fig. 1.4 The normal curvatures R_1 and R_2 of the surface.

Here Γ^j_{ik} ($\Gamma^j_{ik} = \bar{r}^j\bar{r}_{ik}$) are the Christoffel symbols of the second kind. On multiplying subsequently both sides of Eq. (1.22) by $\bar{r}_1$ and $\bar{r}_2$, and making use of Eqs. (1.4) and (1.6), we find

$$\bar{r}_1\bar{r}_{ik} = \Gamma^1_{ik}A_1^2 + \Gamma^2_{ik}A_1A_2\cos\chi,$$
$$\bar{r}_2\bar{r}_{ik} = \Gamma^1_{ik}A_1A_2\cos\chi + \Gamma^2_{ik}A_2^2.$$

Solving the system with respect to Γ^j_{ik} we obtain

$$a\Gamma^1_{ik} = A_2^2(\bar{r}_1\bar{r}_{ik}) - A_1A_2\cos\chi(\bar{r}_2\bar{r}_{ik}),$$
$$a\Gamma^2_{ik} = A_1^2(\bar{r}_2\bar{r}_{ik}) - A_1A_2\cos\chi(\bar{r}_1\bar{r}_{ik}). \tag{1.22a}$$

On differentiating $\bar{r}_1^2 = A_1^2, \bar{r}_2^2 = A_2^2$ and $\bar{r}_1\bar{r}_2 = A_1A_2\cos\chi$ with respect to α_1 and α_2 we find

$$\bar{r}_i\bar{r}_{ii} = A_i\frac{\partial A_i}{\partial\alpha_1}, \qquad \bar{r}_1\bar{r}_{12} = A_1\frac{\partial A_1}{\partial\alpha_2}, \qquad \bar{r}_2\bar{r}_{12} = A_2\frac{\partial A_2}{\partial\alpha_1}, \tag{1.22b}$$

$$\bar{r}_2\bar{r}_{11} + \bar{r}_1\bar{r}_{12} = \frac{\partial a_{12}}{\partial\alpha_1} = a_{12,1},$$
$$\bar{r}_2\bar{r}_{12} + \bar{r}_1\bar{r}_{22} = \frac{\partial a_{12}}{\partial\alpha_2} = a_{12,2}, \tag{1.22c}$$

where $a_{12} = a_{21} = A_1A_2\cos\chi$ and $a_{12,i} = \partial a_{12}/\partial\alpha_1$. From (1.22b) and (1.22c) we have

$$\bar{r}_2\bar{r}_{11} = \frac{\partial a_{12}}{\partial\alpha_1} - A_1\frac{\partial A_1}{\partial\alpha_2},$$
$$\bar{r}_1\bar{r}_{12} = \frac{\partial a_{12}}{\partial\alpha_2} - A_2\frac{\partial A_2}{\partial\alpha_1}. \tag{1.22d}$$

By substituting expressions (1.22b) and (1.22d) into (1.22a) we obtain

$$
\begin{aligned}
a\Gamma_{11}^{1} &= A_1A_2^2\frac{\partial A_1}{\partial \alpha_1} - a_{12}\left(\frac{\partial a_{12}}{\partial \alpha_1} - A_1\frac{\partial A_1}{\partial \alpha_2}\right),\\
a\Gamma_{12}^{1} &= A_1A_2^2\frac{\partial A1}{\partial \alpha_2} - A_2a_{12}\frac{\partial A_2}{\partial \alpha_1},\\
a\Gamma_{22}^{1} &= A_2^2\frac{\partial a_{12}}{\partial \alpha_2} - A_2^3\frac{\partial A_2}{\partial \alpha_1} - A_2a_{12}\frac{\partial A_2}{\partial \alpha_2},\\
a\Gamma_{22}^{2} &= A_2A_1^2\frac{\partial A_2}{\partial \alpha_2} - a_{12}\left(\frac{\partial a_{12}}{\partial \alpha_2} - A_2\frac{\partial A_2}{\partial \alpha_1}\right),\\
a\Gamma_{12}^{2} &= A_2A_1^2\frac{\partial A_2}{\partial \alpha_1} - A_1a_{12}\frac{\partial A_1}{\partial \alpha_2},\\
a\Gamma_{11}^{2} &= A_1^2\frac{\partial a_{12}}{\partial \alpha_1} - A_1^3\frac{\partial A_1}{\partial \alpha_2} - A_1a_{12}\frac{\partial A_1}{\partial \alpha_1}.
\end{aligned}
\tag{1.23}
$$

The derivatives of the normal vector $\bar{m}$ with respect to α_i lie in the tangent plane $(\bar{r}_1, \bar{r}_2)$ of S. By decomposing $\bar{m}_i$ $(i=1, 2)$ along $\bar{r}_i$ we get

$$
\bar{m}_i = -b_i^1\bar{r}_1 - b_i^2\bar{r}_2, \tag{1.24}
$$

where b_i^1 and b_i^2 are the mixed coefficients of the second fundamental form. Taking the scalar product of Eq. (1.24) with $\bar{r}_1$ and $\bar{r}_2$, respectively, yields

$$
\begin{aligned}
\bar{m}_i\bar{r}_1 &= -b_i^1\bar{r}_1^2 - b_i^2\bar{r}_2\bar{r}_1,\\
\bar{m}_i\bar{r}_2 &= -b_i^1\bar{r}_1\bar{r}_2 - b_i^2\bar{r}_2^2.
\end{aligned}
$$

Using Eqs. (1.4) and (1.18), b_{1i} and b_{2i} are found to be

$$
\begin{aligned}
b_{1i} &= A_1^2b_i^1 + a_{12}b_i^2,\\
b_{2i} &= A_2^2b_i^2 + a_{12}b_i^1.
\end{aligned}
\tag{1.25}
$$

It is easy to show that

$$
\begin{aligned}
b_i^1 &= \frac{1}{A_1\sin^2\chi}\left(\frac{b_{i1}}{A_1} - \frac{b_{i2}}{A_2}\cos\chi\right),\\
b_i^2 &= \frac{1}{A_2\sin^2\chi}\left(\frac{b_{i2}}{A_2} - \frac{b_{i1}}{A_1}\cos\chi\right).
\end{aligned}
\tag{1.26}
$$

Assuming that the surface is referred to orthogonal coordinates $\chi=\pi/2$, $a_{12}=a_{12}=0$, $a=(A_1A_2)^2$, Eqs. (1.23) take the form

$$
\begin{aligned}
\Gamma_{11}^{1} &= \frac{1}{A_1}\frac{\partial A_1}{\partial \alpha_1}, & \Gamma_{12}^{1} &= \frac{1}{A_1}\frac{\partial A_1}{\partial \alpha_2}, & \Gamma_{22}^{1} &= -\frac{A_2}{A_1^2}\frac{\partial A_2}{\partial \alpha_1},\\
\Gamma_{22}^{2} &= \frac{1}{A_2}\frac{\partial A_2}{\partial \alpha_2}, & \Gamma_{12}^{2} &= \frac{1}{A_2}\frac{\partial A_2}{\partial \alpha_1}, & \Gamma_{11}^{2} &= -\frac{A_1}{A_2^2}\frac{\partial A_1}{\partial \alpha_2}.
\end{aligned}
\tag{1.27}
$$

On substituting Eqs. (1.27) into Eq. (1.22) and making use of Eqs. (1.5), (1.22a) and (1.22b), the formulas for the derivatives of the unit vectors $\bar{e}_1$ and $\bar{e}_2$ are found to be

$$\begin{aligned}\frac{\partial \bar{e}_1}{\partial \alpha_1} &= -\frac{\bar{e}_2}{A_2}\frac{\partial A_1}{\partial \alpha_2} - A_1 k_{11}\bar{m}, & \frac{\partial \bar{e}_1}{\partial \alpha_2} &= -\frac{\bar{e}_2}{A_1}\frac{\partial A_2}{\partial \alpha_1} - A_2 k_{12}\bar{m},\\ \frac{\partial \bar{e}_2}{\partial \alpha_1} &= -\frac{\bar{e}_1}{A_2}\frac{\partial A_1}{\partial \alpha_2} - A_1 k_{12}\bar{m}, & \frac{\partial \bar{e}_2}{\partial \alpha_2} &= -\frac{\bar{e}_1}{A_1}\frac{\partial A_2}{\partial \alpha_1} - A_2 k_{22}\bar{m}.\end{aligned} \tag{1.28}$$

From Eqs. (1.24) we have

$$A_1^2 b_i^1 = b_{1i}, \qquad A_2^2 b_i^1 = b_{i2},$$

and, using Eqs. (1.18), we get

$$A_1 b_i^1 = -A_i k_{1i}, \qquad A_2 b_i^1 = -A_i k_{2i}.$$

On substituting the above into Eq. (1.24), and using Eq. (1.5), we obtain

$$\bar{m}_i = A_i(k_{1i}\bar{e}_i + k_{2i}\bar{e}_i). \tag{1.29}$$

If the coordinate lines are the lines of principal curvature $(k_{12} = 0, k_{ii} = 1/R_i)$, Eqs. (1.28) and (1.29) can be simplified to take the form

$$\begin{aligned}\frac{\partial \bar{e}_1}{\partial \alpha_1} &= -\frac{\bar{e}_2}{A_2}\frac{\partial A_1}{\partial \alpha_2} - \bar{m}\frac{A_1}{R_1}, & \frac{\partial \bar{e}_1}{\partial \alpha_2} &= \frac{\bar{e}_2}{A_1}\frac{\partial A_2}{\partial \alpha_1},\\ \frac{\partial \bar{e}_2}{\partial \alpha_1} &= \frac{\bar{e}_1}{A_1}\frac{\partial A_1}{\partial \alpha_2}, & \frac{\partial \bar{e}_2}{\partial \alpha_2} &= -\frac{\bar{e}}{A_1}\frac{\partial A_2}{\partial \alpha_1} - \bar{m}\frac{A_2}{R_2},\\ R_i\bar{m}_i &= A_i\bar{e}_i.\end{aligned} \tag{1.30}$$

1.3 The equations of Gauss and Codazzi

The coefficients of the first and second fundamental forms are interdependent and satisfy the three differential Gauss–Codazzi equations. The Gauss formula defines the Gaussian curvature K of the surface,

$$K = \frac{1}{R_1 R_2} = \frac{b_{11} - b_{12}^2}{a} = \frac{A_1 A_2 (k_{11}k_{22} - k_{12}^2)}{a}. \tag{1.31}$$

With the help of Eqs. (1.21a), (1.21b) and (1.26) it can be written in the form

$$\begin{aligned}\frac{b_{11} - b_{12}^2}{A_1 A_2 \sin\chi} = &\frac{\partial^2 \chi}{\partial \alpha_1\, \partial \alpha_2} + \frac{\partial}{\partial \alpha_1}\frac{(\partial A_2/\partial \alpha_1) - \cos\chi(\partial A_1/\partial \alpha_2)}{A_1 \sin\chi}\\ &+ \frac{\partial}{\partial \alpha_2}\frac{(\partial A_1/\partial \alpha_2) - \cos\chi(\partial A_2/\partial \alpha_1)}{A_2 \sin\chi}.\end{aligned} \tag{1.32}$$

To derive the Codazzi equations we proceed from

$$\frac{\partial b_{11}}{\partial \alpha_2} - \frac{\partial b_{12}}{\partial \alpha_1} = -\frac{\partial(\bar{r}_1 \bar{m}_1)}{\partial \alpha_2} + \frac{\partial(\bar{r}_2 \bar{m}_2)}{\partial \alpha_1} = -\bar{m}_1 \bar{r}_{12} + \bar{m}_2 \bar{r}_{22}.$$

On substituting $\bar{r}_{ik}$ given by Eqs. (1.22) the first Codazzi equation is found to be

$$\frac{\partial b_{11}}{\partial \alpha_2} - \frac{\partial b_{12}}{\partial \alpha_1} = -\bar{m}_1 \left(\Gamma^1_{12}\bar{r}_1 + \Gamma^2_{12}\bar{r}_2\right) + \bar{m}_2 \left(\Gamma^1_{11}\bar{r}_1 + \Gamma^2_{11}\bar{r}_2\right)$$
$$= \Gamma^1_{12} b_{11} + (\Gamma^2_{12} - \Gamma^2_{11}) b_{12} - \Gamma^2_{11} b_{22}. \tag{1.33}$$

Here use is made of Eq. (1.18) and the fact that $\bar{m}_i \bar{m} = 0$.

Similarly, proceeding from the difference $\partial b_{22}/\partial \alpha_1 - \partial b_{12}/\partial \alpha_2$ and repeating the steps as above, we obtain the second Codazzi equation

$$\frac{\partial b_{22}}{\partial \alpha_1} - \frac{\partial b_{12}}{\partial \alpha_2} = \Gamma^1_{12} b_{22} + (\Gamma^1_{12} - \Gamma^2_{22}) b_{12} - \Gamma^1_{22} b_{11}. \tag{1.34}$$

The Christoffel symbols Γ^j_{ik} in (1.26) satisfy (1.17).

On substituting Eqs. (1.21a) and (1.21b) into the left-hand side of Eq. (1.34), we get

$$\frac{\partial b_{11}}{\partial \alpha_2} - \frac{\partial b_{12}}{\partial \alpha_1} = A_1 \left(\frac{\partial (A_1 k_{12})}{\partial \alpha_1} - \frac{\partial (A_1 k_{11})}{\partial \alpha_2} \right) + A_2 k_{12} \frac{\partial A_1}{\partial \alpha_1} - A_1 k_{11} \frac{\partial A_1}{\partial \alpha_2}. \tag{1.35}$$

On using Eq. (1.35) in (1.33) and dividing the resultant equation by $-A_1$ and $-A_2$, respectively, we obtain

$$\begin{aligned}
&\frac{\partial (A_1 k_{11})}{\partial \alpha_2} - \frac{\partial (A_2 k_{12})}{\partial \alpha_1} + k_{11} \left(\frac{\partial A_1}{\partial \alpha_2} - A_1 \Gamma^1_{12} \right) \\
&\qquad - A_2 k_{12} \left(\frac{1}{A_1} \frac{\partial A_1}{\partial \alpha_1} - \Gamma^1_{11} + \Gamma^2_{12} \right) - \frac{A_2^2 k_{22} \Gamma^1_{11}}{A_1} = 0, \quad A_2^2 = (A_2)^2, \\
&\frac{\partial (A_2 k_{22})}{\partial \alpha_1} - \frac{\partial (A_1 k_{12})}{\partial \alpha_2} + k_{22} \left(\frac{\partial A_2}{\partial \alpha_1} - A_2 \Gamma^2_{12} \right) \\
&\qquad - A_1 k_{12} \left(\frac{1}{A_2} \frac{\partial A_2}{\partial \alpha_2} - \Gamma^2_{22} + \Gamma^1_{12} \right) - \frac{A_1^2 k_{11} \Gamma^2_{22}}{A_2} = 0.
\end{aligned} \tag{1.36}$$

In the case of orthogonal coordinates Eqs. (1.32) and (1.36) can be written in the form

$$\begin{aligned}
&\frac{\partial}{\partial \alpha_1} \left(\frac{1}{A_1} \frac{\partial A_2}{\partial \alpha_1} \right) + \frac{\partial}{\partial \alpha_2} \left(\frac{1}{A_2} \frac{\partial A_1}{\partial \alpha_2} \right) = A_1 A_2 (k_{12}^2 - k_{11} k_{22}) = -\frac{A_1 A_2}{R_1 R_2}, \\
&\frac{\partial (A_1 k_{11})}{\partial \alpha_2} - \frac{\partial (A_2 k_{12})}{\partial \alpha_1} - k_{12} \frac{\partial A_2}{\partial \alpha_1} - k_{22} \frac{\partial A_1}{\partial \alpha_2} = 0, \\
&\frac{\partial (A_2 k_{22})}{\partial \alpha_1} - \frac{\partial (A_1 k_{12})}{\partial \alpha_2} - k_{12} \frac{\partial A_1}{\partial \alpha_2} - k_{11} \frac{\partial A_2}{\partial \alpha_1} = 0.
\end{aligned} \tag{1.37}$$

Here use is made of Eq. (1.27) for the Christoffel symbols Γ^j_{ik}.

If the coordinate lines are the lines of curvature ($k_{12} = 0, k_{ii} = 1/R_i$), then Eqs. (1.37) take the simplest form

$$\frac{\partial}{\partial \alpha_1}\left(\frac{1}{A_1}\frac{\partial A_2}{\partial \alpha_1}\right) + \frac{\partial}{\partial \alpha_2}\left(\frac{1}{A_2}\frac{\partial A_1}{\partial \alpha_2}\right) = -\frac{A_1 A_2}{R_1 R_2},$$
$$\frac{\partial}{\partial \alpha_2}\left(\frac{A_1}{R_1}\right) = \frac{1}{R_2}\left(\frac{\partial A_1}{\partial \alpha_1}\right), \qquad \frac{\partial}{\partial \alpha_1}\left(\frac{A_2}{R_2}\right) = \frac{1}{R_1}\left(\frac{\partial A_2}{\partial \alpha_1}\right). \tag{1.38}$$

1.4 General curvilinear coordinates

Consider a non-singular surface S_z that is located at a distance z from S and $S_z \parallel S$. Let $\bar{\rho}$ be the position vector of a point $M_z \in S_z$ (Fig. 1.5)

$$\bar{\rho}(\alpha_1, \alpha_2) = \bar{r}(\alpha_1, \alpha_2) + z\bar{m}(\alpha_1, \alpha_2). \tag{1.39}$$

Here z is the normal distance measured from the point $M \in S$ and to M_z. On differentiating Eq. (1.39) with respect to α_i and z, we find

$$\bar{\rho}_i = \frac{\partial \bar{\rho}}{\partial \alpha_i} = \bar{r}_i + \bar{m}_i z, \qquad \bar{\rho}_3 = \frac{\partial \bar{\rho}}{\partial z} = \bar{m}_i. \tag{1.40}$$

The vectors $\bar{\rho}_1$ and $\bar{\rho}_2$ are tangent to the α_1- and α_2-coordinate lines and are linearly independent. Thus, together with $\bar{\rho}_3$, they comprise a local base $\{\bar{\rho}_1, \bar{\rho}_2, \bar{\rho}_3\}$ at M_z.

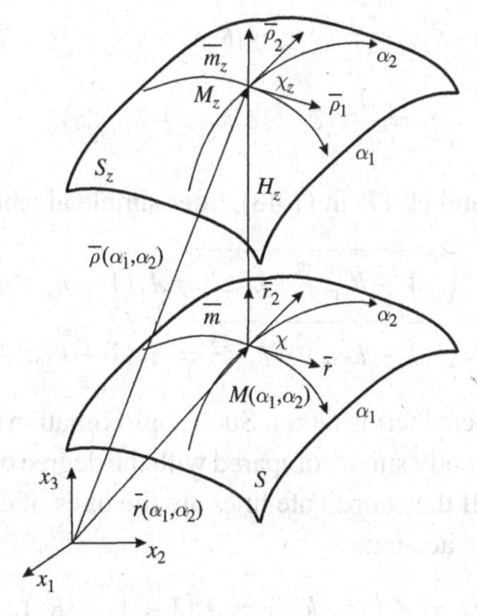

Fig. 1.5 Parameterization of the equidistant surface S_z.

On substituting $\bar{m}_i$ given by Eq. (1.24) into (1.40), we get

$$\bar{\rho}_1 = \bar{r}_1\left(1 - zb_1^1\right) - \bar{r}_1 z b_1^2, \qquad \bar{\rho}_2 = \bar{r}_2\left(1 - zb_2^2\right) - \bar{r}_1 z b_1^1, \qquad \bar{\rho}_3 = \bar{m}. \quad (1.41)$$

If the coordinate system is orthogonal then, using Eq. (1.29) for $\bar{m}_i$, we obtain

$$\bar{\rho}_i = A_i\bar{e}_i + \left(A_i k_{1i}\bar{e}_1 + k_{2i}\bar{e}_2\right)z, \qquad \bar{\rho}_3 = \bar{m}. \quad (1.42)$$

Furthermore, if the coordinate lines are the lines of curvature, Eqs. (1.42) become

$$\bar{\rho}_i = A_i\bar{e}_i + \left(1 + z/R_i\right), \qquad \bar{\rho}_3 = \bar{m}. \quad (1.43)$$

Taking the scalar product of Eqs. (1.41) and (1.42) with $\bar{m}$ yields

$$\bar{\rho}_i\bar{m} = 0. \quad (1.44)$$

Let g_{ik} and H_i be the scalar products of $\bar{\rho}_i$ and $\bar{\rho}_k$ in general and orthogonal curvilinear coordinates, respectively,

$$g_{ik} = \bar{\rho}_i\bar{\rho}_k \quad (i, k = 1, 2), \quad (1.45)$$

$$H_i = |\bar{\rho}_i|. \quad (1.46)$$

H_i are the Lamé coefficients of S_z. The unit vectors $\bar{e}_i^z \in S_z$ are given by

$$\bar{e}_i^z = \frac{\bar{\rho}_i}{H_i}. \quad (1.47)$$

Making use of Eqs. (1.42) in (1.47), for $\bar{e}_1^z$ and $\bar{e}_2^z$ we find

$$\begin{aligned} \bar{e}_1^z &= \frac{A_1}{H_1}\left(\bar{e}_1 + \bar{e}_1 k_{11} z + \bar{e}_2 k_{12} z\right), \\ \bar{e}_2^z &= \frac{A_2}{H_2}\left(\bar{e}_2 + \bar{e}_2 k_{22} z + \bar{e}_1 k_{12} z\right). \end{aligned} \quad (1.48)$$

On use of Eqs. (1.42) and (1.47) in (1.48), after simple algebra, we obtain

$$\begin{aligned} H_1 &= A_1\sqrt{(1 + k_{11}z)^2 + k_{12}z^2} = A_1\left(1 + k_{11}z + \cdots\right), \\ H_2 &= A_2\sqrt{(1 + k_{22}z)^2 + k_{12}z^2} = A_2\left(1 + k_{22}z + \cdots\right). \end{aligned} \quad (1.49)$$

In Eqs. (1.49) we neglected terms $O(z^2)$. Such approximation has been proven to be satisfactory for z sufficiently small compared with the degree of accuracy required in practical applications. If the coordinate lines are the lines of curvature, $k_{12} = 0$, the formulas (1.49) become accurate,

$$H_i = A_i\left(1 + k_{ii}z\right) = A_i\left(1 + k_{11}z/R_{ii}\right). \quad (1.50)$$

Also the unit vectors $\bar{e}_i^z \in S_z$ and $\bar{e}_i \in S$ become identical if the chosen coordinate lines are the lines of curvature. Indeed, by setting $k_{12} = 0$ in Eq. (1.48) and using formula (1.50), we obtain

$$\bar{e}_1^z = \bar{e}_1, \qquad \bar{e}_2^z = \bar{e}_2. \tag{1.51}$$

The length of a line element on S_z is given by

$$\begin{aligned}(\mathrm{d}s^z)^2 = |\mathrm{d}\bar{\rho}|^2 &= |\bar{\rho}_1\,\mathrm{d}\alpha_1 + \bar{\rho}_2\,\mathrm{d}\alpha_2 + \bar{m}\,\mathrm{d}z|^2 \\ &= g_{11}\,\mathrm{d}\alpha_1^2 + 2g_{12}\,\mathrm{d}\alpha_1\,\mathrm{d}\alpha_2 + \mathrm{d}z^2.\end{aligned} \tag{1.52}$$

By substituting Eqs. (1.39) into Eq. (1.45), we get

$$g_{ik} = \bar{r}_i\bar{r}_k + 2\bar{m}_i\bar{r}_k z + \bar{m}_i\bar{m}_k z^2.$$

Furthermore, with the help of Eqs. (1.4) and (1.24), we obtain

$$\begin{aligned}g_{ik} &= A_i(1 + 2k_{ii}z) + (\bar{m}_i z)^2, \\ g_{12} &= A_1A_2(\cos\chi + 2k_{12}z) + \bar{m}_1\bar{m}_2 z^2.\end{aligned} \tag{1.53}$$

Neglecting higher-order terms $O(z^2)$, we find

$$g_{ik} = A_i(1 + 2k_{ii}z), \qquad g_{12} = A_1A_2(\cos\chi + 2k_{12}z). \tag{1.54}$$

Assume the coordinate system is orthogonal $g_{12} = \bar{\rho}_1\bar{\rho}_2 = 0$. Hence, from Eq. (1.54), we have $k_{12}z \approx 0$. Note that the condition (1.51) remains valid in orthogonal coordinates even if the coordinate lines are not the lines of curvature.

Let χ^z be the angle between coordinate lines on S_z. Since

$$g_{12} = \bar{\rho}_1\bar{\rho}_2 = |\bar{\rho}_1| \cdot |\bar{\rho}_2| \cos\chi^z,$$

we have that

$$\cos\chi^z = \frac{g_{12}}{|\bar{\rho}_1| \cdot |\bar{\rho}_2|} = \frac{g_{12}}{\sqrt{g_{11}g_{12}}}.$$

Using Eqs. (1.54), we get

$$\cos\chi^z = \frac{\cos\chi + 2k_{12}z}{\sqrt{(1 + 2k_{11}z)(1 + 2k_{22}z)}}. \tag{1.55}$$

We conclude this section with some useful formulas for the cross product of vectors:

$$\begin{aligned}&H_2(\bar{\rho}_1 \times \bar{m}) = -\bar{\rho}_2 H_1, \qquad H_1(\bar{\rho}_2 \times \bar{m}) = \bar{\rho}_1 H_2, \qquad H_1H_2(\bar{\rho}_1 \times \bar{\rho}_2) = \bar{m}, \\ &A_2(\bar{r}_1 \times \bar{m}) = -A_1\bar{r}_2, \qquad A_1(\bar{r}_2 \times \bar{m}) = A_2\bar{r}_1, \qquad A_1A_2(\bar{r}_1 \times \bar{r}_2) = \bar{m},\end{aligned} \tag{1.56}$$

$$(\bar{e}_1 \times \bar{m}) = \bar{e}_2, \qquad (\bar{e}_2 \times \bar{m}) = \bar{e}_1, \qquad (\bar{e}_1 \times \bar{e}_2) = \bar{m}. \tag{1.57}$$

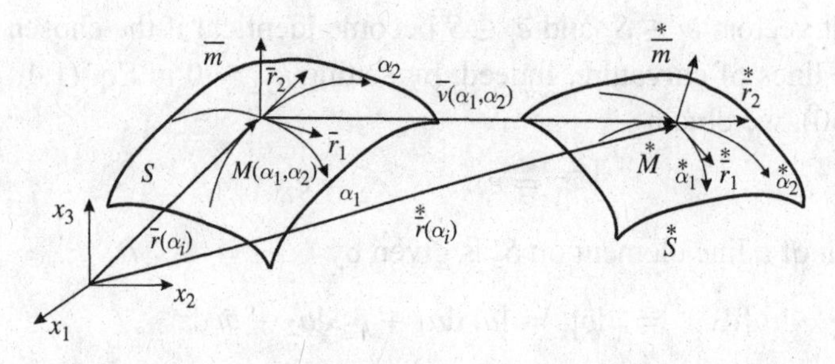

Fig. 1.6 Deformation of the surface.

1.5 Deformation of the surface

As a result of deformation the surface S changes into a new surface $\overset{*}{S}$ with a point $M \in S$ sent into the point $\overset{*}{M} \in \overset{*}{S}$. Henceforth, all quantities that refer to the deformed configuration we shall designate by an asterisk (*) unless specified otherwise. Let $\bar{v}(\alpha_1, \alpha_2)$ be the vector of displacement of point M. Then the position of $\overset{*}{M}$ after deformation (Fig. 1.6) is described by

$$\overset{*}{\bar{r}}(\alpha_1, \alpha_2) = \bar{r}(\alpha_1, \alpha_2) + \bar{v}(\alpha_1, \alpha_2). \tag{1.58}$$

Projections of $\bar{v}$ onto the base $\{\bar{e}_1, \bar{e}_2, \bar{m}\}$ are given by

$$u_1 = \bar{v}\bar{e}_1 = \bar{v}\bar{r}_1/A_1, \qquad u_2 = \bar{v}\bar{e}_2 = \bar{v}\bar{r}_2/A_2, \qquad \omega = \bar{v}\bar{m}. \tag{1.59}$$

Hence

$$\bar{v} = u_1\bar{e}_1 + u_2\bar{e}_2 + \omega\bar{m}, \tag{1.60}$$

where u_1 and u_2 are the tangent, and ω is the normal displacement (deflection), respectively.

On differentiating Eq. (1.58) with respect to α_i and using Eqs. (1.28) and (1.29), we find

$$\begin{aligned} \overset{*}{\bar{r}}_1 &= A_1\{(1 + e_{11})\bar{e}_1 + e_{12}\bar{e}_2 + \bar{m}\varpi_1\}, \\ \overset{*}{\bar{r}}_2 &= A_2\{e_{21}\bar{e}_1 + (1 + e_{22})\bar{e}_2 + \bar{m}\varpi_2\}, \end{aligned} \tag{1.61}$$

where

$$\begin{aligned} e_{11} &= \frac{1}{A_1}\frac{\partial u_1}{\partial \alpha_1} + \frac{u_2}{A_1A_2}\frac{\partial A_1}{\partial \alpha_2} + k_{11}\omega, \qquad e_{22} = \frac{1}{A_2}\frac{\partial u_2}{\partial \alpha_2} + \frac{u_1}{A_1A_2}\frac{\partial A_2}{\partial \alpha_1} + k_{22}\omega, \\ e_{12} &= \frac{1}{A_1}\frac{\partial u_2}{\partial \alpha_1} - \frac{u_1}{A_1A_2}\frac{\partial A_1}{\partial \alpha_2} + k_{12}\omega, \qquad e_{21} = \frac{1}{A_2}\frac{\partial u_1}{\partial \alpha_2} - \frac{u_1}{A_1A_2}\frac{\partial A_2}{\partial \alpha_2} + k_{21}\omega, \\ \varpi_1 &= \frac{1}{A_1}\frac{\partial \omega}{\partial \alpha_1} - k_{11}u_1 - k_{12}u_2, \qquad \varpi_2 = \frac{1}{A_2}\frac{\partial \omega}{\partial \alpha_2} - k_{22}u_1 - k_{21}u_1. \end{aligned} \tag{1.62}$$

The length of a line element on the deformed surface $\overset{*}{S}$ is defined by

$$(\mathrm{d}\overset{*}{s})^2 = (\overset{*}{A}_1)^2\,\mathrm{d}\alpha_1^2 + 2\overset{*}{A}_1\overset{*}{A}_2\,\mathrm{d}\alpha_1\,\mathrm{d}\alpha_2 + (\overset{*}{A}_2)^2\,\mathrm{d}\alpha_2^2 = \overset{*}{a}_{ij}\,\mathrm{d}\alpha_i\,\mathrm{d}\alpha_j, \tag{1.63}$$

where

$$\overset{*}{A}_1 = \left|\overset{*}{\bar{r}}_1\right|, \qquad \overset{*}{A}_2 = \left|\overset{*}{\bar{r}}_2\right|, \qquad \overset{*}{a}_{12} = \overset{*}{\bar{r}}_1\overset{*}{\bar{r}}_2 = \overset{*}{A}_1\overset{*}{A}_2\cos\overset{*}{\chi}, \tag{1.64}$$

and $\overset{*}{\chi}$ is the angle between coordinate lines on $\overset{*}{S}$. On substituting Eqs. (1.61) into (1.64) we find

$$(\overset{*}{A}_1)^2 = A_1^2(1+2\varepsilon_{11}), \qquad (\overset{*}{A}_2)^2 = A_2^2(1+2\varepsilon_{22}), \tag{1.65}$$

$$\cos\overset{*}{\chi} = \frac{2\varepsilon_{12}}{\sqrt{(1+2\varepsilon_{11})(1+2\varepsilon_{22})}}, \tag{1.66}$$

where

$$\begin{aligned} 2\varepsilon_{11} &= 2e_{11} + e_{11}^2 + e_{12}^2 + \varpi_1^2, \\ 2\varepsilon_{22} &= 2e_{22} + e_{22}^2 + e_{21}^2 + \varpi_2^2, \\ 2\varepsilon_{12} &= (1+e_{11})e_{21} + (1+e_{22})e_{12} + \varpi_1\varpi_2. \end{aligned} \tag{1.67}$$

To provide a physically appealing explanation for ε_{ik} ($i, k = 1, 2$), consider the lengths of the same line element before deformation $(\mathrm{d}s)_i$ and after deformation $\left(\mathrm{d}\overset{*}{s}\right)_i$:

$$(\mathrm{d}s)_1 = A_1\,\partial\alpha_1, \qquad (\mathrm{d}s)_2 = A_2\,\partial\alpha_2,$$
$$\left(\mathrm{d}\overset{*}{s}\right)_1 = \overset{*}{A}_1\,\partial\alpha_1, \qquad \left(\mathrm{d}\overset{*}{s}\right)_2 = \overset{*}{A}_2\,\partial\alpha_2.$$

The relative elongations $e_{\alpha 1}, e_{\alpha 2}$ along the α_1 and α_2 lines and the shear angle γ between them are found to be

$$e_{\alpha i} = \frac{\mathrm{d}\overset{*}{s}_i - \mathrm{d}s_i}{\mathrm{d}s_i} = \sqrt{1+2\varepsilon_{ii}} - 1 \quad (i = 1, 2), \tag{1.68}$$

$$\cos\overset{*}{\chi} = \cos\left(\frac{\pi}{2} - \gamma\right) = \sin\gamma = \frac{2\varepsilon_{12}}{\sqrt{(1+2\varepsilon_{11})(1+2\varepsilon_{22})}}. \tag{1.69}$$

It follows from Eqs. (1.68) and (1.69) that ε_{11} and ε_{22} describe tangent deformations along the coordinate lines and ε_{12} is the shear deformation that characterizes the change in γ.

Changes in curvatures, æ_{ii}, and the twist, æ_{12}, of the surface are described by

$$\begin{aligned} \text{æ}_{11} &= \frac{1}{\overset{*}{R}_1} - \frac{1}{R_1} = \overset{*}{k}_{11} - k_{11}, \\ \text{æ}_{22} &= \frac{1}{\overset{*}{R}_2} - \frac{1}{R_2} = \overset{*}{k}_{22} - k_{22}, \\ \text{æ}_{12} &= \text{æ}_{21} = \overset{*}{k}_{12} - k_{12} = \overset{*}{k}_{21} - k_{21}. \end{aligned} \tag{1.70}$$

Here $1/R_i$ and $1/\overset{*}{R}_i$ are curvatures of coordinate lines before and after deformation, and $\overset{*}{k}_{ij} = -\overset{*}{b}_{ij}/\left(\overset{*}{A}_i\overset{*}{A}_j\right)(i,j = 1,2)$.

To express $\overset{*}{\bar{m}}$ in terms of the middle-surface displacements we proceed from the formula

$$\overset{*}{\bar{m}} = \left(\overset{*}{\bar{r}}_1 \times \overset{*}{\bar{r}}_2\right) \Big/ \left(\overset{*}{A}_1\overset{*}{A}_2 \sin\overset{*}{\chi}\right).$$

On substituting $\overset{*}{A}_i$ and $\sin\overset{*}{\chi}$ given by Eqs. (1.65) and (1.66) into the above, we find

$$\overset{*}{\bar{m}} = \left(\overset{*}{\bar{r}}_1 \times \overset{*}{\bar{r}}_2\right) \Big/ \left(\overset{*}{A}_1\overset{*}{A}_2 \sqrt{\mathbf{A}}\right), \qquad \overset{*}{A}_1\overset{*}{A}_2 \sin\overset{*}{\chi} = A_1A_2\sqrt{\mathbf{A}}. \tag{1.71}$$

Here

$$\mathbf{A} = 1 + 2(\varepsilon_{11} + \varepsilon_{22}) + \left(\varepsilon_{11}\varepsilon_{22} - \varepsilon_{12}^2\right). \tag{1.72}$$

On substituting $\overset{*}{\bar{r}}_i$ given by Eqs. (1.61) and (1.57) into (1.71), we find

$$\overset{*}{\bar{m}} = (\bar{e}_1 \mathrm{S}_1 + \bar{e}_2 \mathrm{S}_2 + \bar{m}\mathrm{S}_3)/\sqrt{\mathbf{A}}, \tag{1.73}$$

where

$$\begin{aligned} \mathrm{S}_1 &= e_{12}\varpi_2 - (1 + e_{22})\varpi_1, \\ \mathrm{S}_2 &= e_{21}\varpi_1 - (1 + e_{11})\varpi_2, \\ \mathrm{S}_3 &= (1 + e_{11})(1 + e_{22}) - e_{12}e_{21}. \end{aligned} \tag{1.74}$$

Taking the scalar product of Eqs. (1.61) and (1.73) with $\bar{e}_i$ and $\bar{m}$ yields

$$\cos(\overset{*}{\bar{r}}_i, \bar{e}_k) = (\delta_{ik} + e_{ik})/\sqrt{1 + 2\varepsilon_{ii}}, \qquad \cos(\overset{*}{\bar{r}}_i, \bar{m}) = \varpi_i/\sqrt{1 + 2\varepsilon_{ii}},$$
$$\cos(\bar{m}, \bar{e}_k) = \mathrm{S}_k/\sqrt{\mathbf{A}}, \qquad \cos(\overset{*}{\bar{m}}, \bar{m}) = \mathrm{S}_3/\sqrt{\mathbf{A}}.$$

The quantities e_{ik}, ϖ_i, S_i and S_3 describe the rotations of tangent and normal vectors during deformation.

By differentiating Eq. (1.73) with respect to α_i and making use of Eqs. (1.28) and (1.29), we find

$$\sqrt{\mathbf{A}}\,\overset{*}{\bar{m}}_i = A_i(\bar{e}_1\mathbf{M}_{i1} + \bar{e}_2\mathbf{M}_{i2} + \bar{m}\mathbf{M}_{i3}) - \overset{*}{\bar{m}}\frac{\partial\sqrt{\mathbf{A}}}{\partial\alpha_i}, \tag{1.75}$$

where

$$\begin{aligned}
\mathbf{M}_{11} &= \frac{1}{A_1}\frac{\partial \mathrm{S}_1}{\partial\alpha_1} + \frac{\mathrm{S}_2}{A_1A_2}\frac{\partial A_1}{\partial\alpha_2} + k_{11}\mathrm{S}_3,\\
\mathbf{M}_{21} &= \frac{1}{A_2}\frac{\partial \mathrm{S}_2}{\partial\alpha_2} + \frac{\mathrm{S}_1}{A_2A_1}\frac{\partial A_2}{\partial\alpha_1} + k_{22}\mathrm{S}_3,\\
\mathbf{M}_{12} &= \frac{1}{A_1}\frac{\partial \mathrm{S}_2}{\partial\alpha_1} - \frac{\mathrm{S}_1}{A_1A_2}\frac{\partial A_1}{\partial\alpha_2} + k_{12}\mathrm{S}_3,\\
\mathbf{M}_{22} &= \frac{1}{A_2}\frac{\partial \mathrm{S}_1}{\partial\alpha_2} - \frac{\mathrm{S}_2}{A_1A_2}\frac{\partial A_2}{\partial\alpha_1} + k_{12}\mathrm{S}_3,\\
\mathbf{M}_{13} &= \frac{1}{A_1}\frac{\partial \mathrm{S}_3}{\partial\alpha_1} - k_{11}\mathrm{S}_1 - k_{12}\mathrm{S}_2,\\
\mathbf{M}_{23} &= \frac{1}{A_2}\frac{\partial \mathrm{S}_3}{\partial\alpha_2} - k_{22}\mathrm{S}_2 - k_{12}\mathrm{S}_1.
\end{aligned} \tag{1.76}$$

On substituting Eqs. (1.60) into (1.75), and using the fact that $\overset{*}{\bar{m}}\overset{*}{\bar{r}}_j = 0$, $A_iA_1\overset{*}{k}_{i1} = \overset{*}{\bar{m}}_i\overset{*}{\bar{r}}_1$ and $A_iA_2\overset{*}{k}_{i2} = \overset{*}{\bar{m}}_i\overset{*}{\bar{r}}_2$, we obtain

$$\begin{aligned}
\sqrt{\mathbf{A}}\overset{*}{k}_{i1} &= (1+e_{11})\mathbf{M}_{i1} + e_{12}\mathbf{M}_{i2} + \bar{\varpi}_1\mathbf{M}_{i3},\\
\sqrt{\mathbf{A}}\overset{*}{k}_{i2} &= e_{21}\mathbf{M}_{i1} + (1+e_{22})\mathbf{M}_{i2} + \bar{\varpi}_2\mathbf{M}_{i3}.
\end{aligned} \tag{1.77}$$

$e_{ik}, \varpi_i, \mathrm{S}_i$ and S_3 satisfy the following algebraic equalities:

$$\begin{aligned}
\mathrm{S}_1(1+e_{11}) + e_{12}\mathrm{S}_2 + \mathrm{S}_3\varpi_1 &= 0,\\
\mathrm{S}_2(1+e_{22}) + e_{21}\mathrm{S}_1 + \mathrm{S}_3\varpi_2 &= 0.
\end{aligned} \tag{1.78}$$

Equations (1.78) can be verified by substituting $\overset{*}{\bar{r}}_j$ and $\overset{*}{\bar{m}}$ given by Eqs. (1.61) and (1.73) into the equality $\overset{*}{\bar{m}}\overset{*}{\bar{r}}_j = 0$. We have also the following:

$$\begin{aligned}
\mathrm{S}_1e_{12} - \mathrm{S}_2(1+e_{11}) &= (1+2\varepsilon_{11}) - 2\varepsilon_{12}\varpi_1 = f_{12},\\
\mathrm{S}_3(1+e_{11}) - \mathrm{S}_1\varpi_1 &= (1+e_{22})(1+2\varepsilon_{11}) - 2\varepsilon_{12}e_{12} = g_{12}.\\
\mathrm{S}_3e_{12} - \mathrm{S}_2\varpi_1 &= (1+e_{22})2\varepsilon_{11} - (1+2\varepsilon_{11})e_{12} = h_{12}.
\end{aligned} \tag{1.79}$$

The result can be confirmed by substituting S_i and S_3 given by Eqs. (1.74) and making use of Eqs. (1.67).

On substituting Eqs. (1.76) into Eqs. (1.78), we find

$$\begin{aligned}\sqrt{\mathbf{A}}\overset{*}{k}_{11} =& \frac{1}{A_1}\left[(1+e_{11})\frac{\partial \mathrm{S}_1}{\partial \alpha_1} + e_{12}\frac{\partial \mathrm{S}_2}{\partial \alpha_1} + \bar{\omega}_1\frac{\partial \mathrm{S}_3}{\partial \alpha_1}\right] \\ &- \frac{f_{12}}{A_1A_2}\frac{\partial A_1}{\partial \alpha_2} + k_{11}g_{12} + k_{12}h_{12}, \\ \sqrt{\mathbf{A}}\overset{*}{k}_{22} =& \frac{1}{A_1}\left[e_{21}\frac{\partial \mathrm{S}_1}{\partial \alpha_1} + (1+e_{22})\frac{\partial \mathrm{S}_2}{\partial \alpha_1} + \bar{\omega}_2\frac{\partial \mathrm{S}_3}{\partial \alpha_1}\right] \\ &+ \frac{f_{12}}{A_1A_2}\frac{\partial A_1}{\partial \alpha_2} - k_{11}h_{12} - k_{12}g_{12}.\end{aligned} \tag{1.80}$$

On differentiating Eqs. (1.80) with respect to α_1 we get

$$\begin{aligned}(1+e_{11})\frac{\partial \mathrm{S}_1}{\partial \alpha_1} + e_{12}\frac{\partial \mathrm{S}_2}{\partial \alpha_1} + \bar{\omega}_1\frac{\partial \mathrm{S}_3}{\partial \alpha_1} &= -\mathrm{S}_1\frac{\partial e_{11}}{\partial \alpha_1} - \mathrm{S}_2\frac{\partial e_{12}}{\partial \alpha_1} - \mathrm{S}_3\frac{\partial \bar{\omega}_1}{\partial \alpha_1}, \\ (1+e_{22})\frac{\partial \mathrm{S}_2}{\partial \alpha_1} + e_{21}\frac{\partial \mathrm{S}_1}{\partial \alpha_1} + \bar{\omega}_2\frac{\partial \mathrm{S}_3}{\partial \alpha_1} &= -\mathrm{S}_2\frac{\partial e_{22}}{\partial \alpha_1} - \mathrm{S}_1\frac{\partial e_{21}}{\partial \alpha_1} - \mathrm{S}_3\frac{\partial \bar{\omega}_2}{\partial \alpha_1}.\end{aligned} \tag{1.81}$$

On substituting the right-hand sides of Eqs. (1.81) into (1.80), the final formulas for the curvatures $\overset{*}{k}_{11}$ and $\overset{*}{k}_{22}$ are found to be

$$\begin{aligned}\sqrt{\mathbf{A}}\overset{*}{k}_{11} &= -\frac{1}{A_1}\left(\mathrm{S}_1\frac{\partial e_{11}}{\partial \alpha_1} + \mathrm{S}_2\frac{\partial e_{12}}{\partial \alpha_1} + \mathrm{S}_3\frac{\partial \bar{\omega}_1}{\partial \alpha_1}\right) - \frac{f_{12}}{A_1A_2}\frac{\partial A_1}{\partial \alpha_2} + k_{11}g_{12} + k_{12}h_{12}, \\ \sqrt{\mathbf{A}}\overset{*}{k}_{22} &= -\frac{1}{A_1}\left(\mathrm{S}_2\frac{\partial e_{22}}{\partial \alpha_1} + \mathrm{S}_1\frac{\partial e_{21}}{\partial \alpha_1} + \mathrm{S}_3\frac{\partial \bar{\omega}_2}{\partial \alpha_1}\right) + \frac{f_{12}}{A_1A_2}\frac{\partial A_1}{\partial \alpha_2} - k_{11}h_{12} - k_{12}g_{12}.\end{aligned} \tag{1.82}$$

Formulas (1.82) are obtained under the first Kirchhoff–Love hypothesis. The essence of the hypothesis is that normals to the undeformed middle surface S remain straight and normal to the deformed middle surface and undergo no extension, i.e. $\overset{*}{\bar{m}}\overset{*}{\bar{r}}_j = \overset{*}{\bar{m}}\overset{*}{\bar{\rho}}_j = g_{j3} = 0$ and $\varepsilon^z_{j3} = 0$. The hypotheses were first formulated by Kirchhoff for thin plates and later applied by Love for thin shells. They are known as the Kirchhoff–Love hypotheses and are fundamental in the theory of thin shells. Additional hypotheses will be introduced in the text as needed.

1.6 Equations of compatibility

For the surface to retain continuity during deformation, the parameters ε_{ik} and $æ_{ik}$ (i, k=1, 2) must satisfy the three differential equations called the equations of continuity of deformations. They can be obtained by subtracting the Gauss–Codazzi equations formulated for the undeformed state from the corresponding equations for the deformed configuration. The Gauss formula for the deformed surface $\overset{*}{S}$ is given by (see Eq. (1.32))

$$\frac{\partial^2 \overset{*}{\chi}}{\partial\alpha_1\,\partial\alpha_2}+\frac{\partial}{\partial\alpha_1}\frac{\dfrac{\partial \overset{*}{A}_2}{\partial\alpha_1}-\cos\overset{*}{\chi}\dfrac{\partial\overset{*}{A}_1}{\partial\alpha_2}}{\overset{*}{A}_1\sin\overset{*}{\chi}}+\frac{\partial}{\partial\alpha_2}\frac{\dfrac{\partial\overset{*}{A}_1}{\partial\alpha_2}-\cos\overset{*}{\chi}\dfrac{\partial\overset{*}{A}_2}{\partial\alpha_1}}{\overset{*}{A}_2\sin\overset{*}{\chi}}=\frac{\overset{*}{b}{}_{12}^2-\overset{*}{b}_{11}\overset{*}{b}_{22}}{\overset{*}{A}_1\overset{*}{A}_2\sin\overset{*}{\chi}}. \quad (1.83)$$

On substituting

$$\begin{aligned}
\overset{*}{A}_1\overset{*}{A}_2\sin\overset{*}{\chi}&=A_1A_2\sqrt{\mathbf{A}},\\
\cos\overset{*}{\chi}&=2\varepsilon_{12}/\sqrt{\mathrm{S}_{11}\mathrm{S}_{22}},\\
\sin\overset{*}{\chi}&=\sqrt{\mathbf{A}}/\sqrt{\mathrm{S}_{11}\mathrm{S}_{22}},\\
\frac{\partial\overset{*}{\chi}}{\partial\alpha_i}&=-\frac{1}{\sin\overset{*}{\chi}}\frac{\partial\cos\overset{*}{\chi}}{\partial\alpha_i}=\frac{1}{\sqrt{\mathbf{A}}}\left[-2\frac{\partial\varepsilon_{12}}{\partial\alpha_i}+\varepsilon_{12}\frac{\partial}{\partial\alpha_i}\ln(\mathrm{S}_{11}\mathrm{S}_{22})\right]
\end{aligned}$$

into Eq. (1.83) we get

$$\begin{aligned}
&\frac{\partial}{\partial\alpha_1}\frac{1}{\sqrt{\mathbf{A}}}\left[-\frac{\partial\varepsilon_{12}}{\partial\alpha_2}+\frac{\varepsilon_{12}}{2}\frac{\partial}{\partial\alpha_2}\ln(\mathrm{S}_{11}\mathrm{S}_{22})+\frac{\overset{*}{A}_2}{A_1A_2}\left(\frac{\partial\overset{*}{A}_2}{\partial\alpha_1}-\cos\overset{*}{\chi}\frac{\partial\overset{*}{A}_1}{\partial\alpha_2}\right)\right]\\
&+\frac{\partial}{\partial\alpha_2}\frac{1}{\sqrt{\mathbf{A}}}\left[-\frac{\partial\varepsilon_{12}}{\partial\alpha_1}+\frac{\varepsilon_{12}}{2}\frac{\partial}{\partial\alpha_1}\ln(\mathrm{S}_{11}\mathrm{S}_{22})+\frac{\overset{*}{A}_2}{A_1A_2}\left(\frac{\partial\overset{*}{A}_1}{\partial\alpha_2}-\cos\overset{*}{\chi}\frac{\partial\overset{*}{A}_2}{\partial\alpha_1}\right)\right]\\
&\qquad=\frac{1}{\sqrt{\mathbf{A}}}\frac{\overset{*}{b}{}_{12}^2-\overset{*}{b}_{11}\overset{*}{b}_{22}}{A_1A_2}. \qquad (1.84)
\end{aligned}$$

Further, using Eqs. (1.65) and (1.66) in (1.84), the first Gauss formula is found to be

$$\begin{aligned}
&\frac{\partial}{\partial\alpha_1}\frac{1}{A_1\sqrt{\mathbf{A}}}\left[\frac{A_1\varepsilon_{12}}{2}\frac{\partial}{\partial\alpha_1}\ln\left(\frac{\mathrm{S}_{22}}{\mathrm{S}_{11}}\right)+\frac{\partial(A_2\varepsilon_{22})}{\partial\alpha_1}-\frac{\partial(A_1\varepsilon_{12})}{\partial\alpha_2}-\varepsilon_{12}\frac{\partial A_1}{\partial\alpha_2}+\varepsilon_{22}\frac{\partial A_2}{\partial\alpha_1}+\frac{\partial A_2}{\partial\alpha_1}\right]\\
&+\frac{\partial}{\partial\alpha_2}\frac{1}{A_2\sqrt{\mathbf{A}}}\left[\frac{A_2\varepsilon_{21}}{2}\frac{\partial}{\partial\alpha_2}\ln\left(\frac{\mathrm{S}_{11}}{\mathrm{S}_{22}}\right)+\frac{\partial(A_1\varepsilon_{11})}{\partial\alpha_2}-\frac{\partial(A_2\varepsilon_{21})}{\partial\alpha_1}-\varepsilon_{21}\frac{\partial A_2}{\partial\alpha_1}\right.\\
&\qquad\left.+\varepsilon_{11}\frac{\partial A_1}{\partial\alpha_2}+\frac{\partial A_1}{\partial\alpha_2}\right]=\frac{1}{\sqrt{\mathbf{A}}}\frac{\overset{*}{b}{}_{12}^2-\overset{*}{b}_{11}\overset{*}{b}_{22}}{A_1A_2}.
\end{aligned}$$

It can be written in more concise form as

$$\frac{\partial}{\partial\alpha_1}\frac{1}{A_1\sqrt{\mathbf{A}}}\left[\frac{A_1\varepsilon_{12}}{2}\frac{\partial}{\partial\alpha_2}\ln\left(\frac{\mathrm{S}_{22}}{\mathrm{S}_{11}}\right)+L_2'(\varepsilon_{ik})+(1+\varepsilon_{11}+\varepsilon_{22})\frac{\partial A_2}{\partial\alpha_1}\right]$$
$$+\frac{\partial}{\partial\alpha_2}\frac{1}{A_2\sqrt{\mathbf{A}}}\left[\frac{A_2\varepsilon_{21}}{2}\frac{\partial}{\partial\alpha_2}\ln\left(\frac{\mathrm{S}_{11}}{\mathrm{S}_{22}}\right)+L_1'(\varepsilon_{ik})+(1+\varepsilon_{11}+\varepsilon_{22})\frac{\partial A_1}{\partial\alpha_2}\right]$$
$$=A_1A_2\frac{1}{\sqrt{\mathbf{A}}}\left(\overset{*}{k}{}_{12}^2-\overset{*}{k}_{11}\overset{*}{k}_{22}\right),\qquad(1.85)$$

where the differential operators $L_j'(\varepsilon_{ik})$ are defined by

$$L_1'(\varepsilon_{ik})=\frac{\partial(A_1\varepsilon_{11})}{\partial\alpha_2}-\frac{\partial(A_2\varepsilon_{12})}{\partial\alpha_1}-\varepsilon_{22}\frac{\partial A_1}{\partial\alpha_1}-\varepsilon_{12}\frac{\partial A_2}{\partial\alpha_1},$$
$$L_2'(\varepsilon_{ik})=\frac{\partial(A_2\varepsilon_{22})}{\partial\alpha_1}-\frac{\partial(A_1\varepsilon_{12})}{\partial\alpha_2}-\varepsilon_{11}\frac{\partial A_2}{\partial\alpha_1}-\varepsilon_{12}\frac{\partial A_1}{\partial\alpha_2},$$

Let $\overset{*}{K}$ be the Gaussian curvature of the deformed surface $\overset{*}{S}$

$$\overset{*}{K}=k_{12}^2-k_{11}k_{22}=\frac{1}{A_1A_2}\left[\frac{\partial}{\partial\alpha_1}\left(\frac{1}{A_1}\frac{\partial A_2}{\partial\alpha_1}\right)+\frac{\partial}{\partial\alpha_2}\left(\frac{1}{A_2}\frac{\partial A_1}{\partial\alpha_2}\right)\right].\qquad(1.86)$$

With the help of Eq. (1.86), (1.85) can be written as

$$\frac{\partial}{\partial\alpha_1}\frac{1}{A_1\sqrt{\mathbf{A}}}\left[\frac{A_1\varepsilon_{12}}{2}\frac{\partial}{\partial\alpha_2}\ln\left(\frac{\mathrm{S}_{22}}{\mathrm{S}_{11}}\right)+L'_2(\varepsilon_{ik})+(1+\varepsilon_{11}+\varepsilon_{22})\frac{\partial A_2}{\partial\alpha_1}\right]$$
$$+\frac{\partial}{\partial\alpha_2}\frac{1}{A_2\sqrt{\mathbf{A}}}\left[\frac{A_2\varepsilon_{21}}{2}\frac{\partial}{\partial\alpha_2}\ln\left(\frac{\mathrm{S}_{11}}{\mathrm{S}_{22}}\right)+L'_1(\varepsilon_{ik})+(1+\varepsilon_{11}+\varepsilon_{22})\frac{\partial A_1}{\partial\alpha_2}\right]$$
$$=A_1A_2\frac{1}{\sqrt{\mathbf{A}}}(K-K_0).\qquad(1.87)$$

The increment to the Gaussian curvature is defined by

$$K-\overset{*}{K}=æ_{12}^2-æ_{11}æ_{22}+2kæ_{12}-k_{11}æ_{11}-k_{22}æ_{11}.\qquad(1.88)$$

To obtain the other two equations, recall the Codazzi formulas

$$\frac{\partial\overset{*}{b}_{11}}{\partial\alpha_2}-\frac{\partial\overset{*}{b}_{12}}{\partial\alpha_1}-\overset{*}{\Gamma}{}_{12}^1\overset{*}{b}_{11}-\left(\overset{*}{\Gamma}{}_{12}^2-\overset{*}{\Gamma}{}_{11}^1\right)\overset{*}{b}_{12}+\overset{*}{\Gamma}{}_{11}^2=0.$$

Here $\overset{*}{\Gamma}{}_{ik}^j$ are the Christoffel symbols of $\overset{*}{S}$. On substituting $\overset{*}{b}_{ij}=-A_iA_j\overset{*}{b}_{ij}=A_iA_j(k_{ij}+æ_{ij})$, we get

$$\frac{\partial[A_1A_2(k_{12}+\ae_{12})]}{\partial\alpha_1}-\frac{\partial[A_1^2(k_{11}+\ae_{11})]}{\partial\alpha_2}+A_1^2\overset{*}{k}_{11}\overset{*}{\Gamma}{}^1_{12}$$
$$+A_1A_2\overset{*}{k}_{12}\left(\overset{*}{\Gamma}{}^2_{12}-\overset{*}{\Gamma}{}^1_{11}\right)-A_2^2\overset{*}{k}_{22}\overset{*}{\Gamma}{}^2_{11}=0. \tag{1.89}$$

On subtracting Eqs. (1.37) from this and dividing the resultant equation by $-A_1$, the compatibility equations are found to be

$$\begin{aligned}
&\frac{\partial(A_1\ae_{11})}{\partial\alpha_2}-\frac{\partial(A_2\ae_{12})}{\partial\alpha_1}-\ae_{12}\frac{\partial A_1}{\partial\alpha_2}+A_2\overset{*}{k}_{12}(A^1_{11}-A^2_{12})\\
&\qquad -A_1\overset{*}{k}_{11}A^1_{12}+\frac{A_2^2}{A_1}\overset{*}{k}_{22}A^2_{11}=0,\\
&\frac{\partial(A_2\ae_{22})}{\partial\alpha_1}-\frac{\partial(A_1\ae_{12})}{\partial\alpha_2}-\ae_{12}\frac{\partial A_2}{\partial\alpha_1}+A_1\overset{*}{k}_{12}(A^2_{22}-A^1_{12})\\
&\qquad -A_2\overset{*}{k}_{22}A^2_{12}+\frac{A_1^1}{A_2}\overset{*}{k}_{11}A^1_{22}=0.
\end{aligned} \tag{1.90}$$

In the above we made use of the following formulas for derivatives:

$$\frac{\partial[A_1A_2(k_{12}+\ae_{12})]}{\partial\alpha_1}=A_1\frac{\partial[A_2(k_{12}+\ae_{12})]}{\partial\alpha_1}+A_2(k_{12}+\ae_{12})\frac{\partial A_1}{\partial\alpha_1},$$
$$\frac{\partial[A_1^2(k_{11}+\ae_{11})]}{\partial\alpha_2}=A_1\frac{\partial[A_1(k_{11}+\ae_{11})]}{\partial\alpha_2}+A_1(k_{11}+\ae_{11})\frac{\partial A_1}{\partial\alpha_2}.$$

Let A^j_{ik} be the components of the Christoffel deviator given by

$$A^j_{ik}=\overset{*}{\Gamma}{}^j_{ik}-\Gamma^j_{ik}. \tag{1.91}$$

Defining differential operators $L'_j(\ae_{ik})$ by

$$\begin{aligned}
L'_1(\ae_{ik})&=\frac{\partial(A_1\ae_{11})}{\partial\alpha_2}-\frac{\partial(A_2\ae_{12})}{\partial\alpha_1}-\ae_{12}\frac{\partial A_2}{\partial\alpha_1}-\ae_{22}\frac{\partial A_1}{\partial\alpha_2},\\
L'_2(\ae_{ik})&=\frac{\partial(A_2\ae_{22})}{\partial\alpha_1}-\frac{\partial(A_1\ae_{12})}{\partial\alpha_2}-\ae_{12}\frac{\partial A_1}{\partial\alpha_2}-\ae_{11}\frac{\partial A_2}{\partial\alpha_1},
\end{aligned} \tag{1.92}$$

Eqs. (1.89) can be written in the form

$$\begin{aligned}
&L'_1(\ae_{ik})+A_2\overset{*}{k}_{12}(A^1_{11}-A^2_{12})-A_1\overset{*}{k}_{11}A^1_{12}+\frac{A_2^2}{A_1}\overset{*}{k}_{22}A^2_{11}=0,\\
&L'_2(\ae_{ik})+A_1\overset{*}{k}_{12}(A^2_{22}-A^1_{12})-A_2\overset{*}{k}_{22}A^2_{12}+\frac{A_1^2}{A_2}\overset{*}{k}_{22}A^1_{22}=0.
\end{aligned} \tag{1.93}$$

On substituting $\overset{*}{A}_i, \overset{*}{\chi}$ and $\overset{*}{a}$ into Eqs. (1.23), for $\overset{*}{\Gamma}{}^j_{ik}$ we obtain

$$\begin{aligned}
\overset{*}{a}\overset{*}{\Gamma}{}^{1}_{11} &= \overset{*}{A}_1\overset{*}{A}{}^2_2\frac{\partial\overset{*}{A}_1}{\partial\alpha_1} - \overset{*}{a}_{12}\frac{\partial\overset{*}{a}_{12}}{\partial\alpha_1} - \overset{*}{A}_1\frac{\partial\overset{*}{A}_1}{\partial\alpha_2},\\
\overset{*}{a}\overset{*}{\Gamma}{}^{1}_{12} &= \overset{*}{A}_1\overset{*}{A}{}^2_2\frac{\partial\overset{*}{A}_1}{\partial\alpha_2} - \overset{*}{A}_2\overset{*}{a}_{12}\frac{\partial\overset{*}{A}_1}{\partial\alpha_1},\\
\overset{*}{a}\overset{*}{\Gamma}{}^{2}_{11} &= \overset{*}{A}{}^2_1\frac{\partial\overset{*}{a}_{12}}{\partial\alpha_1} - \overset{*}{A}_1\overset{*}{a}_{12}\frac{\partial\overset{*}{A}_1}{\partial\alpha_1} - \overset{*}{A}{}^3_1\frac{\partial\overset{*}{A}_1}{\partial\alpha_2},\\
\overset{*}{a} &= a\mathbf{A}, \quad \overset{*}{a}_{12} = \overset{*}{A}_1\overset{*}{A}_2\cos\overset{*}{\chi}.
\end{aligned} \tag{1.94}$$

Making use of Eqs. (1.23) and (1.94) in (1.91), the final formulas for the Christoffel deviators become

$$\begin{aligned}
\mathbf{A}A^1_{11} &= \frac{\varepsilon_{12}}{A_1A_2}\frac{\partial(A_1^2\mathrm{S}_{11})}{\partial\alpha_2} - \frac{4\varepsilon_{12}}{A_2}\frac{\partial(A_2\varepsilon_{12})}{\partial\alpha_1} + \mathrm{S}_{22}\frac{\partial\varepsilon_{11}}{\partial\alpha_1},\\
\mathbf{A}A^2_{11} &= \frac{A_1I}{A_2^2}\frac{\partial A_1}{\partial\alpha_2} + -\frac{\mathrm{S}_{11}}{A_2^2}\left(2A_1\frac{\partial(A_2\varepsilon_{12})}{\partial\alpha_1} - \frac{1}{2}\frac{\partial(A_1^2\mathrm{S}_{11})}{\partial\alpha_2}\right) - \frac{2A_2\varepsilon_{12}}{A_2}\frac{\partial\varepsilon_{11}}{\partial\alpha_1},\\
\mathbf{A}A^2_{12} &= \mathrm{S}_{11}\frac{\partial\varepsilon_{22}}{\partial\alpha_1} - \frac{\varepsilon_{12}}{A_1A_2}\frac{\partial(A_1^2\mathrm{S}_{11})}{\partial\alpha_2} + \frac{4\varepsilon_{12}^2}{A_2}\frac{\partial A_2}{\partial\alpha_1},\\
\mathbf{A}A^2_{22} &= \frac{\varepsilon_{12}}{A_1A_2}\frac{\partial(A_2^1\mathrm{S}_{22})}{\partial\alpha_1} - \frac{4\varepsilon_{12}}{A_1}\frac{\partial(A_1\varepsilon_{12})}{\partial\alpha_2} + \mathrm{S}_{11}\frac{\partial\varepsilon_{22}}{\partial\alpha_2},\\
\mathbf{A}A^1_{22} &= \frac{A_2I}{A_1^1}\frac{\partial A_2}{\partial\alpha_1} - \frac{\mathrm{S}_{22}}{A_1^1}\left(2A_2\frac{\partial(A_1\varepsilon_{12})}{\partial\alpha_2} - \frac{1}{2}\frac{\partial(A_2^1\mathrm{S}_{22})}{\partial\alpha_1}\right) - \frac{2A_1\varepsilon_{12}}{A_1}\frac{\partial\varepsilon_{22}}{\partial\alpha_2},\\
\mathbf{A}A^1_{12} &= \mathrm{S}_{22}\frac{\partial\varepsilon_{11}}{\partial\alpha_2} - \frac{\varepsilon_{12}}{A_1A_2}\frac{\partial(A_2^1\mathrm{S}_{22})}{\partial\alpha_1} + \frac{4\varepsilon_{12}^2}{A_1}\frac{\partial A_1}{\partial\alpha_2}.
\end{aligned} \tag{1.95}$$

Exercises

We have developed a general approach for describing the geometry of the surface using orthogonal curvilinear coordinates. However, in the modelling of the small and large intestine cylindrical coordinates become natural, whereas for working with the pregnant uterus, eyeball and urinary bladder spherical coordinates are more practical. All of the exercises which follow are aimed at obtaining specific relations of interest in cylindrical and spherical coordinates, respectively.

1. Find the natural $\{\bar{r}_1, \bar{r}_2, \bar{m}\}$ and reciprocal $\{\bar{r}^1, \bar{r}^2, \bar{m}\}$ bases on the surface S in cylindrical and spherical coordinates.
2. Find the unit base vectors $\bar{e}_i$ and $\bar{e}^i$ ($i = 1, 2$) and the Lamé parameters of the surface.
3. Compute the covariant $\bar{g}_{ij}$ and contravariant $\bar{g}^{ij}$ ($i, j = 1, 2$) metrics of a surface.

4. Derive a formula for the first fundamental form of the surface in cylindrical and spherical coordinates.
5. Find the component of the second fundamental form b_{ij} of S.
6. Find an expression for $\cos \chi^z$.
7. Compute Γ^i_{jk} $(i, j, k = 1, 2)$.
8. Show that $\Gamma^i_{jk} = \Gamma^i_{kj}$.
9. Verify Eqs. (1.31) and (1.32).
10. Verify Eqs. (1.78).

2

Parameterization of shells of complex geometry

2.1 Fictitious deformations

Most biological shells are of complex geometry. This is a result of the considerable anatomical variability of the organs they model. For example, the human stomach resembles a horn or a hook, the pregnant uterus, a pear, and the urine-filled bladder, a prolate or oblate spheroid. Convenient parameterization of such shells is a difficult analytical task and sometimes is even unfeasible. Almost all numerical methods, on the other hand, are based on discretization of the computational domain and hence may appear to be secluded from the problem. However, computational algorithms are most efficient and accurate only when they operate in regular, canonical domains and become computationally demanding and suffer loss of accuracy in complex domains. Therefore, the question of parameterization of shells of complex geometry becomes of utmost importance.

Let an arbitrary point $M(\alpha_1, \alpha_2) \in S$ and its ε_M domain be in one-to-one correspondence with a point $\overset{*}{M}(\overset{*}{\alpha}_1, \overset{*}{\alpha}_2)$ and its $\overset{*}{\varepsilon}_{\overset{*}{M}}$ domain on the deformed middle surface $\overset{*}{S}$ (Fig. 2.1). The transformation is defined analytically by

$$\overset{*}{\alpha}_i = \overset{*}{\alpha}_i(\alpha_1, \alpha_2), \tag{2.1}$$

where (α_1, α_2) and $(\overset{*}{\alpha}_1, \overset{*}{\alpha}_2)$ are the coordinates on S and $\overset{*}{S}$, respectively.

Assuming that Eq. (2.1) is continuously differentiable and $\det(\partial\overset{*}{\alpha}_i/\partial\alpha_k) \neq 0$ $(i, k = 1, 2)$, the inverse transformation is found to be

$$\alpha_i = \alpha_i(\overset{*}{\alpha}_1, \overset{*}{\alpha}_2). \tag{2.2}$$

The coefficients of the direct and inverse transformations are given by

$$C_i^k = \frac{\partial\overset{*}{\alpha}_k}{\partial\alpha_i}, \qquad C = \det(C_i^k), \tag{2.3}$$

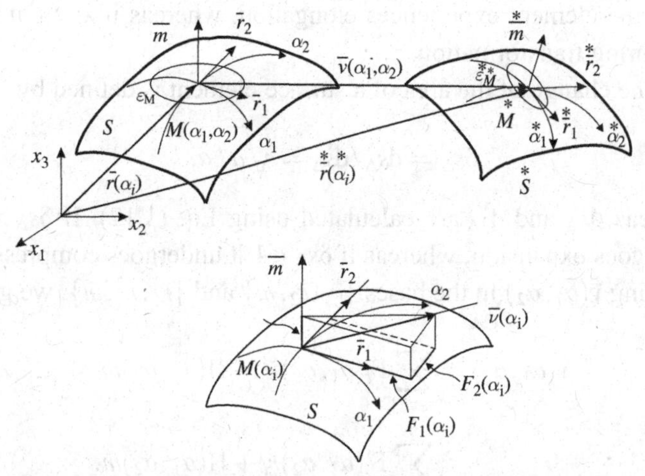

Fig. 2.1 Continuous transformation of an ε_M domain of the surface S. Decomposition of the displacement vector $v(\alpha_1,\alpha_2)$ in the undeformed base $\{\bar{r}_1,\bar{r}_2,\bar{m}\}$.

$$\overset{*}{C}{}_i^k = \frac{\partial \alpha_k}{\partial \overset{*}{\alpha}_i}, \qquad \overset{*}{C} = \det(\overset{*}{C}{}_i^k), \tag{2.4}$$

Assuming that $C \neq 0$ and $\overset{*}{C} \neq 0$, we have

$$\begin{aligned} &\sum_{i=1}^{2} C_i^k \overset{*}{C}{}_k^i = 1, \qquad \sum_{i=1}^{2} C_i^k \overset{*}{C}{}_j^k = 0 \quad (i \neq j), \\ &C\overset{*}{C} = 1, \qquad \overset{*}{C}{}_i^k = [C_k^i]/C. \end{aligned} \tag{2.5}$$

Here $[C_k^i]$ is the cofactor to the element C_k^i of the matrix $(\overset{*}{C}{}_i^k)$.

It follows from the above considerations that the surface $\overset{*}{S}$ can also be referred to the coordinates α_i. Then, the vector equations of S and $\overset{*}{S}$ are given by

$$\begin{aligned} &\bar{r} = \bar{r}(\alpha_1,\alpha_2), \\ &\overset{*}{\bar{r}} = \overset{*}{\bar{r}}(\overset{*}{\alpha}_1,\overset{*}{\alpha}_2) = \overset{*}{\bar{r}}[f_1(\alpha_1,\alpha_2), f_2(\alpha_1,\alpha_2)] = \bar{r}(\alpha_1,\alpha_2) + \bar{v}(\alpha_1,\alpha_2). \end{aligned} \tag{2.6}$$

Here use is made of Eq. (1.58), where $\bar{v}(\alpha_1,\alpha_2)$ is the displacement vector.

The lengths of a linear element before and after transformation, $\mathrm{d}s$ and $\mathrm{d}\overset{*}{s}$, are given by Eqs. (1.11) and (1.63). Let the stretch ratio be defined by

$$\lambda_i = \frac{\mathrm{d}\overset{*}{s}}{\mathrm{d}s} = \sqrt{\frac{\overset{*}{a}_{ik}}{a_{ik}}} \quad (i,k = 1,2). \tag{2.7}$$

If $\lambda_i > 1$ then the element experiences elongation, whereas if $\lambda_i < 1$ it experiences contraction during transformation.

Similarly, the change of the area of a surface element is defined by

$$\delta s_\Delta = ds_\Delta / d\overset{*}{s}_\Delta = \sqrt{a/\overset{*}{a}}, \tag{2.8}$$

where the areas ds_Δ and $d\overset{*}{s}_\Delta$ are calculated using Eq. (1.12). If $\delta s_\Delta > 1$ then the surface undergoes expansion, whereas if $\delta s_\Delta < 1$ it undergoes compression.

On expanding $\bar{v}(\alpha_1, \alpha_2)$ in the bases $\{\bar{r}_1, \bar{r}_2, \bar{m}\}$ and $\{\bar{r}^1, \bar{r}^2, \bar{m}\}$, we get (Fig. 2.1)

$$\begin{aligned} \bar{v}(\alpha_1, \alpha_2) &= \sum_{i=1}^{2} \mathrm{F}_i(\alpha_1, \alpha_2)\bar{r}^i + \mathrm{H}(\alpha_1, \alpha_2)\bar{m} \\ &= \sum_{i=1}^{2} \mathrm{F}^i(\alpha_1, \alpha_2)\bar{r}_i + \mathrm{H}(\alpha_1, \alpha_2)\bar{m}. \end{aligned} \tag{2.9}$$

On substituting Eq. (2.9) into (2.6) we obtain

$$\begin{aligned} \overset{*}{\bar{r}} &= \bar{r}(\alpha_1, \alpha_2) + \sum_{i=1}^{2} \mathrm{F}_i(\alpha_1, \alpha_2)\bar{r}^i + \mathrm{H}(\alpha_1, \alpha_2)\bar{m} \\ &= \bar{r}(\alpha_1, \alpha_2) + \sum_{i=1}^{2} \mathrm{F}^i(\alpha_1, \alpha_2)\bar{r}_i + \mathrm{H}(\alpha_1, \alpha_2)\bar{m}. \end{aligned} \tag{2.10}$$

Evidently, the transformation (2.10) can be achieved by deformation of the middle surface S. For the purpose of parameterization of the surface $\overset{*}{S}$ the transformation could be perfomed fictitiously. Therefore, such a transformation is called fictitious deformation. The problem is to construct the three functions $\mathrm{F}_i(\alpha_i)$ and $\mathrm{H}(\alpha_i) (i = 1, 2)$.

On differentiating Eq. (2.10) with respect to α_i and using Eqs. (1.22) and (1.24), vectors $\overset{*}{\bar{r}}_i$ tangent to coordinate lines $\overset{*}{\alpha}_i$ on $\overset{*}{S}$ are found to be

$$\overset{*}{\bar{r}}_i = \sum_{k=1}^{2} (\delta_{ik} + \underline{e}_i^k)\bar{r}_k + \omega_i\bar{m} = \sum_{k=1}^{2} (a_{ik} + \underline{e}_{ik})\bar{r}^k + \underline{\omega}_i\bar{m}. \tag{2.11}$$

Here

$$\begin{aligned} \underline{e}_i^k &= \frac{\partial \mathrm{F}^k}{\partial \alpha_i} + \sum_{s=1}^{2} \Gamma_{is}^k \mathrm{F}^s(\alpha_i) - \mathrm{H}(\alpha_i)b_i^k := \nabla_i \mathrm{F}^k(\alpha_i) - \mathrm{H}(\alpha_i)b_i^k, \\ \underline{e}_{ki} &= \frac{\partial \mathrm{F}_k}{\partial \alpha_i} - \sum_{s=1}^{2} \Gamma_{ik}^s \mathrm{F}_s(\alpha_i) - \mathrm{H}(\alpha_i)b_{ik} := \nabla_i \mathrm{F}_k(\alpha_i) - \mathrm{H}(\alpha_i)b_{ik}, \\ \underline{\omega}_i &= \frac{\partial \mathrm{H}(\alpha_i)}{\partial \alpha_i} + b_i^k \mathrm{F}_k(\alpha_i), \end{aligned} \tag{2.12}$$

where $\nabla_i(\ldots)$ is the covariant derivative in metric a_{jk}, δ_{ik} is the Kronecker delta and b_{ik} and b_{jk} are the components of the second fundamental form of S. By substituting Eq. (2.11) into (1.4) the components of the metric tensor $\overset{*}{\mathbf{A}}$ on $\overset{*}{S}$ are found to be

$$\overset{*}{a}_{ik} = a_{ik} + 2\underline{\varepsilon}_{ik}, \tag{2.13}$$

where $\underline{\varepsilon}_{ik}$ are the components of the tensor of tangent fictitious deformations given by (compare this with Eqs. (1.67) in Chapter 1)

$$2\underline{\varepsilon}_{ik} = \overset{*}{\bar{r}}_i \overset{*}{\bar{r}}_k - \bar{r}_i \bar{r}_k = \underline{e}_{ik} + \underline{e}_{ki} + a_{js}\underline{e}_{ij}\underline{e}_{ks} + \underline{\varpi}_i \underline{\varpi}_k. \tag{2.14}$$

In just the same way as we introduced bending deformations $æ_{ik}$, we introduce bending fictitious deformations of the surface S:

$$\underline{æ}_{ik} = b_{ik} - \overset{*}{b}_{ik}. \tag{2.15}$$

The components $\underline{\varepsilon}_{ik}$ and $\underline{æ}_{ik}$ are interdependent and satisfy conditions of continuity similar to those given by Eqs. (1.85) and (1.90). Expressing $\underline{\varepsilon}_{ik}$ and $\underline{æ}_{ik}$ in terms of $\mathrm{F}_i(\alpha_i)$ and $\mathrm{H}(\alpha_i)$, we find that the continuity conditions require the existence of continuous derivatives of the functions up to order three at all regular points of the undeformed surface S.

2.2 Parameterization of the equidistant surface

Let a point $M(\alpha_1, \alpha_2) \in S$ be in one-to-one correspondence with point $M_z(\alpha_1, \alpha_2)$ on an equidistant surface S_z ($S_z \| S$). The position vector $\bar{\rho}$ of M_z is given by

$$\bar{\rho}(\alpha_1, \alpha_2) = \bar{r}(\alpha_1, \alpha_2) + \mathrm{H}_z \bar{m}, \tag{2.16}$$

where $\mathrm{H}_z =$ constant. Comparison of Eqs. (2.16) and (2.10) shows that the surface S_z can be obtained from S by fictitious deformation, i.e. by continuous displacement of all points on S by H_z in the direction of the normal vector $\bar{m}$ ($\bar{m} \perp S$) (Fig. 1.5). Since $\mathrm{F}_i(\alpha_1, \alpha_2) = 0$ and $\partial \mathrm{H}_z / \partial \alpha_i = 0$ in (2.16), $\underline{e}_i^k$, $\underline{e}_{ki}$ and $\underline{\omega}_i$ take the forms

$$\underline{e}_i^k = -\mathrm{H}_z b_i^k, \qquad \underline{e}_{ki} = -\mathrm{H}_z b_{ik}, \qquad \underline{\omega}_i = 0. \tag{2.17}$$

Basis vectors $\bar{\rho}_i$ on S_z are defined by

$$\bar{\rho}_i := \frac{\partial \bar{\rho}}{\partial \alpha_i} = \sum_{k=1}^{2} (\delta_i^k - \mathrm{H}_z b_i^k) \bar{r}_k = \sum_{k=1}^{2} \theta_i^k \bar{r}_k. \tag{2.18}$$

Hence, the components a_{ik}^z of the metric tensor on S_z are

$$a^z_{ik} = \bar{\rho}_i \bar{\rho}_k = \sum_{s=1}^{2}\sum_{n=1}^{2} \theta^s_i \theta^n_k a_{sn} = a_{ik} - 2\mathrm{H}_z b_{ik} + (\mathrm{H}_z)^2 \xi_{ik}, \tag{2.19}$$

where ξ_{ik} are the components of the third metric tensor of S and are given by

$$\xi_{ik} = \sum_{n=1}^{2} b^n_i b_{nk} = 2\Gamma b_{ik} - K a_{ik}. \tag{2.20}$$

Here Γ and K are the mean and the Gaussian curvature of S, respectively.

Since $\bar{m} = \bar{m}_z\ (\bar{m}_z \perp S_z)$,

$$\begin{aligned} b^z_{ik} &= -\bar{\rho}_i \bar{m}_z = -\bar{\rho}_i \bar{m} = -\bar{r}_n \bar{m}_k \left(\delta^n_i - \mathrm{H}_z b^n_i\right) \\ &= b_{nk}\left(\delta^n_i - \mathrm{H}_z b^n_i\right) = b_{ik} - \mathrm{H}_z \xi_{ik}. \end{aligned}$$

From the above, with the help of Eq. (2.16), we find

$$b^z_{ik} = \sum_{n=1}^{2} b_{nk} \xi^n_i = b_{ik}(1 - 2\Gamma\mathrm{H}_z) + a_{ik} K \mathrm{H}_z. \tag{2.21}$$

Making use of Eq. (2.18) in the vector product $(\bar{\rho}_1 \times \bar{\rho}_2)$, we obtain

$$\begin{aligned} \bar{\rho}_1 \times \bar{\rho}_2 &= \sum_{k=1}^{2}\sum_{j=1}^{2} (\bar{r}_k \times \bar{r}_j)\theta^k_1 \theta^j_2 = (\bar{r}_1 \times \bar{r}_2)\left(\theta^1_1\theta^2_2 - \theta^2_1\theta^1_2\right) \\ &= (\bar{r}_1 \times \bar{r}_2)\left(1 - 2\Gamma\mathrm{H}_z + K(\mathrm{H}_z)^2\right). \end{aligned} \tag{2.22}$$

Recalling that $(\bar{\rho}_1 \times \bar{\rho}_2)^2 = a^z$ and $(\bar{r}_1 \times \bar{r}_2)^2 = a$, we get, finally,

$$a^z = a\left(1 - 2\Gamma\mathrm{H}_z + K(\mathrm{H}_z)^2\right)^2. \tag{2.23}$$

Thus, the change of a given surface element on S during transformation is found to be

$$\delta s_\Delta = \sqrt{a/a^z} = \left(1 - 2\Gamma\mathrm{H}_z + K(\mathrm{H}_z)^2\right) = \left(\theta^1_1\theta^2_2 - \theta^2_1\theta^1_2\right). \tag{2.24}$$

The Gaussian curvature of S_z is determined by

$$a^z K^z := b^z_{11} b^z_{22} - b^z_{12} b^z_{21} = \left(\theta^1_1\theta^2_2 - \theta^2_1\theta^1_2\right)\left(b^1_1 b^2_2 - b^1_2 b^2_1\right).$$

Using Eq. (2.24) in the above, for the Gaussian curvature of S_z we find

$$a^z K^z = a d^2_s K. \tag{2.25}$$

The coefficients of the first and second fundamental forms of S_z satisfy the Gauss–Codazzi equations

$$a^z K^z = \frac{\partial \Gamma^z{}_{h,ij}}{\partial \alpha_k} - \frac{\partial \Gamma^z{}_{h,ik}}{\partial \alpha_j} + \sum_{s=1}^{2} \left(\Gamma^{(z)s}{}_{ik} \Gamma^z{}_{s,hj} - \Gamma^{(z)s}{}_{ij} \Gamma^z{}_{s,hk} \right), \tag{2.26}$$

$$\frac{\partial b^z_{ij}}{\partial \alpha_k} - \frac{\partial b^z_{ik}}{\partial \alpha_j} + \sum_{s=1}^{2} \left(\Gamma^{(z)s}{}_{ij} b^z_{sk} - \Gamma^{(z)s}{}_{ik} b^z_{sj} \right) = 0. \tag{2.27}$$

Here $\Gamma^z{}_{h,ij}$ and $\Gamma^{(z)s}{}_{ij}$ are the Christoffel symbols of the first and second kind, respectively, given by

$$\Gamma^z{}_{h,ij} = \frac{1}{2} \left(\frac{\partial a^z_{jh}}{\partial \alpha_i} + \frac{\partial a^z_{ih}}{\partial \alpha_j} - \frac{\partial a^z_{ij}}{\partial \alpha_h} \right),$$
$$\Gamma^{(z)s}{}_{ij} = a^{sh} \Gamma^z{}_{h,ij}. \tag{2.28}$$

2.3 A single-function variant of the method of fictitious deformation

Complex biological shells may resemble classical canonical surfaces. For example, with the large intestine the goffered cylinder bears a resemblance to a circular cylinder, with the urinary bladder the prolate spheroid could be viewed as a sphere and with the antropyloric region of the stomach and the ureteropelvic junction of the kidney the distorted funnel is similar to a cone, etc. Let S be a reference canonical surface for $\overset{*}{S}$. Assume that the vector $\bar{m}$ $(\bar{m} \perp S)$ drawn at any point M on S intersects the surface $\overset{*}{S}$ only once (Fig. 2.2). Setting $\mathrm{F}_i(\alpha_1, \alpha_2) = 0$ from Eq. (2.9) for the displacement vector, we have

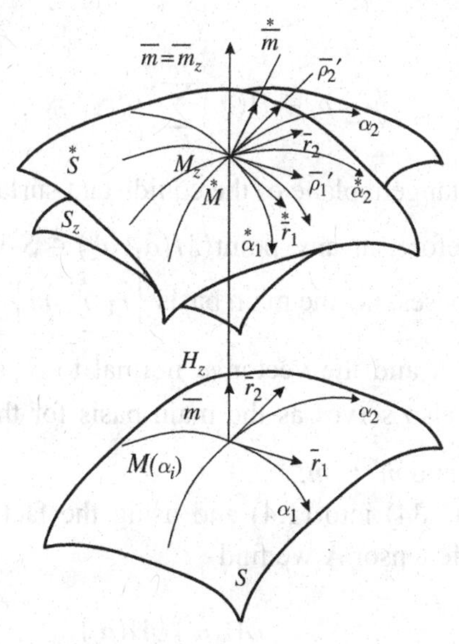

Fig. 2.2 Fictitious deformation of the surface.

$$\bar{v}(\alpha_1, \alpha_2) = \mathrm{H}(\alpha_1, \alpha_2)\bar{m}. \tag{2.29}$$

Here $\mathrm{H}(\alpha_i)$ is the distance measured along $\bar{m}$ from the surface S to $\overset{*}{S}$. Evidently, the vector equation of $\overset{*}{S}$ can be written as

$$\overset{*}{\bar{r}} = \bar{r}(\alpha_1, \alpha_2) + \mathrm{H}(\alpha_1, \alpha_2)\bar{m}. \tag{2.30}$$

Let

$$\mathrm{H}(\alpha_1, \alpha_2) = \mathrm{H}_z + \overset{*}{\mathrm{H}}(\alpha_1, \alpha_2). \tag{2.31}$$

Then, on substituting Eq. (2.31) into (2.30), we get

$$\overset{*}{\bar{r}} = \bar{r}(\alpha_1, \alpha_2) + \left(\mathrm{H}_z + \overset{*}{\mathrm{H}}(\alpha_1, \alpha_2)\right)\bar{m}. \tag{2.32}$$

Equation (2.32) describes the superposition of two consecutive transformations: first, the transformation of the canonical surface S onto an equidistant surface S_z; then the second transformation, of S_z to $\overset{*}{S}$.

On differentiating Eq. (2.30) with respect to α_i and using $\bar{m}_i = -b_i^k \bar{r}_k$, the basis vectors $\overset{*}{\bar{r}}_i$ for the transformation onto $\overset{*}{S}$ are found to be

$$\overset{*}{\bar{r}}_i = \sum_{k=1}^{2} \bar{r}_k (\delta_i^k - \mathrm{H}(\alpha_i) b_i^k) + \frac{\partial \mathrm{H}(\alpha_i)}{\partial \alpha_i}\bar{m}. \tag{2.33}$$

Making use of Eqs. (2.18) and (2.31) in (2.33), we get

$$\overset{*}{\bar{r}}_i = \bar{\rho}'_i + \frac{\partial \overset{*}{\mathrm{H}}(\alpha_i)}{\partial \alpha_i}\bar{m}, \tag{2.34}$$

where

$$\bar{\rho}'_i = \bar{\rho}_i + \overset{*}{\mathrm{H}}(\alpha_i) \sum_{k=1}^{2} b_i^k \bar{r}_k.$$

The vectors $\bar{\rho}'_i$ lie in the tangent plane of the equidistant surface S_z and are collinear to vectors $\bar{r}_i \in S$. Therefore, at any point $\overset{*}{M}(\overset{*}{\alpha}_1, \overset{*}{\alpha}_2) \in \overset{*}{S}$ we can introduce two interrelated orthogonal bases, i.e. the main basis $\left\{\overset{*}{\bar{r}}_1, \overset{*}{\bar{r}}_2, \bar{m}\right\}$ with vectors $\overset{*}{\bar{r}}_i$ tangent to the coordinate lines $\overset{*}{\alpha}_i$ and the vector $\bar{m}$ normal to $\overset{*}{S}$, and an auxiliary basis $\{\bar{\rho}'_1, \bar{\rho}'_2, \bar{m}'\}$. The latter also serves as the main basis for the surface S_z that runs through $\overset{*}{M}$ parallel to S and $\bar{m}' = \bar{m}$.

On substituting Eq. (2.34) into (1.4) and using the fact that $\bar{\rho}'_i\bar{m} = 0$, for the components of the metric tensor $\overset{*}{\mathbf{A}}$ we find

$$\overset{*}{a}_{ik} = a^z_{ik} + \frac{\partial \overset{*}{\mathrm{H}}(\alpha_i)}{\partial \alpha_i}\frac{\partial \overset{*}{\mathrm{H}}(\alpha_i)}{\partial \alpha_k}. \tag{2.35}$$

Here

$$
\begin{aligned}
a^z_{ik} &= \sum_{s=1}^{2}\sum_{n=1}^{2}\left(\delta^s_i - \mathrm{H}(\alpha_i)b^s_i\right)\left(\delta^n_k - \mathrm{H}(\alpha_i)b^n_k\right)a_{sn} \\
&= a_{ik} - 2\mathrm{H}(\alpha_i)b_{ik} + \mathrm{H}(\alpha_i)^2\xi_{ik} \\
&= a_{ik}\left(1 + \overset{*}{\mathrm{H}}(\alpha_i)^2 K\right) + 2\left(\Gamma\,\overset{*}{\mathrm{H}}(\alpha_i) - 1\right)\overset{*}{\mathrm{H}}(\alpha_i)b^i_k,
\end{aligned}
\tag{2.36}
$$

where use is made of Eqs. (2.19) and (2.20). On use of Eqs. (2.35) and (2.36) in (2.13), for the tensor of tangent fictitious deformations $2\underline{\varepsilon}_{ik}$ we obtain

$$
\begin{aligned}
2\underline{\varepsilon}_{ik} &= 2\underline{\varepsilon}^z_{ik} + \frac{\partial\overset{*}{\mathrm{H}}(\alpha_i)}{\partial\alpha_i}\frac{\partial\overset{*}{\mathrm{H}}(\alpha_i)}{\partial\alpha_k}, \\
2\underline{\varepsilon}^z_{ik} &= 2\mathrm{H}(\alpha_i)b_{ik} + \mathrm{H}(\alpha_i)^2\xi_{ik}.
\end{aligned}
\tag{2.37}
$$

From Eq. (2.25) the determinant $\overset{*}{a}$ is found to be

$$
\overset{*}{a} = a^z + a^z_{11}\left(\frac{\partial\overset{*}{\mathrm{H}}(\alpha_i)}{\partial\alpha_2}\right)^2 + a^z_{22}\left(\frac{\partial\overset{*}{\mathrm{H}}(\alpha_i)}{\partial\alpha_1}\right)^2 - 2a^z_{12}\frac{\partial\overset{*}{\mathrm{H}}(\alpha_i)}{\partial\alpha_1}\frac{\partial\overset{*}{\mathrm{H}}(\alpha_i)}{\partial\alpha_2}, \tag{2.38}
$$

where the determinant a^z is given by Eq. (2.23).

On substituting Eq. (2.34) into the formula $\overset{*}{\bar{m}} = (1/2)\overset{*}{c}{}^{ik}\left(\overset{*}{\bar{r}}_i, \overset{*}{\bar{r}}_k\right)$ and using Eqs. (1.8)–(1.10) for the normal vector $\overset{*}{\bar{m}}\left(\overset{*}{\bar{m}} \perp \overset{*}{S}\right)$, we have

$$
\begin{aligned}
\overset{*}{\bar{m}} &= \frac{1}{2}\sqrt{\frac{a_z}{a}}c^{ik}_z\left[(\bar{\rho}'_i, \bar{\rho}'_k) + \frac{\partial\overset{*}{\mathrm{H}}(\alpha_i)}{\partial\alpha_k}(\bar{\rho}'_i, \bar{m}) + \frac{\partial\overset{*}{\mathrm{H}}(\alpha_i)}{\partial\alpha_i}(\bar{m}, \bar{\rho}'_k)\right] \\
&= \frac{\mathrm{H}_0}{2}\sum_{i=1}^{2}\sum_{k=1}^{2}c^{ik}_z\left(\bar{m}c^z_{ik} + \sum_{j=1}^{2}\left(c^z_{ij}\bar{\rho}'^j\frac{\partial\overset{*}{\mathrm{H}}(\alpha_i)}{\partial\alpha_k} - c^z_{jk}\bar{\rho}'^j\frac{\partial\overset{*}{\mathrm{H}}(\alpha_i)}{\partial\alpha_i}\right)\right) \\
&= \mathrm{H}_0\left(\bar{m} - \sum_{i=1}^{2}\bar{\rho}'^i\frac{\partial\overset{*}{\mathrm{H}}(\alpha_i)}{\partial\alpha_i}\right) = \mathrm{H}_0\left(\bar{m} - \sum_{k=1}^{2}\bar{\rho}'_i a^{ik}_z\frac{\partial\overset{*}{\mathrm{H}}(\alpha_i)}{\partial\alpha_k}\right),
\end{aligned}
\tag{2.39}
$$

where

$$
a_z = a\mathrm{H}_0^2, \qquad \mathrm{H}_0 = \left(1 + \sum_{i=1}^{2}\sum_{k=1}^{2}a^{ik}_z\frac{\partial\overset{*}{\mathrm{H}}(\alpha_i)}{\partial\alpha_i}\frac{\partial\overset{*}{\mathrm{H}}(\alpha_i)}{\partial\alpha_k}\right).
$$

Further, on substituting $\bar{\rho}'_i = \bar{\rho}_i + \overset{*}{\mathrm{H}}(\alpha_i)b^k_i\bar{r}_k$ into (2.39) for $\overset{*}{\bar{m}}$ in terms of basis vectors $\bar{r}_k \in S$ we obtain

$$\stackrel{*}{\bar{m}} = \mathrm{H}_0\left(\bar{m} - \bar{r}_k a_z^{ik}\frac{\partial \stackrel{*}{\mathrm{H}}(\alpha_i)}{\partial \alpha_k}(\delta_i^k - \mathrm{H}(\alpha_i) b_i^k)\right). \tag{2.40}$$

By differentiating Eq. (2.39) with respect to α_i we get

$$\stackrel{*}{\bar{m}}_i = \mathrm{H}_0\left(\bar{m} - \sum_{k=1}^{2}\bar{\rho}'^k\frac{\partial \stackrel{*}{\mathrm{H}}(\alpha_i)}{\partial \alpha_k}\right)\frac{\partial \mathrm{H}_0}{\partial \alpha_i} + \mathrm{H}_0\frac{\partial}{\partial \alpha_i}\left(\bar{m} - \sum_{k=1}^{2}\bar{\rho}'^k\frac{\partial \stackrel{*}{\mathrm{H}}(\alpha_i)}{\partial \alpha_k}\right). \tag{2.41}$$

Using Eq. (2.39), the first term in the above can be written in the form

$$\left(\bar{m} - \bar{r}^k\frac{\partial \stackrel{*}{\mathrm{H}}(\alpha_i)}{\partial \alpha_k}\right) = \frac{\stackrel{*}{\bar{m}}_i}{\mathrm{H}_0}\frac{\partial \mathrm{H}_0}{\partial \alpha_i}. \tag{2.42}$$

In the second term we introduce the covariant derivative with respect to a_{ik}^z as (see Eq. (2.12))

$$\begin{aligned}\frac{\partial}{\partial \alpha_i}\left(\bar{m} - \sum_{k=1}^{2}\bar{\rho}'^k\frac{\partial \stackrel{*}{\mathrm{H}}(\alpha_i)}{\partial \alpha_k}\right) &:= \nabla_i^z\left(\bar{m} - \sum_{k=1}^{2}\bar{\rho}'^k\frac{\partial \stackrel{*}{\mathrm{H}}(\alpha_i)}{\partial \alpha_k}\right)\\ &= \bar{m} - \sum_{k=1}^{2}\frac{\partial \stackrel{*}{\mathrm{H}}(\alpha_i)}{\partial \alpha_k}\nabla_i^z\bar{\rho}'^k - \sum_{k=1}^{2}\bar{\rho}'^k\nabla_i^z\left(\frac{\partial \stackrel{*}{\mathrm{H}}(\alpha_i)}{\partial \alpha_k}\right).\end{aligned} \tag{2.43}$$

Derivatives of vectors $\bar{\rho}_i'$ and $\bar{m}$ can be found from Eqs. (1.18) and (1.22) by substituting $\bar{\rho}_i'$ for $\bar{r}_i$. Since the components of the second fundamental form of S_z in terms of b_{ik} are given by

$$\begin{aligned} b_{ik}^z &= \sum_{n=1}^{2} b_{nk}\theta_i^n = b_{ik}(1 - 2\Gamma\mathrm{H}(\alpha_i)) + a_{ik}K\mathrm{H}(\alpha_i),\\ b_i^{(z)k} &= \sum_{s=1}^{2} a_z^{ks} b_{is}^z. \end{aligned} \tag{2.44}$$

Thus, Eq. (2.43) can be written in the form

$$\frac{\partial}{\partial \alpha_i}\left(\bar{m} - \bar{\rho}'^k\frac{\partial \stackrel{*}{\mathrm{H}}(\alpha_i)}{\partial \alpha_k}\right) = \bar{m} - \sum_{k=1}^{2}\frac{\partial \stackrel{*}{\mathrm{H}}(\alpha_i)}{\partial \alpha_k}b_i^{(z)k}\bar{m} - \sum_{k=1}^{2}\bar{\rho}'^k\,\nabla_i^z\frac{\partial \stackrel{*}{\mathrm{H}}(\alpha_i)}{\partial \alpha_k}. \tag{2.45}$$

On substituting Eqs. (2.42) and (2.45) into (2.41) for $\stackrel{*}{\bar{m}}_i$ we get, finally,

$$\stackrel{*}{\bar{m}}_i = \frac{\stackrel{*}{\bar{m}}_i}{\mathrm{H}_0}\frac{\partial \mathrm{H}_0}{\partial \alpha_i} + \mathrm{H}_0\left(\bar{m} - \frac{\partial \stackrel{*}{\mathrm{H}}(\alpha_i)}{\partial \alpha_k}b_i^{(z)k}\bar{m} - \bar{\rho}'^k\,\nabla_i^z\frac{\partial \stackrel{*}{\mathrm{H}}(\alpha_i)}{\partial \alpha_k}\right). \tag{2.46}$$

The coefficients b_{ik} are found from Eq. (1.18) by substituting Eqs. (2.34) and (2.46):

$$\begin{aligned} b_{ik} &= \mathrm{H}_0 \sum_{j=1}^{2} \left(b_{ij}^{(z)} \left(\delta^j_k + \underset{z}{a}^{ji} \frac{\partial \overset{*}{\mathrm{H}}(\alpha_i)}{\partial \alpha_i} \frac{\partial \overset{*}{\mathrm{H}}(\alpha_i)}{\partial \alpha_k} \right) + \nabla^z_i \left(\frac{\partial \overset{*}{\mathrm{H}}(\alpha_i)}{\partial \alpha_k} \right) \right) \\ &= \mathrm{H}_0 \sum_{j=1}^{2} \left(b_i^{(z)j} \left(\underset{z}{a}_{kj} - \frac{\partial \overset{*}{\mathrm{H}}(\alpha_i)}{\partial \alpha_k} \frac{\partial \overset{*}{\mathrm{H}}(\alpha_i)}{\partial \alpha_j} \right) + \nabla^z_i \left(\frac{\partial \overset{*}{\mathrm{H}}(\alpha_i)}{\partial \alpha_k} \right) \right) \\ &= \mathrm{H}_0 \sum_{j=1}^{2} \left(b_i^{(z)j} \overset{*}{a}_{kj} + \nabla^z_i \left(\frac{\partial \overset{*}{\mathrm{H}}(\alpha_i)}{\partial \alpha_k} \right) \right). \end{aligned} \tag{2.47}$$

The Christoffel symbols $\overset{*}{\Gamma}_{ij,k}$ of the first kind are calculated from $\overset{*}{\Gamma}_{ij,k} = \overset{*}{\bar{r}}_{ij}\overset{*}{\bar{r}}_k$. Thus, on differentiating Eq. (2.33) with respect to α_j and multiplying the resultant equation by $\overset{*}{\bar{r}}_k$, we find

$$\overset{*}{\Gamma}_{ij,k} = \Gamma^{(z)}_{ij,k} + \frac{\partial^2 \overset{*}{\mathrm{H}}(\alpha_i)}{\partial \alpha_i\, \partial \alpha_j} \frac{\partial \overset{*}{\mathrm{H}}(\alpha_i)}{\partial \alpha_k}, \tag{2.48}$$

where $\Gamma^{(z)}_{ij,k}$ are calculated using (2.28). The Christoffel symbols of the second kind $\overset{*}{\Gamma}{}^k_{ij}$ are found to be

$$\begin{aligned} \overset{*}{\Gamma}{}^k_{ij} &= \overset{*}{a}{}^{ak} \overset{*}{\Gamma}_{ij,a} \\ &= \left(\Gamma^{(z)}_{ij,a} + \frac{\partial^2 \overset{*}{\mathrm{H}}(\alpha_i)}{\partial \alpha_i\, \partial \alpha_j} \right) \left(\underset{z}{a}^{ak} - \mathrm{H}_0^2 \sum_{m=1}^{2} \sum_{j=1}^{2} \underset{z}{a}^{am} \underset{z}{a}^{jk} \frac{\partial \overset{*}{\mathrm{H}}(\alpha_i)}{\partial \alpha_m} \frac{\partial \overset{*}{\mathrm{H}}(\alpha_i)}{\partial \alpha_j} \right), \end{aligned} \tag{2.49}$$

where

$$\overset{*}{a}{}^{ak} = \underset{z}{a}^{ak} - \mathrm{H}_0^2 \sum_{j=1}^{2} \sum_{m=1}^{2} \underset{z}{a}^{am} \underset{z}{a}^{jk} \frac{\partial \overset{*}{\mathrm{H}}(\alpha_i)}{\partial \alpha_m} \frac{\partial \overset{*}{\mathrm{H}}(\alpha_i)}{\partial \alpha_j}.$$

2.4 Parameterization of a complex surface in preferred coordinates

The problem of parameterization of a complex surface simplifies significantly if the surface S is referred to coordinate lines (α_1, α_2) that are the lines of curvature. On differentiating Eq. (2.30) with respect to α_i with the help of Eq. (1.30), the base vectors $\overset{*}{\bar{r}}_i \in \overset{*}{S}$ are found to be

$$\overset{*}{\bar{r}}_i = A^z_i (\bar{e}_i + y_i \bar{m}). \tag{2.50}$$

Here $A_i^z = A_i\theta_i = A_i[1 + \mathrm{H}(\alpha_i)/R_i]$, where A_i are the Lamé parameters, R_i are the principal radii of curvature and $\bar{e}_i$ are the unit vectors on S. The coefficients y_i are given by

$$y_i = \frac{1}{A_i^z}\frac{\partial \mathrm{H}(\alpha_j)}{\partial \alpha_i} \equiv \frac{1}{A_i[1 + \mathrm{H}(\alpha_j)/R_i]}\frac{\partial \mathrm{H}(\alpha_j)}{\partial \alpha_i}. \tag{2.51}$$

The first term in Eq. (2.50) defines the basis vectors $\bar{\rho}_i'$ on the equidistant surface S_z. Hence, at any point $\overset{*}{M} \in \overset{*}{S}$ the unit vector $\bar{e}_i^z = \bar{\rho}_i/A_i^z$ ($\bar{e}_i^z \in S_z$) equals the unit vector $\bar{e}_i$ defined on the canonical surface S: $\bar{e}_i^z = \bar{e}_i$.

Analogously to Eq. (2.40), for decomposition of the normal vector $\overset{*}{\bar{m}}$ we have

$$\overset{*}{\bar{m}} = \varsigma(\bar{m} - \varsigma_i\bar{e}_i). \tag{2.52}$$

On use of Eqs. (2.50) and (2.52), taking the scalar product of Eq. (2.52) with $\overset{*}{\bar{m}}$ and $\overset{*}{\bar{r}}_i$ yields

$$\varsigma_i = y_i, \qquad \varsigma = 1/\sqrt{1 + y_1^2 + y_2^2}.$$

Hence,

$$\overset{*}{\bar{m}} = \varsigma(\bar{m} - y_i\bar{e}_i) = \frac{1}{\sqrt{1 + y_1^2 + y_2^2}}(\bar{m} - y_1\bar{e}_1 - y_2\bar{e}_2). \tag{2.53}$$

On substituting Eq. (2.50) into $\overset{*}{a}_{ik} = \overset{*}{\bar{r}}_i \overset{*}{rk}_i$ we obtain

$$\overset{*}{a}_{ik} = A_i^z A_k^z(\delta_{ik} + y_i y_k). \tag{2.54}$$

The determinant of the metric tensor $\overset{*}{\mathbf{A}}$ on $\overset{*}{S}$ is found to be

$$\begin{aligned}\overset{*}{a} &= \overset{*}{a}_{11}\overset{*}{a}_{22} - \overset{*}{a}{}_{12}^{2} = a_{11}^z a_{22}^z(1 + y_1^2 + y_2^2 + y_1^2y_2^2) - a_{11}^z a_{22}^z y_1^2 y_2^2 \\ &= a_{11}^z a_{22}^z(1 + y_1^2 + y_2^2) = \frac{a_{11}^z a_{22}^z}{\varsigma^2}.\end{aligned} \tag{2.55}$$

The contravariant components $\overset{*}{a}{}^{ik}$ of $\overset{*}{\mathbf{A}}$ are calculated as

$$\overset{*}{a}{}^{11} = \frac{\overset{*}{a}_{22}}{\overset{*}{a}} = \frac{(1 + y_2^2)\varsigma^2}{a_{11}^z}, \qquad \overset{*}{a}{}^{22} = \frac{(1 + y_1^2)\varsigma^2}{a_{22}^z}, \qquad \overset{*}{a}{}^{12} = -\frac{a_{22}^z}{\overset{*}{a}} = -\frac{y_1y_2\varsigma^2}{A_1^zA_2^z}. \tag{2.56}$$

On differentiating Eq. (2.53) with respect to α_i we get

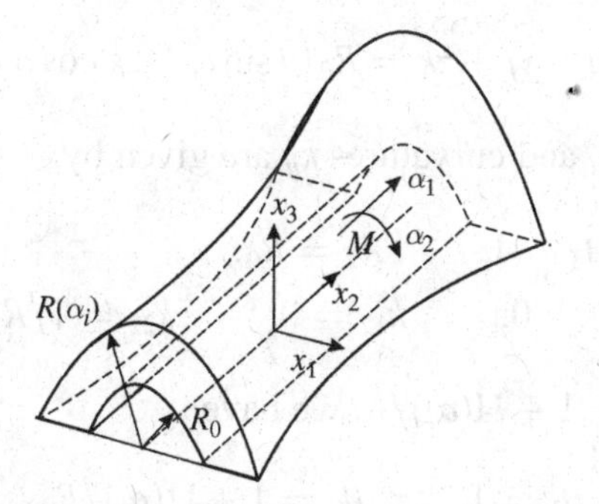

Fig. 2.3 A shell of complex geometry in relation to the canonical cylindrical surface.

$$
\begin{aligned}
\overset{*}{\bar{m}}_i &= \frac{\partial \varsigma}{\partial \alpha_1}(\bar{m} - y_i\bar{e}_i) \\
&+\varsigma\left[\frac{A_1}{R_1}\bar{e}_1 - \frac{\partial y_i}{\partial \alpha_1}\bar{e}_i + \frac{\partial \varsigma}{\partial \alpha_1}\left(\frac{A_1}{R_1}\bar{m} + \frac{\bar{e}_2}{A_2}\frac{\partial A_1}{\partial \alpha_2}\right) - \frac{\partial \varsigma}{\partial \alpha_2}\frac{\bar{e}_1}{A_2}\frac{\partial A_1}{\partial \alpha_2}\right], \\
\overset{*}{\bar{m}}_2 &= \frac{\partial \varsigma}{\partial \alpha_2}(\bar{m} - y_i\bar{e}_i) \\
&+\varsigma\left[\frac{A_2}{R_2}\bar{e}_2 - \frac{\partial y_i}{\partial \alpha_2}\bar{e}_i + \frac{\partial \varsigma}{\partial \alpha_2}\left(\frac{A_2}{R_2}\bar{m} + \frac{\bar{e}_1}{A_1}\frac{\partial A_2}{\partial \alpha_1}\right) - \frac{\partial \varsigma}{\partial \alpha_1}\frac{\bar{e}_2}{A_1}\frac{\partial A_2}{\partial \alpha_1}\right],
\end{aligned}
\tag{2.57}
$$

where use is made of Eqs. (1.30). By substituting Eqs. (2.49) and (2.57) into $\overset{*}{b}_{ik} = -\overset{*}{\bar{m}}_i\overset{*}{\bar{r}}_k$ we obtain

$$
\begin{aligned}
\overset{*}{b}_{11} &= -A_1^z\varsigma\left(\frac{A_1}{R_1}\sqrt{1+y_1^2} - \frac{\partial y_1}{\partial \alpha_1} - \frac{y_2}{A_2}\frac{\partial A_1}{\partial \alpha_2}\right), \\
\overset{*}{b}_{22} &= -A_2^z\varsigma\left(\frac{A_2}{R_2}\sqrt{1+y_2^2} - \frac{\partial y_2}{\partial \alpha_2} - \frac{y_1}{A_1}\frac{\partial A_2}{\partial \alpha_1}\right), \\
\overset{*}{b}_{12} &= -A_2^z\varsigma\left(y_1y_2\frac{A_1}{R_1} - \frac{\partial y_2}{\partial \alpha_1} - \frac{y_1}{A_2}\frac{\partial A_1}{\partial \alpha_2}\right) = \overset{*}{b}_{21}.
\end{aligned}
\tag{2.58}
$$

Formulas (2.48), (2.49) and (2.28) are used to calculate the Christoffel symbols $\overset{*}{\Gamma}{}^k_{ij}$ on $\overset{*}{S}$. $\overset{*}{\Gamma}{}^{(z)k}_{ij}$ are calculated from Eqs. (1.27) by replacing A_i^z and their derivatives for A_i and $\partial A_i/\partial \alpha_{1,2}$, respectively.

For example, consider a shell of complex geometry $\overset{*}{S}$ as shown in Fig. 2.3. Let a cylinder of constant radius R_0 be the reference surface for $\overset{*}{S}$. Its orientation with respect to $\overset{*}{S}$ is such that the function $\mathrm{H}(\alpha_i)$ and its derivatives satisfy the uniqueness of transformation (2.30).

Introduce polar coordinates α_1 and α_2 on S, such that α_1 is the axial and α_2 is the polar angular coordinate. They are related to the global Cartesian coordinates by

$$\bar{r}(\alpha_i) = x\bar{i} + y\bar{j} + z\bar{k} = R_0\left(\bar{i}\sin\alpha_2 + \bar{k}\cos\alpha_2\right) + \alpha_1\bar{j}. \tag{2.59}$$

The Lamé parameters A_i and curvatures k_{ij} are given by

$$\begin{gathered} A_1 = 1, \qquad A_2 = R_0, \\ k_{11} = 1/R_1 = 0, \qquad k_{12} = 0, \qquad k_{22} = 1/R_2 = 1/R_0. \end{gathered} \tag{2.60}$$

For the coefficients $\theta_i = 1 + \mathrm{H}(\alpha_i)/R_i$ we have

$$\theta_1 = 1, \qquad \theta_2 = 1 + \mathrm{H}(\alpha_i)/R_0. \tag{2.61}$$

Hence, from Eq. (2.51) for y_i, we obtain

$$y_1 = \frac{\partial \mathrm{H}(\alpha_i)}{\partial \alpha_1}, \qquad y_2 = \frac{1}{R_0 + \mathrm{H}(\alpha_j)}. \tag{2.62}$$

On substituting Eqs. (2.60)–(2.62) into (2.54) we find

$$\begin{aligned} \overset{*}{a}_{11} &= 1 + y_1^2 = 1 + \left(\frac{\partial \mathrm{H}(\alpha_i)}{\partial \alpha_1}\right)^2, \\ \overset{*}{a}_{22} &= (R_0 + \mathrm{H}(\alpha_i))^2\left(1 + y_2^2\right) = (R_0 + \mathrm{H}(\alpha_i))^2 + \left(\frac{\partial \mathrm{H}(\alpha_i)}{\partial \alpha_2}\right)^2, \\ \overset{*}{a}_{12} &= \overset{*}{a}_{21} = (R_0 + \mathrm{H}(\alpha_i))^2 y_1 y_2 = \frac{\partial \mathrm{H}(\alpha_i)}{\partial \alpha_1}\frac{\partial \mathrm{H}(\alpha_i)}{\partial \alpha_2}. \end{aligned} \tag{2.63}$$

The coefficients of the second fundamental form are found to be

$$\begin{aligned} \overset{*}{b}_{11} &= \varsigma\frac{\partial y_1}{\partial \alpha_1} = \varsigma\frac{\partial^2 \mathrm{H}(\alpha_i)}{\partial \alpha_1^2}, \\ \overset{*}{b}_{12} &= \overset{*}{b}_{21} = \varsigma\left(\frac{\partial y_1}{\partial \alpha_2} - y_1 y_2\right) \\ &= \varsigma\left(\frac{\partial^2 \mathrm{H}(\alpha_i)}{\partial \alpha_1\,\partial \alpha_2} - \frac{1}{R_0 + \mathrm{H}(\alpha_i)}\frac{\partial \mathrm{H}(\alpha_i)}{\partial \alpha_1}\frac{\partial \mathrm{H}(\alpha_i)}{\partial \alpha_2}\right), \\ \overset{*}{b}_{22} &= -\frac{1}{R_0 + \mathrm{H}(\alpha_i)}\varsigma\left(1 - \frac{1}{R_0 + \mathrm{H}(\alpha_i)}\frac{\partial^2 \mathrm{H}(\alpha_i)}{\partial \alpha_1\,\partial \alpha_2} + \left(\frac{\partial \mathrm{H}(\alpha_i)/\partial \alpha_2}{R_0 + \mathrm{H}(\alpha_i)}\right)^2\right). \end{aligned} \tag{2.64}$$

Here

$$\varsigma = \sqrt{1 + \frac{\partial^2 \mathrm{H}(\alpha_i)}{\partial \alpha_1^2} + \left(\frac{\partial \mathrm{H}(\alpha_i)/\partial \alpha_2}{R_0 + \mathrm{H}(\alpha_i)}\right)^2}.$$

On substituting Eq. (2.63) into (2.47), after some algebra for $\overset{*}{\Gamma}{}^k_{ij}$, we have

$$\overset{*}{a}\Gamma_{11}^{1} = \frac{\partial \mathrm{H}(\alpha_i)}{\partial \alpha_1}\left[(R_0+\mathrm{H}(\alpha_i))^2\frac{\partial^2 \mathrm{H}(\alpha_i)}{\partial \alpha_1^2} - 2\frac{\partial \mathrm{H}(\alpha_i)}{\partial \alpha_1}\frac{\partial \mathrm{H}(\alpha_i)}{\partial \alpha_2}\frac{\partial^2 \mathrm{H}(\alpha_i)}{\partial \alpha_1\,\partial \alpha_2}\right],$$

$$\overset{*}{a}\Gamma_{11}^{2} = \frac{\partial \mathrm{H}(\alpha_i)}{\partial \alpha_2}\frac{\partial^2 \mathrm{H}(\alpha_i)}{\partial \alpha_1^2},$$

$$\overset{*}{a}\Gamma_{12}^{1} = (R_0+\mathrm{H}(\alpha_i))\frac{\partial \mathrm{H}(\alpha_i)}{\partial \alpha_1}\left[(R_0+\mathrm{H}(\alpha_i))\frac{\partial^2 \mathrm{H}(\alpha_i)}{\partial \alpha_1\,\partial \alpha_2} - \frac{\partial \mathrm{H}(\alpha_i)}{\partial \alpha_1}\frac{\partial \mathrm{H}(\alpha_i)}{\partial \alpha_2}\right],$$

$$\overset{*}{a}\Gamma_{22}^{1} = (R_0+\mathrm{H}(\alpha_i))^3\frac{\partial \mathrm{H}(\alpha_i)}{\partial \alpha_1}\left[\frac{1}{(R_0+\mathrm{H}(\alpha_i))}\frac{\partial^2 \mathrm{H}(\alpha_i)}{\partial \alpha_2^2} -2\left(\frac{1}{(R_0+\mathrm{H}(\alpha_i))}\frac{\partial \mathrm{H}(\alpha_i)}{\partial \alpha_2}\right)^2 - 1\right], \tag{2.65}$$

$$\overset{*}{a}\Gamma_{12}^{2} = \frac{\partial^2 \mathrm{H}(\alpha_i)}{\partial \alpha_1\,\partial \alpha_2}\frac{\partial \mathrm{H}(\alpha_i)}{\partial \alpha_2} + (R_0+\mathrm{H}(\alpha_i))\frac{\partial \mathrm{H}(\alpha_i)}{\partial \alpha_1}\left(1+\left(\frac{\partial \mathrm{H}(\alpha_i)}{\partial \alpha_1}\right)^2\right),$$

$$\overset{*}{a}\Gamma_{22}^{2} = \frac{\partial \mathrm{H}(\alpha_i)}{\partial \alpha_2}\left[R_0+\mathrm{H}(\alpha_i)+\frac{\partial^2 \mathrm{H}(\alpha_i)}{\partial \alpha_2^2} - 2\frac{\partial \mathrm{H}(\alpha_i)}{\partial \alpha_1}\frac{\partial \mathrm{H}(\alpha_i)}{\partial \alpha_2}\frac{\partial^2 \mathrm{H}(\alpha_i)}{\partial \alpha_1\,\partial \alpha_2}\right],$$

where $\overset{*}{a} = (R_0+\mathrm{H}(\alpha_i))^2/\varsigma^2$, and use is made of Eq. (2.63).

To construct the unkown function $\mathrm{H}(\alpha_i)$ let the surface of revolution $\overset{*}{S}$ (Fig. 2.4) be defined by

$$\overset{*}{R} = \overset{*}{R}(x) = \overset{*}{R}(\alpha_i).$$

From geometrical analysis of the triangle $\mathrm{OO_1B}$ we have

$$\begin{aligned} R_0 &= \mathrm{H}(\alpha_i) + \overline{O_1B}, \\ (\overline{OB})^2 &= (\overline{O_1B})^2 + (\overline{O_1O})^2 + 2\overline{O_1B}\cdot\overline{OB_1}\sin\alpha_2. \end{aligned} \tag{2.66}$$

Since $\overline{OB} = \overset{*}{R}(\alpha_1)$ and $\overline{OO_1} = \tilde{R}$ the last equation can be written as

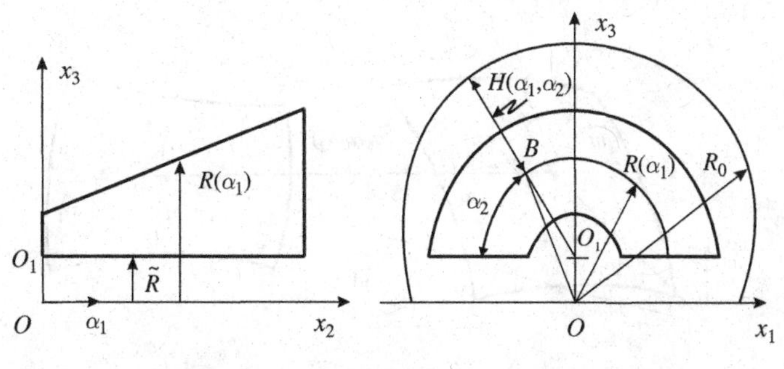

Fig. 2.4 Parameterization of the shell of complex geometry and construction of the $\mathrm{H}(\alpha_1,\alpha_2)$ function.

$$\left(\overline{O_1B}\right)^2 + 2\overline{O_1B}\cdot\tilde{R}\sin\alpha_2 + \tilde{R}^2 - \overset{*}{R}{}^2(\alpha_1) = 0.$$

On use of Eqs. (2.66) in the above, we obtain

$$\mathrm{H}(\alpha_1,\alpha_2) = R_0 + \tilde{R}\sin\alpha_2 \pm \sqrt{\overset{*}{R}{}^2(\alpha_1) - \tilde{R}^2\cos^2\alpha_2}. \tag{2.67a}$$

For $\tilde{R} = 0$ Eq. (2.67a) should satisfy the equality $\overline{O_1B} = \overset{*}{R}(\alpha_1)$. Therefore the final form for $\mathrm{H}(\alpha_i)$ is found to be

$$\mathrm{H}(\alpha_1,\alpha_2) = R_0 + \tilde{R}\sin\alpha_2 - \sqrt{\overset{*}{R}{}^2(\alpha_1) - \tilde{R}^2\cos^2\alpha_2}. \tag{2.67b}$$

Knowing $\mathrm{H}(\alpha_i)$ from Eqs. (2.55)–(2.65), we find coefficients and parameters that characterize the geometry of the surface $\overset{*}{S}$. It is noteworthy that Eq. (2.30) can also be used to parameterize composite shells of complex geometry with variable thicknesses of layers.

2.5 Parameterization of complex surfaces on a plane

Consider a complex planar surface $\overset{*}{S}$ that is referred to coordinates $\overset{*}{\alpha}_1$ and $\overset{*}{\alpha}_2$. Let S be the reference canonical surface to $\overset{*}{S}$ parameterized by orthogonal coordinates α_1 and α_2 (Fig. 2.5). Assume that the coordinate lines are oriented along the contour C of S. Setting $\mathrm{H}(\alpha_1,\alpha_2) = 0$ in Eq. (2.10) and using Eq. (2.9), the position of a point $\overset{*}{M}(\alpha_1,\alpha_2) \in \overset{*}{S}$ is given by

$$\overset{*}{\bar{r}} = \bar{r}(\alpha_1,\alpha_2) + \bar{v}(\alpha_1,\alpha_2) = \bar{r}(\alpha_1,\alpha_2) + \sum_{i=1}^{2} \mathrm{F}_i(\alpha_1,\alpha_2)\bar{e}_i. \tag{2.68}$$

Here the meaning of the parameters is as described by Eqs. (1.4), (1.5), (2.9) and (2.10). Evidently, Eq. (2.68) describes the fictitious tangent deformation of the surface S onto $\overset{*}{S}$.

On differentiating Eq. (2.68) with respect to α_i, for tangent vectors $\overset{*}{\bar{r}}_i$ we get

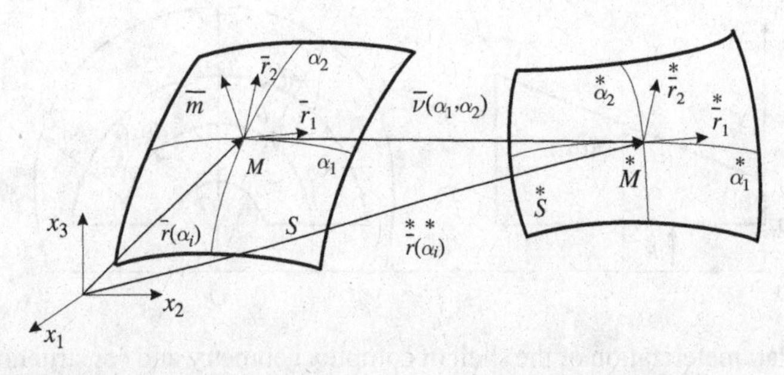

Fig. 2.5 Parameterization of a planar surface of complex geometry.

$$\overset{*}{\bar{r}}_i = A_i \sum_{k=1}^{2} (\delta_{ik} + \underline{e}_{ik})\bar{e}_k, \tag{2.69}$$

where the rotation parameters are given by

$$\underline{e}_{11} = \frac{1}{A_1}\frac{\partial \mathrm{F}_1(\alpha_i)}{\partial \alpha_1} + \frac{\partial A_1}{\partial \alpha_2}\frac{\mathrm{F}_2(\alpha_i)}{A_1 A_2}, \qquad \underline{e}_{22} = \frac{1}{A_2}\frac{\partial \mathrm{F}_2(\alpha_i)}{\partial \alpha_2} + \frac{\partial A_2}{\partial \alpha_1}\frac{\mathrm{F}_1(\alpha_i)}{A_1 A_2},$$
$$\underline{e}_{12} = \frac{1}{A_1}\frac{\partial \mathrm{F}_2(\alpha_i)}{\partial \alpha_1} - \frac{\partial A_1}{\partial \alpha_1}\frac{\mathrm{F}_1(\alpha_i)}{A_1 A_2}, \qquad \underline{e}_{21} = \frac{1}{A_2}\frac{\partial \mathrm{F}_1(\alpha_i)}{\partial \alpha_2} - \frac{\partial A_2}{\partial \alpha_1}\frac{\mathrm{F}_2(\alpha_i)}{A_1 A_2}.$$

Using Eq. (2.69), the components of the tensor $\overset{*}{\mathbf{A}}$ at $\overset{*}{M}(\alpha_i) \in \overset{*}{S}$ are found to be

$$\overset{*}{a}_{ik} = A_i A_k(\delta_{ik} + 2\underline{\varepsilon}_{ik}), \tag{2.70}$$

where $\underline{\varepsilon}_{ik}$ are the physical components of the tensor of fictitious tangent deformation of S given by

$$2\underline{\varepsilon}_{ik} = \underline{e}_{ik} + \underline{e}_{ki} + \underline{e}_{is}\underline{e}_{ks}. \tag{2.71}$$

From Eq. (2.70) the determinant of $\overset{*}{\mathbf{A}}$ is calculated as

$$\overset{*}{a} = (A_1 A_2)^2[(1 + \underline{e}_{11})(1 + \underline{e}_{22}) - \underline{e}_{12}\underline{e}_{21}]. \tag{2.72}$$

By making use of Eqs. (2.70) and (2.72) in the formulas $\overset{*}{a}{}^{ii} = \overset{*}{a}_{ii}/\overset{*}{a}$ and $\overset{*}{a}{}^{12} = \overset{*}{a}{}^{21} = -\overset{*}{a}_{12}/\overset{*}{a}$, one can find the contravariant components of $\overset{*}{\mathbf{A}}$. On use of $\overset{*}{a}{}^{ik}$ and Eqs. (2.69) and (2.70) for vectors of the reciprocal basis $\overset{*}{\bar{r}}{}^i$ we obtain

$$\overset{*}{\bar{r}}{}^i = \sum_{k=1}^{2} \overset{*}{a}{}^{ik}\overset{*}{\bar{r}}_k = \frac{1}{\overset{*}{a}}\sum_{s=1}^{2}\sum_{k=1}^{2} \mathrm{sgn}(A_i A_k)(\delta_{ik} + 2\underline{\varepsilon}_{ik})(\delta_{is} + \underline{e}_{is})\bar{e}_s. \tag{2.73}$$

Finally, with the help of Eqs. (2.69) and (2.73), we calculate the Christoffel symbols of the first and second kind.

In the above derivations we assumed that coordinates α_1 and α_2 are linearly independent. For them to remain such after transformation, the tangent vectors $\overset{*}{\bar{r}}_{\mathrm{i}}$ should remain non-collinear for all $\overset{*}{M}(\alpha_1, \alpha_2) \in \overset{*}{S}$. If at any point $\overset{*}{\bar{r}}_1 \times \overset{*}{\bar{r}}_2 = 0$ then transformation (2.68) at this point experiences singularity. On substituting Eq. (2.65) and using the fact that $\bar{e}_1 \times \bar{e}_2 = 0$, after some algebra, we obtain

$$A_i A_k[(1 + \underline{e}_{11})(1 + \underline{e}_{22}) - \underline{e}_{12}\underline{e}_{21}](e_1 \times e_2) = 0. \tag{2.74}$$

The condition of singularity then becomes

$$A_i A_k[(1 + \underline{e}_{11})(1 + \underline{e}_{22}) - \underline{e}_{12}\underline{e}_{21}] = \sqrt{\overset{*}{a}} = 0, \tag{2.75}$$

where use is made of Eq. (2.72).

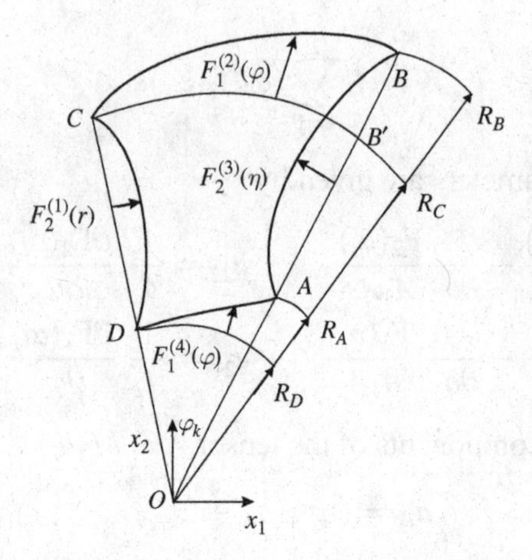

Fig. 2.6 A complex surface bounded by four continuous lines and its parameterization.

As a final point of our discussion, consider a problem of parameterization of a complex surface bounded by four smooth continuous lines (Fig. 2.6). Let the straight lines connecting corner points A, B and C, D intersect at point O.

Assume that the contour lines in polar coordinates (r, φ) $(r = \alpha_1, \varphi = \alpha_2)$ are given analytically by

$$C_1\colon \mathrm{F}_2^{(1)}(r), \quad C_2\colon R_c + \mathrm{F}_1^{(2)}(\varphi), \quad C_3\colon \mathrm{F}_2^{(3)}(\eta), \quad C_4\colon R_D + \mathrm{F}_1^{(4)}(\varphi)$$
$$(R_D \leq r \leq R_C, 0 \leq \varphi \leq \varphi_k, 0 \leq \eta \leq R_B - R_C), \tag{2.76}$$

where R_A, R_B, R_C and R_D are the radii, φ_k is the angle between the rays $\overline{OB}$ and $\overline{OC}$, and $\mathrm{F}_2^{(1)}(r)$and $\mathrm{F}_2^{(3)}(\eta)$ are the normal distances measured from $\overline{OC}$ and $\overline{OB}$ to the contour lines $C_1$and C_3, respectively.

Affine transformation of the line segment $\overline{AB}$ with coordinates $u=0, u=R_B - R_A$, onto the line segment $\overline{A'B'}$ is given by

$$u = \xi_1 + \xi_2 r. \tag{2.77}$$

Making use of boundary conditions

$$r = R_D\colon u = 0; \qquad r = R_C\colon u = R_B - R_A, \tag{2.78}$$

the coefficients ξ_1 and ξ_2 are found to be

$$\xi_1 = -R_D\xi_2, \qquad \xi_2 = (R_B - R_A)/(R_C - R_D).$$

Therefore, $u = \xi_2(r - R_D)$. This allows us to write the equation for C_3 in the form

$$\mathrm{F}_2^{(3)}(u) = \mathrm{F}_2^{(3)}[\xi_2(r - R_D)].$$

By expanding functions $\mathrm{F}_i(r, \varphi)$ $(i=1, 2)$ in the form

$$\begin{aligned} \mathrm{F}_1(r,\varphi) &= \varsigma_1(\varphi) + r\varsigma_2(r), \\ \mathrm{F}_2(r,\varphi) &= \varsigma_3(r) + \varphi\varsigma_4(r), \end{aligned} \tag{2.79}$$

and using boundary conditions given by Eqs. (2.76), the coefficients of expansion ς_j are found to be

$$\begin{aligned} r = R_D&: \mathrm{F}_1(R_D,\varphi) = \mathrm{F}_1^{(4)}(\varphi), \\ r = R_C&: \mathrm{F}_1(R_C,\varphi) = \mathrm{F}_1^{(2)}(\varphi), \\ \varphi = 0&: \mathrm{F}_2(r,0) = \mathrm{F}_2^{(1)}(r), \\ \varphi = \varphi_k&: \mathrm{F}_2(r,\varphi_k) = \mathrm{F}_2^{(3)}[\xi_2(r - R_D)]. \end{aligned} \tag{2.80}$$

Here $\varphi_k = \angle COB$. From Eqs. (2.79) and (2.80) we obtain

$$\begin{aligned} \varsigma_1(\varphi) &= \mathrm{F}_1^{(4)}(\varphi) - \frac{R_D}{R_D - R_C}\left(\mathrm{F}_1^{(4)}(\varphi) - \mathrm{F}_1^{(2)}(\varphi)\right), \\ \varsigma_2(\varphi) &= \frac{1}{R_D - R_C}\left(\mathrm{F}_1^{(4)}(\varphi) - \mathrm{F}_1^{(2)}(\varphi)\right), \\ \varsigma_3(r) &= \mathrm{F}_2^{(1)}(r), \\ \varsigma_4(r) &= \frac{1}{\varphi_k}\left(\mathrm{F}_2^{(3)}[\xi_2(r - R_D)] - \mathrm{F}_2^{(1)}(r)\right). \end{aligned}$$

On substituting the above into (2.79) we get

$$\begin{aligned} \mathrm{F}_1(r,\varphi) &= \mathrm{F}_1^{(4)}(\varphi) + \frac{r - R_D}{R_D - R_C}\left(\mathrm{F}_1^{(4)}(\varphi) - \mathrm{F}_1^{(2)}(\varphi)\right), \\ \mathrm{F}_2(r,\varphi) &= \mathrm{F}_2^{(1)}(r) + \frac{\varphi}{\varphi_k}\left(\mathrm{F}_2^{(3)}[\xi_2(r - R_D)] - \mathrm{F}_2^{(1)}(r)\right). \end{aligned} \tag{2.81}$$

The unit base vectors $\bar{e}_{ik}$ are given by

$$\begin{aligned} \bar{e}_{11} &= \frac{R_D}{R_D - R_C}\left(\mathrm{F}_1^{(4)}(\varphi) - \mathrm{F}_1^{(2)}(\varphi)\right), \\ \bar{e}_{12} &= \frac{\varphi}{\varphi_k}\frac{\mathrm{d}}{\mathrm{d}r}\left(\mathrm{F}_2^{(3)}[\xi_2(r - R_D)] - \mathrm{F}_2^{(1)}(r)\right), \\ \bar{e}_{21} &= \frac{1}{r}\left(\frac{\mathrm{dF}_1^{(4)}(\varphi)}{\mathrm{d}\varphi} + \frac{r - R_D}{R_D - R_C}\frac{\mathrm{d}}{\mathrm{d}\varphi}\left(\mathrm{F}_1^{(4)}(\varphi) - \mathrm{F}_1^{(2)}(\varphi)\right)\right. \\ &\quad \left. - \mathrm{F}_2^{(1)}(r) - \frac{\varphi}{\varphi_k}\left(\mathrm{F}_2^{(3)}[\xi_2(r - R_D)] - \mathrm{F}_2^{(1)}(r)\right)\right), \\ \bar{e}_{22} &= \frac{1}{r}\left(\frac{\mathrm{F}_2^{(3)}[\xi_2(r - R_D)] - \mathrm{F}_2^{(1)}(r)}{\varphi_k} + \mathrm{F}_1^{(4)}(\varphi)\right. \\ &\quad \left. + \frac{r - R_D}{R_D - R_C}\left(\mathrm{F}_1^{(4)}(\varphi) - \mathrm{F}_1^{(2)}(\varphi)\right)\right). \end{aligned} \tag{2.82}$$

Making use of Eqs. (2.82) in (2.71), we find the components of fictitious deformation $2\underline{\varepsilon}_{ik}$ and the Christoffel symbols $\overset{*}{\Gamma}{}^{k}_{ij}$ on $\overset{*}{S}$.

If the complex surface $\overset{*}{S}$ had three corner points, such that $R_D \to 0, R_A \to 0, \left(\mathrm{F}_1^{(4)}(\varphi) \equiv 0\right)$, then the corner points A and D would merge to a single point O. In this case the transformation would have had a singularity at O, as discussed above.

Exercises

1. Compute the stretch ratio λ_i $(i = 1, 2)$ (Eq. (2.7)) and the change of the area of the surface element δs_Δ (Eq. (2.8)) in cylindrical coordinates.
2. Compute $\underline{e}_i^k$ and $\underline{e}_{ki}$ (Eq. (2.12)) in orthogonal Cartesian and cylindrical coordinates.
3. Compute the components $\overset{*}{a}_{ik}$ of the metric tensor $\overset{*}{\mathbf{A}}$ (Eq. (2.13)) in cylindrical and spherical coordinates.
4. Compute the components of the third metric tensor ξ_{ik} (Eq. (2.20)) in cylindrical coordinates.
5. Verify the formula for $\overset{*}{\bar{r}}_i$ $(i = 1, 2)$ given by Eq. (2.33).
6. Verify formulas (2.35) and (2.39).
7. Verify the formula (2.47) for the coefficients b_{ik} $(i, k = 1, 2)$ of the second fundamental form.
8. Using Eqs. (2.70) and (2.74), find expressions for $\Gamma_{ij,k}$ *and* Γ^k_{ij}.
9. Using Eqs. (2.82) in (2.71), find the components of fictitious deformation $2\underline{\varepsilon}_{ik}$.
10. Provide a parameterization of a shell in the shape of a parallelogram.

3

Nonlinear theory of thin shells

3.1 Deformation of the shell

We shall base all our discussion of the deformation of thin shells upon the first Kirchhoff hypothesis. Let the middle surface S of the undeformed thin shell be associated with the orthogonal curvilinear coordinates α_1 and α_2. The position vector $\bar{\rho}$ of an arbitrary point M_z on the equidistant surface $S_z\,(S_z \parallel S)$ is given by Eq. (1.39), where $z \in [-h/2, +h/2]$ and h is the thickness of the shell. The coordinate vectors and the Lamé coefficients satisfy Eqs. (1.42) and (1.49).

The length of a line element on S_z is given by

$$(\mathrm{d}s_z)^2 = H_1^2\,\mathrm{d}\alpha_1^2 + H_2^2\,\mathrm{d}\alpha_2^2 + \mathrm{d}z^2, \tag{3.1}$$

where

$$(\mathrm{d}s_z)_1 = H_1\,\mathrm{d}\alpha_1, \qquad (\mathrm{d}s_z)_2 = H_2\,\mathrm{d}\alpha_2, \tag{3.2}$$

are the lengths of line elements in the direction of the coordinates α_1 and α_2.

In the deformed configuration the position vector $\overset{*}{\bar{\rho}}$ of point $\overset{*}{M}_z \in \overset{*}{S}_z$ is given by

$$\overset{*}{\bar{\rho}} = \bar{\rho}(\alpha_1, \alpha_2) + \bar{v}_z(\alpha_1, \alpha_2, z), \tag{3.3}$$

where $\bar{v}_z$ is the displacement vector. Since for thin shells $z \ll 1$ we assume $\overset{*}{z} \approx z$. The first fundamental form of $\overset{*}{S}_z$ is

$$(\mathrm{d}\overset{*}{s}_z)^2 = \overset{*}{g}_{11}\,\mathrm{d}\alpha_1^2 + 2\overset{*}{g}_{12}\,\mathrm{d}\alpha_1\ \mathrm{d}\alpha_2 + \overset{*}{g}_{22}\,\mathrm{d}\alpha_2^2 + \mathrm{d}z^2, \tag{3.4}$$

$$(\mathrm{d}\overset{*}{s}_z)_1 = \sqrt{\overset{*}{g}_{11}}\ \mathrm{d}\alpha_1, \qquad (\mathrm{d}\overset{*}{s}_z)_2 = \sqrt{\overset{*}{g}_{22}}\ \mathrm{d}\alpha_2. \tag{3.5}$$

Here $\overset{*}{g}_{ik} = \overset{*}{\bar{\rho}}_i \overset{*}{\bar{\rho}}_k$ $(i, k = 1, 2)$ and the vectors $\overset{*}{\bar{\rho}}_1$ and $\overset{*}{\bar{\rho}}_2$ tangent to coordinate lines are obtained by differentiating Eq. (3.3) with respect to α_1 and α_2:

$$\overset{*}{\bar{\rho}}_1 = \overset{*}{\bar{r}}_1(1 - z\overset{*}{b}{}_1^1) - \overset{*}{\bar{r}}_2 z\overset{*}{b}{}_1^2,$$
$$\overset{*}{\bar{\rho}}_2 = \overset{*}{\bar{r}}_2(1 - z\overset{*}{b}{}_2^2) - \overset{*}{\bar{r}}_1 z\overset{*}{b}{}_2^1, \tag{3.6}$$
$$\overset{*}{\bar{\rho}}_3 = \overset{*}{\bar{m}}.$$

Let ε_{11}^z and ε_{22}^z be deformations through point $M_z \in S_z$ in the direction of α_1- and α_2-coordinates defined by

$$\varepsilon_{11}^z = \frac{(\mathrm{d}\overset{*}{s}_z)_1 - (\mathrm{d}s_z)_1}{(\mathrm{d}s_z)_1}, \qquad \varepsilon_{22}^z = \frac{(\mathrm{d}\overset{*}{s}_z)_2 - (\mathrm{d}s_z)_2}{(\mathrm{d}s_z)_2}. \tag{3.7}$$

By substituting Eqs. (3.1) and (3.4) into (3.7), we obtain

$$\varepsilon_{11}^z = \left(\sqrt{\overset{*}{g}_{11}} - H_1\right)\Big/ H_1, \qquad \varepsilon_{22}^z = \left(\sqrt{\overset{*}{g}_{22}} - H_2\right)\Big/ H_2. \tag{3.8}$$

Using Eqs. (1.45) and (1.53), the angle $\overset{*}{\chi}_z$ between the vectors $\overset{*}{\bar{\rho}}_1$ and $\overset{*}{\bar{\rho}}_2$ is found to be

$$\cos\overset{*}{\chi}_z = \frac{\overset{*}{\bar{\rho}}_1\overset{*}{\bar{\rho}}_2}{\left|\overset{*}{\bar{\rho}}_1\right|\left|\overset{*}{\bar{\rho}}_2\right|} = \frac{\overset{*}{g}_{12}}{\sqrt{\overset{*}{g}_{11}\overset{*}{g}_{22}}}, \tag{3.9}$$

where

$$\overset{*}{g}_{ii} = \overset{*}{A}{}_1^2\left(1 + 2\overset{*}{k}_{ii}z\right) + \left(\overset{*}{m}_i z\right)^2,$$
$$\overset{*}{g}_{12} = \overset{*}{A}_1\overset{*}{A}_2\left(\cos\overset{*}{\chi} + 2\overset{*}{k}_{12}z\right) + \overset{*}{m}_1\overset{*}{m}_2 z^2. \tag{3.10}$$

Let ε_{12}^z be the shear deformation, i.e. the change in the angle between initially orthogonal coordinate lines $2\varepsilon_{12}^z = \pi/2 - \overset{*}{\chi}_z$. Evidently,

$$\cos\overset{*}{\chi}_z := \sin(2\varepsilon_{12}^z) = \frac{\overset{*}{g}_{12}}{\sqrt{\overset{*}{g}_{22}\overset{*}{g}_{12}}}. \tag{3.11}$$

Making use of Eq. (3.6) and the fact $\overset{*}{\bar{r}}_1\overset{*}{\bar{m}} = 0$, we find

$$\overset{*}{g}_{13} = \overset{*}{\bar{\rho}}_1\overset{*}{\bar{m}} = 0, \qquad \overset{*}{g}_{23} = \overset{*}{\bar{\rho}}_2\overset{*}{\bar{m}} = 0, \qquad \overset{*}{g}_{33} = \overset{*}{\bar{m}}\overset{*}{\bar{m}} = 1.$$

It follows from the above that the deformation over the thickness of the shell equals zero

$$2\varepsilon_{i3}^z = \frac{\overset{*}{g}_{i3}}{\sqrt{\overset{*}{g}_{ii}\overset{*}{g}_{33}}} = 0, \qquad \varepsilon_{33}^z = \left(\sqrt{\overset{*}{g}_{33}} - H_3\right)\Big/ H_3 = 0 \quad (\overset{*}{g}_{i3} = 0, H_3 = 1).$$

In applications it is more convenient to use deformations ε_{ik}^z expressed in terms of the undeformed middle surface S. Thus, neglecting terms $O(z^2)$ in Eq. (2.10), we have

$$\overset{*}{g}_{ii} = \overset{*}{A}{}_i^2\left(1 + 2\overset{*}{k}_{ii}z\right), \qquad \overset{*}{g}_{12} = \overset{*}{A}_1\overset{*}{A}_2\left(\cos\overset{*}{\chi} + 2\overset{*}{k}_{12}z\right). \tag{3.12}$$

Taking the square root of both sides,

$$\sqrt{\overset{*}{g}_{ii}} \approx \overset{*}{A}_i\left(1 + \overset{*}{k}_{ii}z\right) + \cdots; \tag{3.13}$$

and substituting Eqs. (3.13) and (1.49) into (3.8), ε_{ii}^z are found to be

$$\varepsilon_{ii}^z = \left[\overset{*}{A}_i - A_i + \left(\overset{*}{A}_i\overset{*}{k}_{ii} - A_ik_{ii}\right)z\right]\Big/ A_i(1 + k_{ii}z) \quad (i = 1, 2)$$

Applying Eqs. (1.65) and (1.69), the term in parentheses inside the square brackets can be written as

$$\begin{aligned}
\overset{*}{A}_i\overset{*}{k}_{ii} - A_ik_{ii} &= A_i\sqrt{(1 + 2\varepsilon_{ii})}\overset{*}{k}_{ii} - A_ik_{ii} \\
&= A_i(\sqrt{(1 + 2\varepsilon_{ii})}\overset{*}{k}_{ii} - k_{ii}) \\
&= A_i\text{œ}_{ii}\sqrt{1 + 2\varepsilon_{ii}} + A_ik_{ii}\sqrt{1 + 2\varepsilon_{ii}}.
\end{aligned}$$

Reverse substitution yields

$$\varepsilon_{ii}^z = \left[A_i\left(\sqrt{1 + 2\varepsilon_{ii}} - 1\right) + A_i\text{œ}_{ii}z\sqrt{1 + 2\varepsilon_{ii}} + A_ik_{ii}z\sqrt{1 + 2\varepsilon_{ii}}\right]\Big/ A_i(1 + k_{ii}z).$$

Hence, deformation of the equidistant surface S_z of the shell is fully determined in terms of tangent deformations and curvatures of the middle surface S. Similarly, on introducing $\overset{*}{g}_{12}$ and $\sqrt{\overset{*}{g}_{ii}}$ given by Eqs. (3.12) and (3.13) into (3.11), we obtain

$$\sin\left(2\varepsilon_{12}^z\right) = \frac{\cos\overset{*}{\chi} + 2\overset{*}{k}_{12}z}{\left(1 + \overset{*}{k}_{11}z\right)\left(1 + \overset{*}{k}_{22}z\right)}.$$

Since $\overset{*}{k}_{12}z \approx \text{œ}_{12}z$, we have

$$\sin\left(2\varepsilon_{12}^z\right) = \frac{\cos\overset{*}{\chi} + 2\text{œ}_{12}z}{\left(1 + \overset{*}{k}_{11}z\right)\left(1 + \overset{*}{k}_{22}z\right)}.$$

Because for thin shells $1 + k_{ii}z \approx 1$ and $1 + \overset{*}{k}_{ii}z \approx 1$, the final formulas for tangent and shear deformations of S_z take the forms

$$\varepsilon_{ii}^{z} = \sqrt{1+2\varepsilon_{ii}}\left(1+z\overset{*}{k}_{ii}\right)-1,$$
$$\sin\left(2\varepsilon_{12}^{z}\right) = \frac{1}{2}\left(\cos\overset{*}{\chi}+2æ_{12}z\right). \tag{3.14}$$

3.2 Forces and moments

Consider a differential element in the deformed shell bounded by the surfaces $\alpha_i =$ constant, $\alpha_i + \mathrm{d}\alpha_i =$ constant and $\overset{*}{z} = \pm 0.5h$ (Fig. 3.1). Internal forces acting upon the element are given by $\bar{p}_1\overset{*}{H}_2\,\mathrm{d}\alpha_2\,\mathrm{d}z$ and $\bar{p}_2\overset{*}{H}_1\,\mathrm{d}\alpha_1\,\mathrm{d}z$. Here $\bar{p}_i$ are stress vectors, and $\overset{*}{H}_2\,\mathrm{d}\alpha_2\,\mathrm{d}z$ and $\overset{*}{H}_2\,\mathrm{d}\alpha_1\,\mathrm{d}z$ are the surface areas of differential boundary elements at $z =$ constant. By integrating the internal forces over the thickness of the shell, we obtain the resultant force vectors, $\overline{R}_1$ and $\overline{R}_2$, in the forms

$$\overline{R}_1 = \int_{z_1}^{z_2} \bar{p}_1\overset{*}{H}_2\,\mathrm{d}\alpha_2\,\mathrm{d}z,$$

$$\overline{R}_2 = \int_{z_1}^{z_2} \bar{p}_2\overset{*}{H}_2\,\mathrm{d}\alpha_1\,\mathrm{d}z \quad (z_1 = -h/2,\ z_2 = +h/2).$$

On dividing $\overline{R}_i$ by the length of linear elements, $\overset{*}{A}_i\ \mathrm{d}\alpha_i = A_i\sqrt{1+2\varepsilon_{ii}}\ \mathrm{d}\alpha_i$ $(i = 1, 2)$, we find

$$\overline{R}_1 = \frac{\int_{z_1}^{z_2} \bar{p}_1\overset{*}{H}_2\ \mathrm{d}z}{A_2\sqrt{1+2\varepsilon_{22}}}, \quad \overline{R}_2 = \frac{\int_{z_1}^{z_2} \bar{p}_2\overset{*}{H}_1\ \mathrm{d}z}{A_1\sqrt{1+2\varepsilon_{11}}}. \tag{3.15}$$

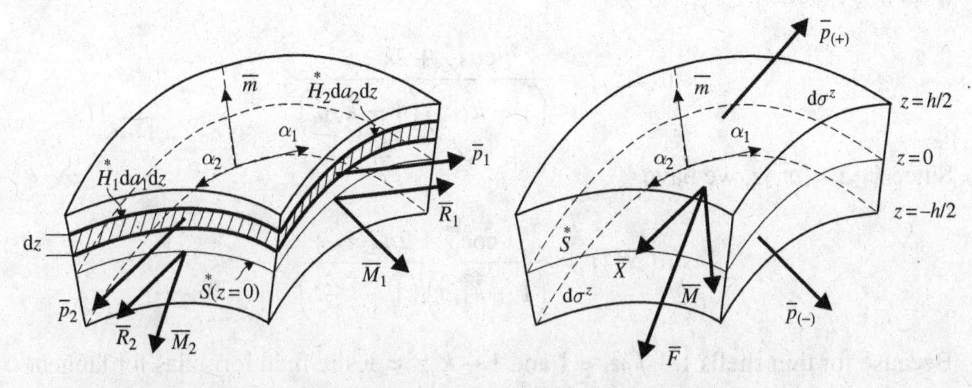

Fig. 3.1 Forces and moments acting upon a three-dimensional solid.

Similar reasoning leads to the definition of the resultant internal moment vectors. The moment of the force $\bar{p}_1$, acting on the face $\alpha_1 =$ constant, about the centre of the middle surface $\overset{*}{S}$ is $\overset{*}{\bar{m}}\, z \times \overset{*}{\bar{p}}_1 H_2\, d\alpha_2\, dz$. Here $\overset{*}{\bar{m}}\, z$ is the radius vector of $\bar{p}_1$. Hence, the resultant moment vector of internal forces $\overline{M}_1$ is given by

$$\overline{M}_1 = \int_{z_1}^{z_2} \overset{*}{\bar{m}}\, z \times \bar{p}_1 \overset{*}{H}_2\, d\alpha_2\, dz.$$

On dividing the above by the length of a line segment in the direction of the α_2-coordinate, $\overset{*}{A}_2\, d\alpha_2 = A_2\sqrt{(1 + 2\varepsilon_{22})}\, d\alpha_2$, we obtain

$$\overline{M}_1 = \frac{1}{A_2\sqrt{(1 + 2\varepsilon_{22})}} \int_{z_1}^{z_2} \left(\overset{*}{\bar{m}} z \times \bar{p}_1\right) \overset{*}{H}_2\, dz. \tag{3.16a}$$

Similarly, we define the resultant moment vector $\overline{M}_2$ as

$$\overline{M}_2 = \frac{1}{A_1\sqrt{1 + 2\varepsilon_{11}}} \int_{z_1}^{z_2} \left(\overset{*}{\bar{m}} z \times \bar{p}_2\right) \overset{*}{H}_1\, dz. \tag{3.16b}$$

The above discussion implies that the internal forces acting on the differential element are statically equivalent to the resultant force and moment vectors, $\overline{R}_i$ and $\overline{M}_i$.

Consider external forces acting on the free surfaces $z = \pm 0.5h$ of the shell. Let

(i) $\bar{p}_{(+)}$ and $\bar{p}_{(-)}$, be the external forces applied on the surface of area $d\sigma^z$,

$$\begin{aligned} d\sigma^z &= \overset{*}{H}_1 \overset{*}{H}_2\, d\alpha_1\, d\alpha_2 \\ &\approx A_1 A_2 (1 + (k_{11} + æ_{11})z)(1 + (k_{22} + æ_{22})z)\sqrt{1 + 2\varepsilon_{11}}\sqrt{1 + 2\varepsilon_{22}}\, d\alpha_1\, d\alpha_2, \end{aligned}$$

and

(ii) $\overline{F}$ be the vector of mass forces per unit volume $d\Omega$ of the deformed element,

$$\begin{aligned} d\Omega &= \overset{*}{H}_1 \overset{*}{H}_2\, d\alpha_1\, d\alpha_2\, d\overset{*}{z} \\ &\approx A_1 A_2 (1 + (k_{11} + æ_{11})z)(1 + (k_{22} + æ_{22})z)\sqrt{1 + 2\varepsilon_{11}}\sqrt{1 + 2\varepsilon_{22}}\, d\alpha_1\, d\alpha_2\, d\overset{*}{z}, \end{aligned}$$

where $H_1^{(+)}$ and $H_2^{(-)}$ are the values of $\overset{*}{H}_i$ at $z = \pm 0.5h$, respectively.

Then, the resultant external force vectors are defined by

$$\bar{p}_{(+)} H_1^{(+)} H_2^{(+)}\, d\alpha_1\, d\alpha_2 \qquad \text{and} \qquad \bar{p}_{(-)} H_1^{(-)} H_2^{(-)}\, d\alpha_1\, d\alpha_2.$$

Their sum divided by the surface area of a deformed element, $\overset{*}{A}_1 \overset{*}{A}_2\, d\alpha_1\, d\alpha_2$, yields

$$\begin{aligned} &\frac{\bar{p}_{(+)} H_1^{(+)} H_2^{(+)}\, d\alpha_1\, d\alpha_2 + \bar{p}_{(-)} H_1^{(-)} H_2^{(-)}\, d\alpha_1\, d\alpha_2}{A_1 A_2 \sqrt{(1 + 2\varepsilon_{11})(1 + 2\varepsilon_{22})}\, d\alpha_1\, d\alpha_2} \\ &\qquad = \frac{\bar{p}_z \overset{*}{H}_1 \overset{*}{H}_2}{A_1 A_2 \sqrt{(1 + 2\varepsilon_{11})(1 + 2\varepsilon_{22})}}. \end{aligned} \tag{3.17}$$

Here $(\bar{p}_z)_{z=0.5h} = \bar{p}_{(+)}, (\bar{p}_z)_{z=-0.5h} = -\bar{p}_{(-)}, \bar{p}_{-z} = -\bar{p}_z$.

Similarly, dividing the resultant of the mass force $\overline{F}$ given by

$$\int_{z_1}^{z_2} \overline{F}\, \mathrm{d}\sigma^{z*}\, \mathrm{d}z = \int_{z_1}^{z_2} \overline{F} \overset{*}{H}_1 \overset{*}{H}_2\, \mathrm{d}\alpha_1\, \mathrm{d}\alpha_2\, \mathrm{d}z,$$

by $\overset{*}{A}_1 \overset{*}{A}_2\, \mathrm{d}\alpha_1\, \mathrm{d}\alpha_2$ and taking the sum of the resultant with Eq. (3.17), we obtain

$$\overline{X} A_1 A_2 = \frac{\bar{p}_z \overset{*}{H}_1 \overset{*}{H}_2}{\sqrt{(1+2\varepsilon_{11})(1+2\varepsilon_{22})}} + \int_{z_1}^{z_2} \frac{\overline{F} \overset{*}{H}_1 \overset{*}{H}_2\, \mathrm{d}z}{\sqrt{(1+2\varepsilon_{11})(1+2\varepsilon_{22})}}. \tag{3.18}$$

$\overline{X}$ is the resultant external force vector referred to the deformed middle surface $\overset{*}{S}$ of the shell.

Moments of external forces about an arbitrary point on $\overset{*}{S}$ are given by

$$\frac{\overset{*}{\bar{m}}\, h}{2} \times \bar{p}_{(+)} H_1^{(+)} H_2^{(+)}\, \mathrm{d}\alpha_1\, \mathrm{d}\alpha_2 \qquad \text{and} \qquad \frac{\overset{*}{\bar{m}}\, h}{2} \times \bar{p}_{(-)} H_1^{(-)} H_2^{(-)}\, \mathrm{d}\alpha_1\, \mathrm{d}\alpha_2.$$

Their sum divided by the surface area $\overset{*}{A}_1 \overset{*}{A}_2\, \mathrm{d}\alpha_1\, \mathrm{d}\alpha_2$ yields

$$\left(\overset{*}{\bar{m}} z \times \bar{p}_{(z)} \frac{\overset{*}{H}_1 \overset{*}{H}_2}{A_1 A_2 \sqrt{(1+2\varepsilon_{11})(1+2\varepsilon_{22})}} \right)_{z_1}^{z_2}.$$

Analogously, the moment of $\overline{F}$ per unit area of the element $\overset{*}{S}$ is given by

$$\int_{z_1}^{z_2} \left(\overset{*}{\bar{m}} z \times \overline{F} \right) \frac{\overset{*}{H}_1 \overset{*}{H}_2\, \mathrm{d}\alpha_1\, \mathrm{d}\alpha_2\, \mathrm{d}z}{A_1 A_2 \sqrt{(1+2\varepsilon_{11})(1+2\varepsilon_{22})}\, \mathrm{d}\alpha_1\, \mathrm{d}\alpha_2}$$

$$= \int_{z_1}^{z_2} \left(\overset{*}{\bar{m}} z \times \overline{F} \right) \frac{\overset{*}{H}_1 \overset{*}{H}_2\, \mathrm{d}z}{A_1 A_2 \sqrt{(1+2\varepsilon_{11})(1+2\varepsilon_{22})}}.$$

Hence, the resultant external moment vector $\overline{M}$ of external forces is found to be

$$\overline{M} A_1 A_2 = \left(\overset{*}{\bar{m}} z \times \bar{p}_{(z)} \frac{\overset{*}{H}_1 \overset{*}{H}_2}{\sqrt{(1+2\varepsilon_{11})(1+2\varepsilon_{22})}} \right)_{z_1}^{z_2}$$

$$+ \int_{z_1}^{z_2} \left(\overset{*}{\bar{m}} z \times \overline{F} \right) \frac{\overset{*}{H}_1 \overset{*}{H}_2\, \mathrm{d}z}{\sqrt{(1+2\varepsilon_{11})(1+2\varepsilon_{22})}}. \tag{3.19}$$

Again, we have arrived at the conclusion that the external forces acting upon the differential element are statically equivalent to the resultants of the external force and moment vectors, $\overline{X}$ and $\overline{M}$.

Decomposing $\bar{p}_i, \bar{p}_z$ and $\overline{F}$ in the direction of unit vectors $\overset{*}{\bar{e}}{}_1^z = \overset{*}{\bar{\rho}}_1/\overset{*}{H}_1, \overset{*}{\bar{e}}{}_2^z = \overset{*}{\bar{\rho}}_2/\overset{*}{H}_2$ and $\overset{*}{\bar{e}}{}_3^z = \overset{*}{\bar{m}}$, we have

$$\begin{aligned}
\bar{p}_1 &= \sigma_{11}\overset{*}{\bar{e}}{}_1^z + \sigma_{12}\overset{*}{\bar{e}}{}_2^z + \sigma_{13}\,\overset{*}{\bar{m}},\\
\bar{p}_2 &= \sigma_{21}\overset{*}{\bar{e}}{}_1^z + \sigma_{22}\overset{*}{\bar{e}}{}_2^z + \sigma_{23}\,\overset{*}{\bar{m}},\\
\bar{p}_z &= \sigma_{31}\overset{*}{\bar{e}}{}_1^z + \sigma_{32}\overset{*}{\bar{e}}{}_2^z + \sigma_{33}\,\overset{*}{\bar{m}},\\
\overline{F} &= \overset{*}{F}_1\overset{*}{\bar{e}}{}_1^z + \overset{*}{F}_2\overset{*}{\bar{e}}{}_2^z + \overset{*}{F}_3\,\overset{*}{\bar{m}},
\end{aligned} \tag{3.20}$$

where σ_{ij} $(i,\ j = 1,\ 2)$ are the internal stresses $(\sigma_{ij} = \sigma_{ji})$ and $\overset{*}{F}_j$ are the projections of $\overline{F}$ on the base $\overset{*}{\bar{e}}{}_1^z, \overset{*}{\bar{e}}{}_2^z, \overset{*}{\bar{m}}$ (Fig. 3.2). On substituting Eqs. (2.20) into (3.15) and (3.16), for $\overline{R}_i$ and $\overline{M}_i$, we find

$$\begin{aligned}
A_2\sqrt{(1+2\varepsilon_{22})}\,\overline{R}_1 &= \int_{z_1}^{z_2}\left(\sigma_{11}\overset{*}{\bar{e}}{}_1^z + \sigma_{12}\overset{*}{\bar{e}}{}_2^z + \sigma_{13}\,\overset{*}{\bar{m}}\right)\overset{*}{H}_2\,\mathrm{dz},\\
A_1\sqrt{(1+2\varepsilon_{11})}\,\overline{R}_2 &= \int_{z_1}^{z_2}\left(\sigma_{21}\overset{*}{\bar{e}}{}_1^z + \sigma_{22}\overset{*}{\bar{e}}{}_2^z + \sigma_{23}\,\overset{*}{\bar{m}}\right)\overset{*}{H}_1\,\mathrm{dz},\\
A_2\sqrt{(1+2\varepsilon_{22})}\,\overline{M}_1 &= \int_{z_1}^{z_2}\left(\overset{*}{\bar{m}}z \times (\sigma_{11}\overset{*}{\bar{e}}{}_1^z + \sigma_{12}\overset{*}{\bar{e}}{}_2^z)\right)\overset{*}{H}_2\,\mathrm{dz},\\
A_1\sqrt{(1+2\varepsilon_{11})}\,\overline{M}_2 &= \int_{z_1}^{z_2}\left(\overset{*}{\bar{m}}z \times (\sigma_{21}\overset{*}{\bar{e}}{}_1^z + \sigma_{22}\overset{*}{\bar{e}}{}_2^z)\right)\overset{*}{H}_1\,\mathrm{dz}.
\end{aligned}$$

Since $\overset{*}{\bar{e}}{}_1^z = \overset{*}{\bar{e}}_1, \overset{*}{\bar{e}}{}_2^z = \overset{*}{\bar{e}}_2$, (Eq. (1.51)), we have

$$\begin{aligned}
A_2\sqrt{(1+2\varepsilon_{22})}\,\overline{R}_1 &= \int_{z_1}^{z_2}\left(\sigma_{11}\overset{*}{\bar{e}}_1 + \sigma_{12}\overset{*}{\bar{e}}_2 + \sigma_{13}\,\overset{*}{\bar{m}}\right)\overset{*}{H}_2\,\mathrm{dz},\\
A_1\sqrt{(1+2\varepsilon_{11})}\,\overline{R}_2 &= \int_{z_1}^{z_2}\left(\sigma_{21}\overset{*}{\bar{e}}_1 + \sigma_{22}\overset{*}{\bar{e}}_2 + \sigma_{23}\,\overset{*}{\bar{m}}\right)\overset{*}{H}_1\,\mathrm{dz},\\
A_2\sqrt{(1+2\varepsilon_{22})}\,\overline{M}_1 &= \int_{z_1}^{z_2}\left(\overset{*}{\bar{m}}z \times (\sigma_{11}\overset{*}{\bar{e}}_1 + \sigma_{12}\overset{*}{\bar{e}}_2)\right)\overset{*}{H}_2\,\mathrm{dz}\\
A_1\sqrt{(1+2\varepsilon_{11})}\,\overline{M}_2 &= \int_{z_1}^{z_2}\left(\overset{*}{\bar{m}}z \times (\sigma_{21}\overset{*}{\bar{e}}_1 + \sigma_{22}\overset{*}{\bar{e}}_2)\right)\overset{*}{H}_1\,\mathrm{dz}
\end{aligned}$$

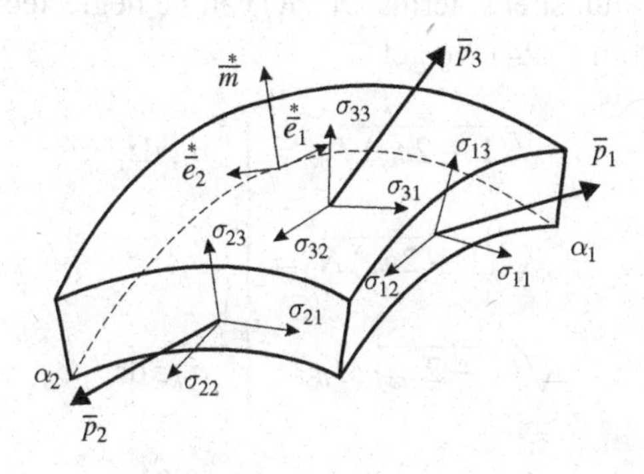

Fig. 3.2 Internal stresses in the shell.

where (see Eqs. (1.61) and (1.72))

$$\overset{*}{\bar{e}}_1 = [\bar{e}_1(1 + e_{11}) + \bar{e}_2 e_{22} + \bar{m}\varpi_1]/(1 + 2\varepsilon_{11}),$$

$$\overset{*}{\bar{e}}_2 = [\bar{e}_1 e_{11} + \bar{e}_2(1 + e_{22}) + \bar{m}\varpi_2]/(1 + 2\varepsilon_{22}), \tag{3.21}$$

$$\overset{*}{\bar{m}} = (\bar{e}_1 E1 + \bar{e}_2 E2 + mE3)/\sqrt{\mathbf{A}}.$$

On use of Eqs. (1.57), $\overline{R}_i$ and $\overline{M}_i$ can be written in the form

$$\overline{R}_1 = \overset{*}{T}_{11}\overset{*}{\bar{e}}_1 + \overset{*}{T}_{12}\overset{*}{\bar{e}}_2 + \overset{*}{N}_1\overset{*}{\bar{m}}, \qquad \overline{R}_2 = \overset{*}{T}_{21}\overset{*}{\bar{e}}_1 + \overset{*}{T}_{22}\overset{*}{\bar{e}}_2 + \overset{*}{N}_2\overset{*}{\bar{m}}, \tag{3.22}$$

$$\overline{M}_1 = \overset{*}{M}_{11}\overset{*}{\bar{e}}_2 - \overset{*}{M}_{12}\overset{*}{\bar{e}}_1, \qquad \overline{M}_2 = \overset{*}{M}_{21}\overset{*}{\bar{e}}_2 - \overset{*}{M}_{22}\overset{*}{\bar{e}}_1, \tag{3.23}$$

where

$$A_2\sqrt{(1+2\varepsilon_{22})}\,\overset{*}{T}_{11} = \int_{z_1}^{z_2} \sigma_{11}\overset{*}{H}_2\,\mathrm{dz}, \qquad A_2\sqrt{(1+2\varepsilon_{22})}\,\overset{*}{T}_{12} = \int_{z_1}^{z_2} \sigma_{12}\overset{*}{H}_2\,\mathrm{dz},$$

$$A_2\sqrt{(1+2\varepsilon_{22})}\,\overset{*}{N}_1 = \int_{z_1}^{z_2} \sigma_{13}\overset{*}{H}_2\,\mathrm{dz}, \qquad A_1\sqrt{(1+2\varepsilon_{11})}\,\overset{*}{T}_{21} = \int_{z_1}^{z_2} \sigma_{21}\overset{*}{H}_1\,\mathrm{dz},$$

$$A_1\sqrt{(1+2\varepsilon_{11})}\,\overset{*}{T}_{22} = \int_{z_1}^{z_2} \sigma_{22}\overset{*}{H}_1\,\mathrm{dz}, \qquad A_1\sqrt{(1+2\varepsilon_{11})}\,\overset{*}{N}_2 = \int_{z_1}^{z_2} \sigma_{23}\overset{*}{H}_1\,\mathrm{dz},$$

$$A_2\sqrt{(1+2\varepsilon_{22})}\,\overset{*}{M}_{11} = \int_{z_1}^{z_2} \sigma_{11}\overset{*}{H}_2 z\,\mathrm{dz}, \qquad A_2\sqrt{(1+2\varepsilon_{22})}\,\overset{*}{M}_{12} = \int_{z_1}^{z_2} \sigma_{12}\overset{*}{H}_2 z\,\mathrm{dz},$$

$$A_1\sqrt{(1+2\varepsilon_{11})}\,\overset{*}{M}_{21} = \int_{z_1}^{z_2} \sigma_{21}\overset{*}{H}_1 z\,\mathrm{dz}, \qquad A_1\sqrt{(1+2\varepsilon_{11})}\,\overset{*}{M}_{22} = \int_{z_1}^{z_2} \sigma_{22}\overset{*}{H}_1 z\,\mathrm{dz}. \tag{3.24}$$

The vectors $\overset{*}{T}_{i1}\overset{*}{\bar{e}}_1 + \overset{*}{T}_{i2}\overset{*}{\bar{e}}_2$ lie in the tangent plane of the deformed middle surface $\overset{*}{S}$. They are called the in-plane forces, namely $\overset{*}{T}_{11}$ and $\overset{*}{T}_{22}$ are normal forces, $\overset{*}{T}_{12}$ and $\overset{*}{T}_{21}$ are shear forces, and $\overset{*}{N}_i$ are lateral (cut) forces. The moments $\overset{*}{M}_{11}$ and $\overset{*}{M}_{22}$ are bending moments and $\overset{*}{M}_{12}$, and $\overset{*}{M}_{21}$ are twisting moments (Fig. 3.3). Since, for thin shells, terms $O(h/R)$ can be neglected without loss of the accuracy required, from (3.24) we get

$$\sqrt{(1+2\varepsilon_{ik})}\,\overset{*}{T}_{ik} = \int_{z_1}^{z_2} \sigma_{ik}\,\mathrm{dz},$$

$$\sqrt{(1+2\varepsilon_{ik})}\,\overset{*}{N}_i = \int_{z_1}^{z_2} \sigma_{i3}\,\mathrm{dz}, \tag{3.25}$$

$$\sqrt{(1+2\varepsilon_{ik})}\,\overset{*}{M}_{ik} = \int_{z_1}^{z_2} \sigma_{ik} z\,\mathrm{dz}.$$

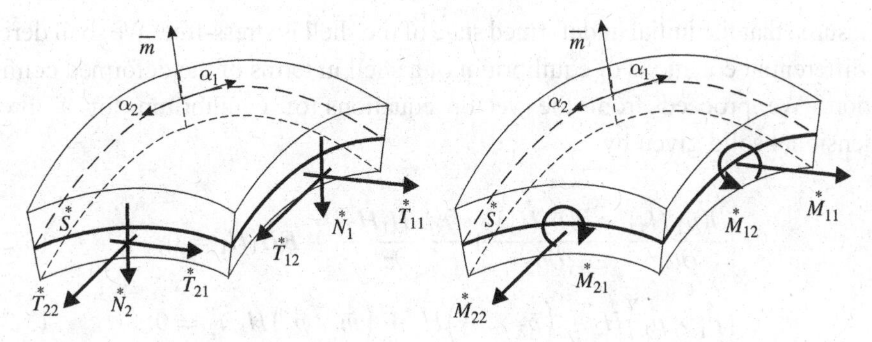

Fig. 3.3 Forces and moments in a thin shell.

On use of Eqs. (3.20) in (3.18) and (3.19) and neglecting terms $\overset{*}{k}_{ii}z \ll 1$, we find

$$\begin{aligned} \overline{X}A_1A_2 &= \overset{*}{X}_1\overset{*}{\bar{e}}_1 + \overset{*}{X}_2\overset{*}{\bar{e}}_2 + \overset{*}{X}_3\overset{*}{\bar{m}}, \\ \overline{M} &= \overset{*}{M}_1\overset{*}{\bar{e}}_2 - \overset{*}{M}_2\overset{*}{\bar{e}}_1, \end{aligned} \tag{3.26}$$

where

$$\overset{*}{X}_i = \frac{\sigma_{i3}}{\sqrt{(1+2\varepsilon_{11})(1+2\varepsilon_{22})}} + \int_{z_1}^{z_2} \frac{\overset{*}{F}_i\,\mathrm{d}z}{\sqrt{(1+2\varepsilon_{11})(1+2\varepsilon_{22})}},$$

$$\overset{*}{X}_3 = \frac{\sigma_{33}}{\sqrt{(1+2\varepsilon_{11})(1+2\varepsilon_{22})}} + \int_{z_1}^{z_2} \frac{\overset{*}{F}_3\,\mathrm{d}z}{\sqrt{(1+2\varepsilon_{11})(1+2\varepsilon_{22})}},$$

$$\overset{*}{M}_i = \frac{\sigma_{i3}z}{\sqrt{(1+2\varepsilon_{11})(1+2\varepsilon_{22})}} + \int_{z_1}^{z_2} \frac{\overset{*}{F}_i z\,\mathrm{d}z}{\sqrt{(1+2\varepsilon_{11})(1+2\varepsilon_{22})}}.$$

Here $\overset{*}{X}_i$ and $\overset{*}{M}_i$ are the projections of the external force and moment vectors on the base $\overset{*}{\bar{e}}_1, \overset{*}{\bar{e}}_2, \overset{*}{\bar{m}} \in \overset{*}{S}$.

3.3 Equations of equilibrium

From the modelling perspective, a thin shell can be treated as a three-dimensional solid. However, the complexity of the problem would be reduced significantly if its dimensionality could be reduced from three to two. To achieve this reduction, we introduce the second Kirchhoff–Love hypothesis. It states that 'the transverse normal stress is significantly smaller than other stresses in the shell, $\sigma_{33} \ll \sigma_{ik}$ $(i, k = 1, 2)$ and thus may be neglected'. In addition, recalling that the deformed state of the shell is completely defined in terms of deformations and curvatures of its middle surface, the shell can be regarded as a two-dimensional solid. Thus the equilibrium-conditions analysis can be based on the study of the resultant forces and moments taken over the thickness of the shell.

Assume that the initial undeformed state of the shell is stress-free. We shall derive the differential equations of equilibrium of a shell in terms of the deformed configuration. We proceed from the vector equations of equilibrium for a three-dimensional solid given by

$$\frac{\partial(\bar{p}_1 \overset{*}{H}_2)}{\partial\alpha_1} + \frac{\partial(\bar{p}_2 \overset{*}{H}_1)}{\partial\alpha_2} + \frac{\partial(\bar{p}_z \overset{*}{H}_1 \overset{*}{H}_2)}{\partial z} + \overline{F} \overset{*}{H}_1 \overset{*}{H}_1 = 0, \tag{3.27}$$

$$\left(\overset{*}{\bar{\rho}}_1 \times \bar{p}_1\right)\overset{*}{H}_2 + \left(\overset{*}{\bar{\rho}}_2 \times \bar{p}_2\right)\overset{*}{H}_1 + \left(\overset{*}{\bar{m}} \times \bar{p}_z\right)\overset{*}{H}_1 \overset{*}{H}_2 = 0. \tag{3.28}$$

On multiplying Eq. (3.27) by dz and integrating over the thickness of the shell, $z \in [z_1, z_2]$, we obtain

$$\frac{\partial(A_2 \overline{R}_1)}{\partial\alpha_1} + \frac{\partial(A_1 \overline{R}_2)}{\partial\alpha_2} + A_1 A_2 \overline{X} = 0. \tag{3.29}$$

Here $\overline{R}_i$ and $\overline{X}$ satisfy Eqs. (3.15) and (3.18).

By integrating the vector product of Eq. (3.27) and $\overset{*}{\bar{m}} z$ over z, we find

$$\int_{z_1}^{z_2} \left(\overset{*}{\bar{m}} z \times \left(\frac{\partial(\bar{p}_1 \overset{*}{H}_2)}{\partial\alpha_1} + \frac{\partial(\bar{p}_2 \overset{*}{H}_1)}{\partial\alpha_2} + \frac{\partial(\bar{p}_z \overset{*}{H}_1 \overset{*}{H}_2)}{\partial z} + \overline{F} \overset{*}{H}_1 \overset{*}{H}_2 \right)\right) \mathrm{d}z = 0. \tag{3.30}$$

Since

$$\begin{aligned}
\int_{z_1}^{z_2} \left(\overset{*}{\bar{m}} z \times \frac{\partial(\bar{p}_1 \overset{*}{H}_2)}{\partial\alpha_1} \right) \mathrm{d}z &= \int_{z_1}^{z_2} \frac{\partial}{\partial\alpha_1} \left(\overset{*}{\bar{m}} z \times \bar{p}_1 \overset{*}{H}_2 \right) \mathrm{d}z - \int_{z_1}^{z_2} \frac{\partial}{\partial\alpha_1} \left(\overset{*}{\bar{m}}_1 z \times \bar{p}_1 \overset{*}{H}_2 \right) \mathrm{d}z \\
&= \frac{\partial(A_2 \overline{M}_1)}{\partial\alpha_1} - \int_{z_1}^{z_2} \left(\left(\overset{*}{\bar{\rho}}_1 - \overset{*}{\bar{r}}_1 \right) \times \frac{\partial(\bar{p}_1 \overset{*}{H}_2)}{\partial\alpha_1} \right) \mathrm{d}z \\
&= \frac{\partial(A_2 \overline{M}_1)}{\partial\alpha_1} + \left(\overset{*}{\bar{r}}_1 \times \overline{R}_1 \right) A_2 - \int_{z_1}^{z_2} \left(\overset{*}{\bar{\rho}}_1 \times \bar{p}_1 \right) \overset{*}{H}_2 \, \mathrm{d}z,
\end{aligned}$$

$$\int_{z_1}^{z_2} \left(\overset{*}{\bar{m}} z \times \frac{\partial(\bar{p}_2 \overset{*}{H}_1)}{\partial\alpha_2} \right) \mathrm{d}z = \frac{\partial(A_1 \overline{M}_2)}{\partial\alpha_2} + \left(\overset{*}{\bar{r}}_2 \times \overline{R}_2 \right) A_1 - \int_{z_1}^{z_2} \left(\overset{*}{\bar{\rho}}_2 \times \bar{p}_2 \right) \overset{*}{H}_1 \, \mathrm{d}z,$$

$$\overset{*}{\bar{m}}_1 z = \overset{*}{\bar{\rho}}_1 - \overset{*}{\bar{r}}_1, \qquad \overset{*}{\bar{m}}_2 z = \overset{*}{\bar{\rho}}_2 - \overset{*}{\bar{r}}_2, \tag{3.31}$$

On substituting the left-hand sides of (3.31) into (3.30), with the help of Eq. (3.28), we get

$$\frac{\partial(A_2\overline{M}_1)}{\partial\alpha_1}+\frac{\partial(A_1\overline{M}_2)}{\partial\alpha_2}+A_1\left(\overset{*}{\bar{r}}_1\times\overline{R}_2\right)+A_2\left(\overset{*}{\bar{r}}_2\times\overline{R}_1\right)+A_1A_2\overline{M}=0. \tag{3.32}$$

Here $\overline{M}_i$ and $\overline{M}$ satisfy Eqs. (3.16) and (3.19). Although Eqs. (3.29)and (3.32) are derived under the assumption $h=$constant, they are also valid for shells of variable thickness $h=h(\alpha_1,\alpha_2)$.

On substituting $\overline{R}_i$, and $\overline{X}$ given by Eqs. (3.22) and (3.27) into (3.29), for the equilibrium equations we have

$$\begin{aligned}&\frac{\partial(A_2\overset{*}{T}_{11})}{\partial\alpha_1}+\frac{\partial(A_1\overset{*}{T}_{21})}{\partial\alpha_2}+\overset{*}{T}_{12}\frac{A_2}{\overset{*}{A}_2}\frac{\partial\overset{*}{A}_1}{\partial\alpha_2}-\overset{*}{T}_{22}\frac{A_1}{\overset{*}{A}_1}\frac{\partial\overset{*}{A}_2}{\partial\alpha_1}\\&\qquad+\overset{*}{A}_1A_2\overset{*}{N}_1\overset{*}{k}_{11}+A_1\overset{*}{A}_2\overset{*}{N}_2\overset{*}{k}_{12}+A_1A_2\overset{*}{X}_1=0,\\&\frac{\partial(A_1\overset{*}{T}_{22})}{\partial\alpha_2}+\frac{\partial(A_2\overset{*}{T}_{12})}{\partial\alpha_1}+\overset{*}{T}_{21}\frac{A_1}{\overset{*}{A}_1}\frac{\partial\overset{*}{A}_2}{\partial\alpha_1}-\overset{*}{T}_{11}\frac{A_2}{\overset{*}{A}_2}\frac{\partial\overset{*}{A}_1}{\partial\alpha_2}\\&\qquad+A_2\overset{*}{A}_1\overset{*}{N}_1\overset{*}{k}_{21}+\overset{*}{A}_2A_1\overset{*}{N}_2\overset{*}{k}_{22}+A_1A_2\overset{*}{X}_2=0\end{aligned} \tag{3.33}$$

and

$$\begin{aligned}&\frac{\partial(A_2\overset{*}{N}_1)}{\partial\alpha_1}+\frac{\partial(A_1\overset{*}{N}_2)}{\partial\alpha_2}-\overset{*}{A}_1A_2\overset{*}{T}_{11}\overset{*}{k}_{11}-A_1\overset{*}{A}_2\overset{*}{T}_{22}\overset{*}{k}_{22}\\&\qquad-A_1\overset{*}{A}_2\overset{*}{T}_{12}\overset{*}{k}_{12}-A_1\overset{*}{A}_2\overset{*}{T}_{21}\overset{*}{k}_{12}+A_1A_2\overset{*}{X}_3=0.\end{aligned} \tag{3.34}$$

In deriving Eqs. (3.33) and (3.34) use is made of formulas for $\partial\overset{*}{\bar{e}}_i/\partial\alpha_k$ and $\overset{*}{\bar{m}}_i$, namely

$$\frac{\partial\overset{*}{\bar{e}}_1}{\partial\alpha_1}=-\frac{\overset{*}{\bar{e}}_2}{\overset{*}{A}_2}\frac{\partial\overset{*}{A}_1}{\partial\alpha_2}-\overset{*}{A}_1\overset{*}{k}_{11}\overset{*}{\bar{m}},\qquad \frac{\partial\overset{*}{\bar{e}}_1}{\partial\alpha_2}=\frac{\overset{*}{\bar{e}}_2}{\overset{*}{A}_1}\frac{\partial\overset{*}{A}_2}{\partial\alpha_1}-\overset{*}{A}_2\overset{*}{k}_{12}\overset{*}{\bar{m}},$$

$$\frac{\partial\overset{*}{\bar{e}}_2}{\partial\alpha_1}=\frac{\overset{*}{\bar{e}}_1}{\overset{*}{A}_2}\frac{\partial\overset{*}{A}_1}{\partial\alpha_2}-\overset{*}{A}_1\overset{*}{A}_{12}\overset{*}{\bar{m}},\qquad \frac{\partial\overset{*}{\bar{e}}_2}{\partial\alpha_2}=-\frac{\overset{*}{\bar{e}}_1}{\overset{*}{A}_1}\frac{\partial\overset{*}{A}_2}{\partial\alpha_1}-\overset{*}{A}_2\overset{*}{k}_{22}\overset{*}{\bar{m}},$$

$$\overset{*}{\bar{m}}_i=\overset{*}{A}_i\left(\overset{*}{k}_{1i}\overset{*}{\bar{e}}_1+\overset{*}{k}_{2i}\overset{*}{\bar{e}}_2\right),$$

Analogously, on substituting $\overline{M}_i$ and $\overline{M}$ given by Eqs. (3.23) and (3.27) into (3.32) with the help of Eqs. (3.22) and the formulas given above, the equilibrium equations of moments take the form

$$\frac{\partial(A_2\overset{*}{M}_{11})}{\partial\alpha_1}+\frac{\partial(A_1\overset{*}{M}_{21})}{\partial\alpha_2}+\overset{*}{M}_{12}\frac{A_2}{\overset{*}{A}_2}\frac{\partial\overset{*}{A}_1}{\partial\alpha_2}-\overset{*}{M}_{22}\frac{A_1}{\overset{*}{A}_1}\frac{\partial\overset{*}{A}_2}{\partial\alpha_1}$$
$$+\overset{*}{A}_1A_2\overset{*}{M}_1-A_1\overset{*}{A}_2\overset{*}{N}_1=0,$$
$$\frac{\partial(A_2\overset{*}{M}_{12})}{\partial\alpha_1}+\frac{\partial(A_1\overset{*}{M}_{22})}{\partial\alpha_2}+\overset{*}{M}_{21}\frac{A_1}{\overset{*}{A}_1}\frac{\partial\overset{*}{A}_2}{\partial\alpha_1}-\overset{*}{M}_{11}\frac{A_2}{\overset{*}{A}_2}\frac{\partial\overset{*}{A}_1}{\partial\alpha_2}$$
$$+\overset{*}{A}_2A_1\overset{*}{M}_2-A_2\overset{*}{A}_1\overset{*}{N}_2=0 \tag{3.35}$$

and

$$A_1\overset{*}{A}_2\overset{*}{T}_{12}-\overset{*}{A}_1A_2\overset{*}{T}_{21}+A_1\overset{*}{A}_2\overset{*}{M}_{12}\overset{*}{k}_{11}-\overset{*}{A}_1A_2\overset{*}{M}_{21}\overset{*}{k}_{22}$$
$$+\overset{*}{A}_1A_2\overset{*}{M}_{22}\overset{*}{k}_{12}-A_1\overset{*}{A}_2\overset{*}{M}_{11}\overset{*}{k}_{12}=0. \tag{3.36}$$

The system of Eqs. (3.33)–(3.36) contains ten unknowns: $\overset{*}{T}_{ik}$, $\overset{*}{N}_i$ and $\overset{*}{M}_{ik}$ ($i, k =$ 1, 2). Introducing differential operators defined by

$$L_1\left(\overset{*}{T}_{ik}\right)=\frac{\partial(A_2\overset{*}{T}_{11})}{\partial\alpha_1}+\frac{\partial(A_1\overset{*}{T}_{21})}{\partial\alpha_2}+\overset{*}{T}_{12}\frac{A_2}{\overset{*}{A}_2}\frac{\partial\overset{*}{A}_1}{\partial\alpha_2}-\overset{*}{T}_{22}\frac{A_1}{\overset{*}{A}_1}\frac{\partial\overset{*}{A}_2}{\partial\alpha_1},$$
$$L_2\left(\overset{*}{T}_{ik}\right)=\frac{\partial(A_2\overset{*}{T}_{12})}{\partial\alpha_1}+\frac{\partial(A_1\overset{*}{T}_{22})}{\partial\alpha_2}+\overset{*}{T}_{21}\frac{A_1}{\overset{*}{A}_1}\frac{\partial\overset{*}{A}_2}{\partial\alpha_1}-\overset{*}{T}_{11}\frac{A_2}{\overset{*}{A}_2}\frac{\partial\overset{*}{A}_1}{\partial\alpha_2}, \tag{3.37}$$

and

$$L_1\left(\overset{*}{M}_{ik}\right)=\frac{\partial(A_2\overset{*}{M}_{11})}{\partial\alpha_1}+\frac{\partial(A_1\overset{*}{M}_{21})}{\partial\alpha_2}+\overset{*}{M}_{12}\frac{A_2}{\overset{*}{A}_2}\frac{\partial\overset{*}{A}_1}{\partial\alpha_2}-\overset{*}{M}_{22}\frac{A_1}{\overset{*}{A}_1}\frac{\partial\overset{*}{A}_2}{\partial\alpha_1},$$
$$L_2\left(\overset{*}{M}_{ik}\right)=\frac{\partial(A_2\overset{*}{M}_{12})}{\partial\alpha_1}+\frac{\partial(A_1\overset{*}{M}_{22})}{\partial\alpha_2}+\overset{*}{M}_{21}\frac{A_1}{\overset{*}{A}_1}\frac{\partial\overset{*}{A}_2}{\partial\alpha_1}-\overset{*}{M}_{11}\frac{A_2}{\overset{*}{A}_2}\frac{\partial\overset{*}{A}_1}{\partial\alpha_2}, \tag{3.38}$$

the first two equations in (3.33) and (3.35) can be written in the form

$$L_1\left(\overset{*}{T}_{ik}\right)\overset{*}{A}_1A_2\overset{*}{N}_1\overset{*}{k}_{11}+A_1\overset{*}{A}_2\overset{*}{N}_2\overset{*}{k}_{12}+A_1A_2\overset{*}{X}_1=0,$$
$$L_2\left(\overset{*}{T}_{ik}\right)+\overset{*}{A}_2A_1\overset{*}{N}_2\overset{*}{k}_{22}+A_2\overset{*}{A}_1\overset{*}{N}_1\overset{*}{k}_{21}+A_1A_2\overset{*}{X}_2=0, \tag{3.39}$$

$$\frac{\partial(A_2\overset{*}{N}_1)}{\partial\alpha_1}+\frac{\partial(A_1\overset{*}{N}_2)}{\partial\alpha_2}-\overset{*}{A}_1A_2\overset{*}{T}_{11}\overset{*}{k}_{11}-A_1\overset{*}{A}_2\overset{*}{T}_{22}\overset{*}{k}_{22}$$
$$-A_1\overset{*}{A}_2\overset{*}{T}_{12}\overset{*}{k}_{12}-A_1\overset{*}{A}_2\overset{*}{T}_{21}\overset{*}{k}_{12}+A_1A_2\overset{*}{X}_3=0, \tag{3.40}$$

$$L_1\left(\overset{*}{M}_{ik}\right)+\overset{*}{A}_1\overset{*}{A}_2\overset{*}{M}_1-A_1\overset{*}{A}_2\overset{*}{N}_1=0,$$
$$L_2\left(\overset{*}{M}_{ik}\right)+\overset{*}{A}_2A_1\overset{*}{M}_1-A_2\overset{*}{A}_1\overset{*}{N}_1=0, \tag{3.41}$$

$$\overset{*}{A}_1\overset{*}{A}_2\overset{*}{T}_{12}-\overset{*}{A}_1A_2\overset{*}{T}_{21}+A_1\overset{*}{A}_2\overset{*}{M}_{12}\overset{*}{k}_{11}-\overset{*}{A}_1A_2\overset{*}{M}_{21}\overset{*}{k}_{22}$$
$$+\overset{*}{A}_1A_2\overset{*}{M}_{22}\overset{*}{k}_{12}-A_1\overset{*}{A}_2\overset{*}{M}_{11}\overset{*}{k}_{12}=0. \tag{3.42}$$

Equations (3.39)–(3.42) are nonlinear. The nonlinearity is introduced by the curvatures of the surface, $\overset{*}{k}_{ij}=k_{ij}+æ_{ij}$, projections of the forces and moments $\overset{*}{X}_i,\overset{*}{X}_3,\overset{*}{M}_i,\overset{*}{T}_{ik},\overset{*}{N}_i$ and $\overset{*}{M}_{ik}$ on the deformed axes, and, additionally, may be brought in by constitutive relations for the constructive material of a shell.

Proceeding from the second equation of equilibrium (3.29) and projecting it onto the orthogonal base $\bar{e}_1,\bar{e}_2,\bar{m}$, for the tangent T_{ik} and lateral forces N_i, we find

$$T_{11}=\overline{R}_1\bar{e}_1,\qquad T_{12}=\overline{R}_1\bar{e}_2,\qquad T_{21}=\overline{R}_2\bar{e}_1,\qquad T_{22}=\overline{R}_2\bar{e}_2,$$
$$N_i=\overline{R}_i\overline{m},\quad X_i=\overline{X}\overline{e}_i,\quad X_3=\overline{X}\overline{m}.$$

On substituting expressions for $\overline{R}_i,\overline{X}$ and $\overset{*}{\bar{e}}_1,\overset{*}{\bar{e}}_2,\overset{*}{\bar{m}}$ given by Eqs. (3.21), (3.23) and (3.27), we obtain

$$T_{11}=\left(\overset{*}{T}_{11}\overset{*}{\bar{e}}_1+\overset{*}{T}_{12}\overset{*}{\bar{e}}_2+\overset{*}{N}_1\overset{*}{\bar{m}}\right)\bar{e}_1=\overset{*}{T}_{11}(1+e_{11})/(1+2\varepsilon_{11})$$
$$+\overset{*}{T}_{12}e_{12}/(1+2\varepsilon_{22})+\overset{*}{N}_1\mathrm{S}_1/\sqrt{\mathbf{A}},$$
$$T_{12}=\overset{*}{T}_{11}e_{12}/(1+2\varepsilon_{11})+\overset{*}{T}_{12}(1+e_{22})/(1+2\varepsilon_{22})+\overset{*}{N}_1\mathrm{S}_2/\sqrt{\mathbf{A}}, \tag{3.43}$$
$$T_{22}=\overset{*}{T}_{21}e_{12}/(1+2\varepsilon_{11})+\overset{*}{T}_{22}(1+e_{22})/(1+2\varepsilon_{22})+\overset{*}{N}_2\mathrm{S}_2/\sqrt{\mathbf{A}},$$
$$T_{21}=\overset{*}{T}_{21}(1+e_{11})/(1+2\varepsilon_{11})+\overset{*}{T}_{22}e_{21}/(1+2\varepsilon_{22})+\overset{*}{N}_2\mathrm{S}_1/\sqrt{\mathbf{A}},$$

$$N_1=\overset{*}{T}_{11}\varpi_1/(1+2\varepsilon_{11})+\overset{*}{T}_{12}\varpi_2/(1+2\varepsilon_{22})+\overset{*}{N}_1\mathrm{S}_3/\sqrt{\mathbf{A}}, \tag{3.44}$$
$$N_2=\overset{*}{T}_{22}\varpi_2/(1+2\varepsilon_{22})+\overset{*}{T}_{21}\varpi_1/(1+2\varepsilon_{11})+\overset{*}{N}_2\mathrm{S}_3/\sqrt{\mathbf{A}},$$

$$X_1=\overset{*}{X}_1(1+e_{11})/(1+2\varepsilon_{11})+\overset{*}{X}_2e_{12}/(1+2\varepsilon_{22})+\overset{*}{X}_3\varpi_1/\sqrt{\mathbf{A}},$$
$$X_2=\overset{*}{X}_1e_{21}/(1+2\varepsilon_{11})+\overset{*}{X}_2(1+e_{22})/(1+2\varepsilon_{22})+\overset{*}{X}_3\varpi_2/\sqrt{\mathbf{A}},$$
$$X_3=\left(\overset{*}{X}_1\mathrm{S}_1+\overset{*}{X}_2\mathrm{S}_2+\overset{*}{X}_3\mathrm{S}_3\right)\Big/\sqrt{\mathbf{A}}.$$

Analogously, by substituting Eqs. (1.19) and (1.23) into (3.29) we obtain the equilibrium equations for the thin shell in terms of the undeformed configuration:

$$\begin{aligned} &L_1(T_{ik}) + A_1A_2N_1k_{11} + A_1A_2N_2k_{12} + A_1A_2X_1 = 0, \\ &L_2(T_{ik}) + A_1A_2N_2k_{22} + A_1A_2N_1k_{21} + A_1A_2X_2 = 0, \end{aligned} \tag{3.45}$$

$$\begin{aligned} &\frac{\partial(A_2N_1)}{\partial\alpha_1} + \frac{\partial(A_1N_2)}{\partial\alpha_2} - A_1A_2T_{11}k_{11} - A_1A_2T_{22}k_{22} \\ &\qquad - A_1A_2T_{12}k_{12} - A_1A_2T_{21}k_{12} + A_1A_2X_2 = 0. \end{aligned} \tag{3.46}$$

The resultant internal and external force vectors $\overline{R}_i$ and $\overline{X}$ are given by

$$\begin{aligned} \overline{R}_1 &= T_{11}\bar{e}_1 + T_{12}\bar{e}_2 + N_1\bar{m}, \\ \overline{R}_2 &= T_{21}\bar{e}_1 + T_{22}\bar{e}_2 + N_2\bar{m}, \\ \overline{X} &= X_1\bar{e}_1 + X_2\bar{e}_2 + X_3\bar{m}. \end{aligned} \tag{3.47}$$

The equilibrium equations for moments Eqs. (3.41) and (3.42) can be recast in a similar way. The resultant equations are very bulky and are not given here. They are left as an exercise.

Exercises

1. Discuss the advantages and limitations of the Kirchhoff–Love hypotheses.
2. Verify Eqs. (3.29) and (3.32).
3. Verify Eqs. (3.33) and (3.34).
4. Verify Eqs. (3.35) and (3.36).
5. Verify Eqs. (3.45) and (3.46).
6. Derive the equation of equilibrium of a shell for moments in terms of the undeformed configuration.
7. Derive the final form of the equations of equilibrium (3.39)–(3.42) of a thin shell assuming that the constructive material is *Fung-elastic* (Fung, 1993).

4

The continuum model of the biological tissue

4.1 Structure of the tissue

A fundamental goal in constitutive modelling is to predict the mechanical behaviour of a material under various loading states. A biological tissue is a collection of cells, and extracellular matrices, that perform various specialized functions. There are four basic primary tissues types: muscular, nervous, epithelial and connective. Muscle tissue produces mechanical work through contraction–relaxation. For example, skeletal muscles are responsible for locomotion primarily through voluntary muscle contraction, cardiac muscles provide the active pumping of blood from the heart, and smooth muscles, which are part of the organs of the digestive tract, facilitate peristalsis, propulsion, microcirculation, etc.

Nervous tissue provides communication among organs and systems predominantly by electrical signals. Neurons are responsible for the production and propagation of the waves of depolarization in the myelinated and unmyelinated nerve fibres, smooth muscle syncytia and other cell aggregates. Neuroglial cells are a diverse group of morphoelements that play a supportive, mainly trophic, role.

Epithelial tissue covers the outer and inner surfaces of most of the organs. Various types of cells line the digestive, reproductive and urinary tracts, blood vessels, ducts, etc. They act as a protective barrier and are instrumental in selective regulation of the transport of specific agents and substrates.

Connective tissue includes a diverse set of cells surrounded by a large amount of extracellular matrix. Its main function is to provide a level of mechanical support to the organ.

The wall of the abdominal viscera is a biological composite formed of three histologically identified layers: mucosa, connective stroma, with embedded smooth muscle fibres, and serosa. The mucosa is the innermost layer, consisting of sheets of epithelial cells that line the outer surface of the gastrointestinal tract and have primarily secretory, digestive, absorptive and protective functions. The cells are tightly packed, with little extracellular substance between them, and are supported on a basal membrane. One of the characteristics of the cells is a stable apico-basal polarity, which is expressed in morphological, electrophysiological and transport properties. The ionic channels are predominantly located within the apical membrane, while ion-pumps are situated in the baso-lateral domain. This arrangement is of morphomechanical importance. It endows a required amount of turgor pressure in the subepithelial cavities. The epithelial cells are bound together by tight junctions to form an integrated net that gives the cell sheet mechanical strength and makes it impermeable to passive diffusion of small molecules and solutes.

The fibrillary connective tissue consists of insoluble high-relative-molecular-mass polymers of proteins, called collagen and elastin. It functions as a supportive stroma to multiple cellular elements. The fibres are loosely woven and packed in an ordered way to form a net for smooth muscle bundles. Such structural organization of the tissue provides stability to the wall and allows organs to undergo reversible changes in length and diameter, while offering remarkable properties in terms of stiffness and elasticity.

The muscle coat, depending on the organ, is made of two or three distinct, smooth muscle layers arranged into syncytia. The outer layer is composed of smooth muscle fibres with their axes oriented along the principal structural axis of the organ. The inner layer is made of cells that run circumferentially relative to the outer layer. Although syncytia within the tissue are morphologically distinct, there are intermediate muscle bundles that pass from one layer to the other. Additionally, fibres form multiple cell–stroma junctions with the connective tissue net. They are of mechanical significance, i.e. they provide an even stress–strain distribution during the contraction–relaxation reaction. The thickness of the muscle layers can vary greatly between individuals and according to the anatomical part of the organ.

Multiple physiological functions of the organs of the gastrointestinal tract are under precise control of the autonomic nervous system. It is composed of various types of nerve cells identified as sensory, inter-, motor neurons, etc. Each neuron has a soma, a number of branching dendrites and the unmyelinated axon. They are arranged in planar neuronal networks via neuronal junctions called synapses. Their main function is to transfer electrically coded information among neurons and from neurons to muscles.

The outermost layer, the serosa, is composed of a thin sheet of epithelial cells and connective tissue.

4.2 Biocomposite as a mechanochemical continuum

The connection between the geometrical and statical quantities studied in the previous chapters must be complemented by equations establishing relationships among the stresses and deformations, their rates, temperature and structural changes

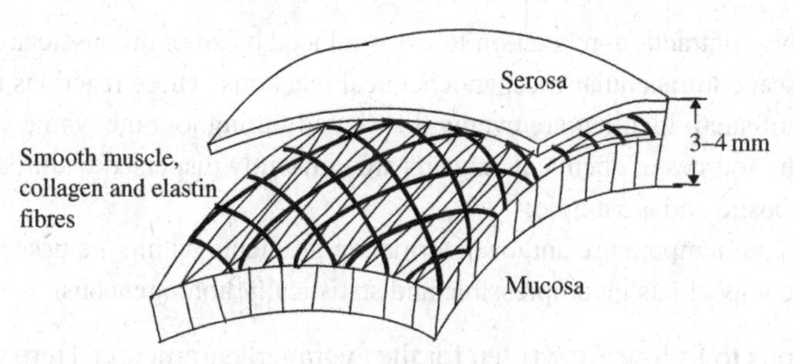

Fig. 4.1 The wall of the gastrointestinal-tract organs as a biological composite.

of constructive materials, e.g. the tissue that forms the wall of a biological shell. The complete theoretical formulation is best achieved by application of the principles of thermodynamics supported by extensive experimentation, including in-plane and complex loading testing. The advantage of such an approach is that it employs generalized quantities such as entropy, free energy and Gibbs potential as fundamental descriptors. Specific problems are encountered due to the discrete morphological structure of the biological tissue and the continuum scale, which is typically ~1 μm. For example, because of existing anisotropy, multidimensional strain data from the uniaxial experiments are not enough to extrapolate to the fully three-dimensional constitutive equations. Further, small specimen sizes, tethering effects, heterogeneity of deformation and difficulty in maintaining constant force distribution along specimen edges make experiments on soft tissues very difficult. Also, being heterogeneous, anisotropic, nonlinear, viscoelastic, incompressible composites, soft biomaterials defy simple material models. Accounting for these particulars both in constitutive models and in experimental evaluations remains a great challenge.

There is no strict method for deciding on a specific choice of form of the model. The process of formulation is somewhat arbitrary and depends on the investigator's needs and preferences. In this book we shall adopt the phenomenological approach (Usik, 1973; Nikitin, 1980) to model a soft biocomposite – the wall of gastrointestinal organs. Our derivations will be based on the following histological assumptions.

(i) The biomaterial is a three-phase, multicomponent, mechanochemically active, anisotropic medium; phase 1 comprises the connective tissue net, phase 2 mechanochemically active smooth muscle fibres and phase 3 inert myofibrils (Fig. 4.1).
(ii) The phase interfaces are semipermeable to certain substrates.
(iii) Smooth muscle, collagen and elastin fibres are the main weight-bearing and active force-generating elements.
(iv) The biocomposite endows properties of general curvilinear anisotropy and viscoelasticity; the viscous properties are due to the mechanics of smooth muscle fibre and the elastic properties depend mainly on the collagen and elastin fibres.

(v) The active contraction–relaxation forces produced by smooth muscle are the result of multicascade intracellular mechanochemical reactions. These reactions run in a large number of small loci that are evenly distributed throughout the whole volume of the tissue; the sources of chemical reagents are uniformly dispersed within the volume of the composite and are ample.

(vi) There are no temperature and/or deformation gradients within the tissue.

(vii) The biocomposite is incompressible and statistically homogeneous.

All equations to follow are written for the averaged parameters. Here we adopt the following notation: the quantities obtained by averaging over the volume of a particular phase are contained in angle brackets, whereas those free of angle brackets are attained by averaging over the entire volume.

Let ρ be the mean density of the tissue. The partial density of the ζth substrate ($\zeta = 1-n$) in phase β ($\beta = 1, 2, 3$) is defined as

$$\rho_\zeta^\beta = m_\zeta^\beta / \upsilon,$$

where m_ζ^β is the mass of the ζth substrate and υ is the total elementary volume of the tissue $\upsilon = \sum_{\beta=1}^{3} \upsilon^\beta$. The mass and the effective concentrations of substrates are

$$c_\zeta^\beta = \rho_\zeta^\beta / \rho, \quad \langle c^\beta \rangle = m_\zeta^\beta / (\upsilon^\beta \rho^\beta). \tag{4.1}$$

Assuming $\rho = \langle \rho^\beta \rangle = \text{constant}$, we have

$$\langle \rho^\beta \rangle = \sum_{\zeta=1}^{n} \frac{m_\zeta^\beta}{\upsilon^\beta}. \tag{4.2}$$

Setting $\beta = 1$, we find

$$\langle \rho^1 \rangle = \sum_{\zeta=1}^{n} \frac{m_\zeta^1}{\upsilon^1} = \sum_{\zeta=1}^{n} \frac{\rho_\zeta^1 \upsilon}{\upsilon^1} = \sum_{\zeta=1}^{n} \frac{c_\zeta^1 \rho \upsilon}{\upsilon^1} = \rho \sum_{\zeta=1}^{n} \frac{c_\zeta^1}{\eta},$$

where η is the porosity of phase β ($\eta = \upsilon / \upsilon^\beta$). It is easy to show that

$$\eta = \sum_{\zeta=1}^{n} c_\zeta^1 \equiv c^1. \tag{4.3}$$

Hence the mass c_ζ^β and the effective $\left\langle c_\zeta^\beta \right\rangle$ concentrations are interrelated by $c_\zeta^\beta = \eta \left\langle c_\zeta^\beta \right\rangle$. The sum of all concentrations c_ζ^β in the medium equals unity $\left(\sum_{\zeta=1}^{n} c_\zeta^\beta = 1 \right)$.

Change in the concentration of constituents in different phases is due to the exchange of matter among phases, external fluxes, chemical reactions and diffusion. Since chemical reactions run only in phase 2 and the substrates move at the same velocity, there is no diffusion within phases. Hence, the equations of conservation of mass of the ζth substrate in the medium are given by

$$\rho \frac{\mathrm{d}c_\zeta^1}{\mathrm{d}t} = Q_\zeta^1, \qquad \rho \frac{\mathrm{d}c_\zeta^2}{\mathrm{d}t} = Q_\zeta^2 + \sum_{j=1}^{r} \nu_{\zeta j} J_j, \qquad \rho \frac{\mathrm{d}c_\zeta^3}{\mathrm{d}t} = Q_\zeta^3. \tag{4.4}$$

Here Q_ς^β is the velocity of influx of the ζth substrate into phase α, $\nu_{\zeta j} J_j$ is the rate of formation of the ζth substrate in the jth chemical reaction ($j = 1-r$). The quantity $\nu_{\zeta j}$ is related to the relative molecular mass M_ζ of the substrate ζ and is analogous to the stoichiometric coefficient in the jth reaction. $\nu_{\zeta j}$ takes positive values if the substrate is formed and becomes negative if the substrate disassociates. Since the mass of reacting components is conserved in each chemical reaction, we have

$$\sum_{\zeta=1}^{n} \nu_{\zeta j} = 0.$$

Assume that there is a flux Q_ς^β of the matter into (i) phase 1 from external sources and phase 3, (ii) phase 2 from phases 1 and 3 only, and (iii) phase 3 from phase 2. Hence, we have

$$Q_\zeta^1 = -Q_\zeta + Q_\zeta^{\mathrm{e}}, \qquad Q_\zeta^2 = Q_\zeta + Q_\zeta^{\mathrm{m}}, \qquad Q_\zeta^3 = -Q_\zeta^{\mathrm{m}}, \tag{4.5}$$

where Q_ζ^{e} is the flux of externally distributed sources, Q_ζ is the exchange flux between phases and Q_ζ^{m} is the flux of matter from phase 3 into phase 2 (Fig. 4.2). Applying the incompressibility condition to (4.5), we have

$$\sum_{\zeta=1}^{n} Q_\zeta^{\mathrm{e}} = 0.$$

Let also $\sum_{\zeta=1}^{n} Q_\zeta = 0$. Assuming that the effective concentration of substrates in phase 3 remains constant, throughout, $\left\langle c_\zeta^3 \right\rangle$ = constant, and using Eqs. (4.5) in (4.4), we obtain

$$\rho \frac{\mathrm{d}c_\zeta^1}{\mathrm{d}t} = -Q_\zeta + Q_\zeta^{\mathrm{e}}, \qquad \rho \frac{\mathrm{d}c_\zeta^2}{\mathrm{d}t} = Q_\zeta - Q_\zeta^3 + \sum_{j=1}^{r} \nu_{\zeta j} J_j, \qquad \rho \frac{\mathrm{d}c_\zeta^3}{\mathrm{d}t} = -Q_\zeta. \tag{4.6}$$

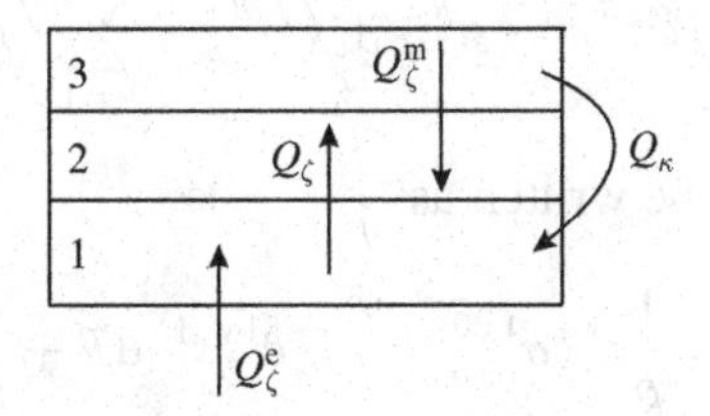

Fig. 4.2 Flux exchanges in a three-phase biocomposite.

In the above, we neglected the convective transport of the matter within phases.

The equations of continuity and the conservation of momentum for the tissue treated as a three-dimensional solid in a fixed Cartesian coordinate system (x_1, x_2, x_3) are given by

$$\frac{\partial u}{\partial x_1} + \frac{\partial v}{\partial x_2} + \frac{\partial w}{\partial x_3} = 0, \tag{4.7}$$

$$\rho \frac{\partial^2 u_i}{\partial t^2} = \frac{\partial \sigma_{ij}}{\partial x_j} + \rho f_i. \tag{4.8}$$

Here u, v and w are the components of the displacement vector, f_i is the mass force, and σ_{ij} $(i, j = x_1, x_2, x_3)$ are stresses (see Chapter 2).

Let $U^{(\beta)}$, $s^{(\beta)}$ and σ^{α}_{ij} be the free energy, entropy and stresses of each phase. Hence the Gibbs relations for each phase are defined by

$$c^1\langle U^1\rangle = U_0^1(c_\zeta^1, T) + \frac{1}{2\rho} E_{ijlm}\varepsilon_{ij}\varepsilon_{lm}, \tag{4.9}$$

$$c^\beta\langle U^\beta\rangle = U_0^\beta(c_\zeta^\beta, T) + \frac{1}{2\rho} Y_{ijlm}\varsigma_{ij}^\beta\varsigma_{lm}^\beta, \tag{4.10}$$

$$\begin{aligned} \mathrm{d}(c^1\langle U^1\rangle) &= \mathrm{d}U^1 \\ &= \frac{1}{\rho} c^1\langle\sigma^1\rangle_{ij}\mathrm{d}\varepsilon_{ij}^1 - c^1\langle s^1\rangle \mathrm{d}T + \sum_{\varsigma=1}^{n}\left\langle\mu_\zeta^1\right\rangle \mathrm{d}\left\langle c_\zeta^1\right\rangle, \end{aligned} \tag{4.11}$$

$$\begin{aligned} \mathrm{d}(c^\beta\langle U^\beta\rangle) &= \mathrm{d}U^\beta \\ &= \frac{1}{\rho} c^\beta\langle\sigma^\beta\rangle_{ij}\mathrm{d}\varsigma_{ij}^\beta - c^\beta\langle s^\beta\rangle \mathrm{d}T + \sum_{\varsigma=1}^{n}\left\langle\mu_\zeta^\beta\right\rangle \mathrm{d}\left\langle c_\zeta^\beta\right\rangle, \end{aligned} \tag{4.12}$$

$$\left\langle\mu_\zeta^\beta\right\rangle = \partial\langle U^\beta\rangle/\partial\left\langle c_\zeta^\beta\right\rangle, \quad \langle s^\beta\rangle = \partial\langle U^\beta\rangle/\partial T \quad (\beta = 2, 3),$$

where T is the temperature, μ_ζ^β is the chemical potential of the ζth substrate in phase β, $\mu_\zeta^\beta = \partial c^\beta\langle U^\beta\rangle/\partial c_\zeta^\beta$, ς_{ij} is the elastic and Δ_{ij} is the viscous part of deformation ($\varepsilon_{ij}^\beta = \varsigma_{ij}^\beta + \Delta_{ij}^\beta$, $\beta = 2, 3$). Making use of the equality

$$\partial(c^\beta\langle U^\beta\rangle)/\partial c_\zeta^\beta = \left\langle\mu_\zeta^\beta\right\rangle + \langle U^\beta\rangle - \sum_{\varsigma=1}^{n}\left\langle\mu_\zeta^\beta\right\rangle\left\langle c_\zeta^\beta\right\rangle,$$

Eqs. (4.11) and (4.12) can be written as

$$\mathrm{d}(c^1\langle U^1\rangle) = \frac{1}{\rho} c^1\langle\sigma^1\rangle_{ij}\,\mathrm{d}\varepsilon_{ij}^1 - c^1\langle s^1\rangle \mathrm{d}T + \sum_{\varsigma=1}^{n}\mu_\zeta^1\,\mathrm{d}c_\zeta^1, \tag{4.13}$$

$$\mathrm{d}\left(c^\beta\langle U^\beta\rangle\right) = \frac{1}{\rho}c^\beta\langle\sigma^\beta\rangle_{ij}\,\mathrm{d}\varsigma_{ij}^\beta - c^\beta\langle s^\beta\rangle\mathrm{d}T + \sum_{\varsigma=1}^{n}\mu_\zeta^\beta\,\mathrm{d}c_\zeta^\beta \quad (\beta = 2,3).$$

Assuming that the mass sources are present only in phases 1 and 2, the general heat flux and the second law of thermodynamics for the tissue are described by

$$\mathrm{d}U = \frac{1}{\rho}\sigma_{ij}\,\mathrm{d}\varepsilon_{ij} - s\,\mathrm{d}T - \mathrm{d}q' + \frac{1}{\rho}\sum_{\varsigma=1}^{n}\frac{\partial F}{\partial c_\varsigma^1}Q_\zeta^{\mathrm{e}}\,\mathrm{d}t, \tag{4.14}$$

$$T\,\mathrm{d}s = \mathrm{d}q^{\mathrm{e}} + \mathrm{d}q' + \sum_{\varsigma=1}^{n}TS_\varsigma^1\frac{Q_\zeta^{\mathrm{e}}}{\rho}\,\mathrm{d}t, \tag{4.15}$$

$$S_\zeta^1 = \left(\frac{\partial s}{\partial c_\zeta^1}\right)_{T,c_\vartheta^i(\vartheta\neq\zeta),\varsigma_{ij},\varepsilon_{ij}} = \frac{\partial^2 F}{\partial T\,\partial c_\zeta^1}.$$

Here $U = \sum_{\beta=1}^{3}c^\beta\langle U^\beta\rangle$, $s = \sum_{\beta=1}^{3}c^\beta\langle s^\beta\rangle$, $\sigma_{ij} = \sum_{\alpha=1}^{3}c^\beta\langle\sigma^\beta\rangle_{ij}$ and S_ζ^1 is the partial entropy of the biocomposite.

To complete the formulation of the model we need to specify the thermodynamic fluxes Q_ζ^{e}, Q_ζ and J_j and stresses $\sigma_{ij}^{(\beta)}$.

Let the heat flux $\overline{q}$ be given by

$$\rho\,\mathrm{d}q^{(\mathrm{e})} = -\left(\frac{\partial q_x}{\partial x} + \frac{\partial q_y}{\partial y} + \frac{\partial q_z}{\partial z}\right)\mathrm{d}t \equiv -\mathrm{div}\,\overline{q}\,\mathrm{d}t.$$

On use of Eq. (4.13) in (4.14) and (4.15), the equation of the balance of entropy of the composite takes the form

$$\rho\frac{\mathrm{d}s}{\mathrm{d}t} - \sum_{\zeta=1}^{n}\frac{\partial s}{\partial c_\zeta^1}Q_\zeta^{\mathrm{e}} = -\mathrm{div}\left(\frac{\overline{q}}{T}\right) + \frac{\boldsymbol{R}}{T}, \tag{4.16}$$

where

$$\begin{aligned}\boldsymbol{R} = &-\frac{\overline{q}}{T}\left(\frac{\partial T}{\partial x} + \frac{\partial T}{\partial y} + \frac{\partial T}{\partial z}\right) + \sigma_{ij}^2\frac{\mathrm{d}\Delta_{ij}^2}{\mathrm{d}t} + \sigma_{ij}^3\frac{\mathrm{d}\Delta_{ij}^3}{\mathrm{d}t}\\ &+ \sum_{\varsigma=1}^{n}\left(\mu_\zeta^1 - \mu_\zeta^2\right)Q_\zeta^{\mathrm{e}} + \sum_{\varsigma=1}^{n}\left(\mu_\zeta^3 - \mu_\zeta^2\right)Q_\zeta^{\mathrm{e}} - \sum_{j=1}^{r}J_j\Lambda_j\end{aligned} \tag{4.17}$$

and

$$\Lambda_j = \sum_{\zeta=1}^{n}\nu_{\varsigma j}\mu_\zeta^2. \tag{4.18}$$

Here $\boldsymbol{R}$ is the dissipative function and Λ_j is the affinity constant of the jth chemical reaction.

Let the thermodynamic forces acting in the system be

$$-\frac{1}{T^2}\left(\frac{\partial T}{\partial x}+\frac{\partial T}{\partial y}+\frac{\partial T}{\partial z}\right),\quad \frac{1}{T}\frac{\mathrm{d}\Delta_{ij}^2}{\mathrm{d}t},\quad \frac{1}{T}\frac{\mathrm{d}\Delta_{ij}^3}{\mathrm{d}t},$$

$$\frac{\mu_\varsigma^1-\mu_\varsigma^2}{T},\quad \frac{\mu_\varsigma^3-\mu_\varsigma^2}{T},\quad -\frac{\Lambda_j}{T}. \tag{4.19}$$

Assuming a linear relationship among thermodynamic fluxes $\overline{q}$, $\sigma_{ij}^{(\alpha)}$, Q_ζ^{e}, Q_ζ and J_j and thermodynamic forces, we have

$$q_i = -W_{ij}\frac{\partial T}{\partial x_j}, \tag{4.20}$$

$$\sigma_{kl}^1 = E_{ijkl}\varepsilon_{ij}, \tag{4.21}$$

$$\sigma_{kl}^2 = B_{ijkl}\frac{\mathrm{d}\Delta_{ij}^2}{\mathrm{d}t}+B_{klij}^*\frac{\mathrm{d}\Delta_{ij}^3}{\mathrm{d}t}+\sum_{\beta=1}^{r}D_{\beta kl}^*\Lambda_\beta$$
$$-\sum_{\alpha=1}^{n}O_{\alpha kl}^*\left(\mu_\alpha^1-\mu_\alpha^2\right)-\sum_{\alpha=1}^{n}Y_{\alpha kl}^*\left(\mu_\alpha^3-\mu_\alpha^2\right), \tag{4.22}$$

$$\sigma_{ij}^3 = B_{ijkl}^*\frac{\mathrm{d}\Delta_{ij}^2}{\mathrm{d}t}+B_{ijkl}\frac{\mathrm{d}\Delta_{ij}^3}{\mathrm{d}t}+\sum_{\beta=1}^{r}D_{\beta ij}^*\Lambda_\beta$$
$$-\sum_{\alpha=1}^{n}O_{\alpha kl}^*\left(\mu_\alpha^1-\mu_\alpha^2\right)-\sum_{\alpha=1}^{n}V_{\alpha kl}^*\left(\mu_\alpha^3-\mu_\alpha^2\right), \tag{4.23}$$

$$J_\beta = D_{\beta ij}\frac{\mathrm{d}\Delta_{ij}^2}{\mathrm{d}t}+D_{\beta kl}^*\frac{\mathrm{d}\Delta_{ij}^2}{\mathrm{d}t}-\sum_{\gamma=1}^{r}{}^1l_{\beta\gamma}\Lambda_\beta$$
$$+\sum_{\alpha=1}^{n}{}^2l_{\alpha\beta}\left(\mu_\alpha^1-\mu_\alpha^2\right)+\sum_{\alpha=1}^{n}{}^3l_{\alpha\beta}\left(\mu_\alpha^3-\mu_\alpha^2\right), \tag{4.24}$$

$$Q_\alpha = O_{\alpha ij}\frac{\mathrm{d}\Delta_{ij}^2}{\mathrm{d}t}+O_{\alpha kl}^*\frac{\mathrm{d}\Delta_{ij}^3}{\mathrm{d}t}-\sum_{\beta=1}^{r}{}^2l_{\alpha\beta}\Lambda_\beta$$
$$+\sum_{\beta=1}^{n}{}^4l_{\alpha\beta}\left(\mu_\beta^1-\mu_\beta^2\right)+\sum_{\gamma=1}^{n}{}^5l_{\beta_\gamma}\left(\mu_\gamma^3-\mu_\gamma^2\right), \tag{4.25}$$

$$Q_n^{\mathrm{e}} = Y_{nij}\frac{\mathrm{d}\Delta_{ij}^2}{\mathrm{d}t}+Y_{nkl}^*\frac{\mathrm{d}\Delta_{ij}^3}{\mathrm{d}t}-\sum_{\beta=1}^{r}{}^3l_{n\beta}\Lambda_\beta$$
$$+\sum_{\alpha=1}^{n}{}^5l_{n\alpha}\left({\mu_\alpha}^1-\mu_\alpha^2\right)-\sum_{\alpha=1}^{n}{}^6l_{n\alpha}\left({\mu_\alpha}^3-\mu_\alpha^2\right). \tag{4.26}$$

Here ${}^{m}l_{n\alpha,n\beta,\alpha\beta,\beta\gamma}$ ($m = 1$–6) are scalars, whereas $B^{(*)}_{ijkl}, E_{ijkl}, D^{(*)}_{nij}, Y^{(*)}_{\alpha ij}, O^{(*)}_{\alpha ij}$ and W_{ij} are the parameters of tensorial nature. They satisfy the Onsager reciprocal relations

$$B^{(*)}_{ijkl} = B^{(*)}_{klij}, \qquad D^{*}_{nij} = -D_{nij}, \qquad W_{ji} = W_{ij},$$
$$V^{*}_{\alpha ij} = -V_{\alpha ij}, \qquad O^{*}_{\alpha ij} = -O_{\alpha ij}, \qquad {}^{m}l_{n\alpha,n\beta,\alpha\beta,\beta\gamma} = {}^{m}l_{\alpha n,\beta n,\beta\alpha,\gamma\beta}.$$

For example, assuming that the tissue is transversely anisotropic, B_{ijkl} and D_{nij} are defined by

$$\begin{aligned}
B_{ikjl} =& \lambda_1\left(\delta_{ik}\delta_{jl} + \delta_{il}\delta_{jk} - \frac{2}{3}\delta_{ij}\delta_{kl}\right) + \lambda_2\left(\delta_{ij}b_{kl} + \delta_{kl}b_{ij} - \frac{1}{3}\delta_{ij}\delta_{kl} - 3b_{ij}b_{kl}\right)\\
&+ \lambda_3(\delta_{ik}b_{jl} + \delta_{jk}b_{il} + \delta_{il}b_{jk} + \delta_{jl}b_{ik} - 4b_{ij}b_{kl}),\\
D_{nij} =& D_n(\delta_{ij} - 3b_{ij}) \quad (n = 1, 2, \ldots, r).
\end{aligned}$$

Numerous experimental data on uniaxial and biaxial loading show that collagen and elastin fibres possess nonlinear elastic characteristics, whereas muscle tissue exhibits viscoelastic characteristics. Hence, for stresses we have

$$\sigma_{ij} = \sum_{\alpha=1}^{3} \sigma^{\alpha}_{ij} = c^1 E_{ijkl}\varepsilon_{kl} + c^2 E^{\mathrm{ve}}_{ijkl}(\varepsilon_{kl} - \Delta^2_{kl}) + c^3 E^{\mathrm{ve}}_{ijkl}(\varepsilon_{kl} - \Delta^3_{kl}), \tag{4.27}$$

where E^{ve}_{ijkl} is the tensor of viscous characteristics ($E^{\mathrm{ve}}_{ijkl} = E^{\mathrm{ve}}_{klij}$). By differentiating Eq. (4.27) with respect to time, the constitutive relations of the mechanochemically active biological tissue are found to be

$$\begin{aligned}
B^{*}_{klij}\underline{E}^{\mathrm{ve}}_{ijmn}\frac{\mathrm{d}\sigma_{kl}}{\mathrm{d}t} &+ \left(\mathbf{I} - \frac{1}{c^3}B^{*}_{klij}\underline{E}^{\mathrm{ve}}_{ijmn}\frac{\mathrm{d}c^3}{\mathrm{d}t}\right)\sigma_{kl}\\
&= c^1 E_{ijmn}\varepsilon_{ij} - \left(\frac{c^1}{c^3}B^{*}_{klij}\underline{E}^{\mathrm{ve}}_{ijmn}E_{mnkl}\frac{\mathrm{d}c^3}{\mathrm{d}t} + B^{*}_{klij}\underline{E}^{\mathrm{ve}}_{ijmn}E_{mnkl}\frac{\mathrm{d}c^1}{\mathrm{d}t}\right)\varepsilon_{mn}\\
&\quad - \frac{c^2}{c^3}B^{*}_{klij}\frac{dc^3}{\mathrm{d}t}\varepsilon_{mn} + B^{*}_{ijmn}\left(c^1\underline{E}^{\mathrm{ve}}_{ijkl}E_{klmn} + c^2 + c^3\right)\frac{\mathrm{d}\varepsilon_{ij}}{\mathrm{d}t}\\
&\quad - B^{*}_{ijmn}\left(\frac{\mathrm{d}c^2}{\mathrm{d}t} - \frac{c^2}{c^3}\frac{\mathrm{d}c^3}{\mathrm{d}t}\right)\frac{\mathrm{d}\Delta^2_{ij}}{\mathrm{d}t} + c^2 Z_{mn} + c^3 Z^{*}_{mn},
\end{aligned}$$

$$B^{*}_{ijmn}\underline{E}^{\mathrm{ve}}_{ijkl}\frac{\mathrm{d}\Delta^2_{kl}}{\mathrm{d}t} - \varsigma^2_{mn} + \underline{E}^{\mathrm{ve}}_{ijmn}Z_{ij} = 0,$$

$$B^{*}_{ijmn}\underline{E}^{\mathrm{ve}}_{ijkl}\frac{\mathrm{d}\Delta^3_{kl}}{\mathrm{d}t} - \varsigma^3_{mn} + \underline{E}^{\mathrm{ve}}_{ijmn}Z^{*}_{ij} = 0,$$

$$c^2\frac{\mathrm{d}Z_{mn}}{\mathrm{d}t}=B^{\mathrm{T}}_{mnij}\frac{\mathrm{d}\Delta^3_{ij}}{\mathrm{d}t}+\sum_{\beta=1}^{r}D^*_{\beta mn}\Lambda_\beta-\sum_{\alpha=1}^{n}O^*_{\alpha mn}\left(\mu^1_\alpha-\mu^2_\alpha\right)$$
$$-\sum_{\alpha=1}^{n}Y^*_{\alpha mn}\left(\mu^3_\alpha-\mu^2_\alpha\right), \tag{4.28}$$
$$c^3\frac{\mathrm{d}Z^*_{mn}}{\mathrm{d}t}=B_{mnij}\frac{\mathrm{d}\Delta^2_{ij}}{\mathrm{d}t}+\sum_{\beta=1}^{r}D_{\beta mn}\Lambda_\beta-\sum_{\alpha=1}^{n}O_{\alpha mn}\left(\mu^1_\alpha-\mu^2_\alpha\right)$$
$$-\sum_{\alpha=1}^{n}Y_{\alpha mn}\left(\mu^3_\alpha-\mu^2_\alpha\right)$$
$$\left(\mathbf{B}^{\mathrm{T}}=\mathbf{B}\right).$$

Here $\underline{\mathbf{E}}$ is the tensor inverse to $\mathbf{E}$ ($\underline{\mathbf{E}}\mathbf{E}=\mathbf{I}$), $\mathbf{I}$ is the identity tensor and Z^*_{ij} is the 'biofactor' that accounts for various biological phenomena including electromechanical, chemical, remodelling and ageing processes in the tissue.

Although the system of Eqs. (4.28) describes the mechanics of biocomposites, it does not provide the required relationships between in-plane forces, moments and deformations in the thin shell. To establish the missing link, consider the distribution of ε^z_{ik} and stresses σ_{ik} $(i, k=1, 2, 3)$ in the shell. Recalling the first Kirchhoff–Love geometrical hypothesis, $\varepsilon^z_{13}=\varepsilon^z_{23}=0$, it would be appealing to exclude the shear stresses and lateral forces from the equilibrium equations by neglecting the terms $\sigma_{13}=\sigma_{23}=0$ *and* $\overset{*}{N}_1=\overset{*}{N}_2=0$. However, that would strongly violate the equilibrium conditions. Accepting the second Kirchhoff–Love hypothesis, which states that the normal stress σ_{33} is significantly smaller than σ_{ij} $(i, k=1, 2)$, we can eliminate only the terms containing σ_{33}. Then, Eqs. (4.28) take the form

$$B^*_{klij}\underline{E}^{\mathrm{ve}}_{ijmn}\frac{\mathrm{d}\sigma_{kl}}{\mathrm{d}t}+\left(\mathbf{I}-\frac{1}{c^3}B^*_{klij}\underline{E}^{\mathrm{ve}}_{ijmn}\frac{\mathrm{d}c^3}{\mathrm{d}t}\right)\sigma_{kl}$$
$$=c^1\underline{E}^{\mathrm{ve}}_{ijmn}\varepsilon^z_{ij}-B^*_{klij}\underline{E}^{\mathrm{ve}}_{ijmn}E_{mnkl}\left(\frac{c^1}{c^3}\frac{\mathrm{d}c^3}{\mathrm{d}t}+\frac{\mathrm{d}c^1}{\mathrm{d}t}\right)\varepsilon^z_{mn}$$
$$-\frac{c^2}{c^3}B^*_{klij}\frac{\mathrm{d}c^3}{\mathrm{d}t}\varepsilon^z_{mn}+B^*_{ijmn}\left(c^1\underline{E}^{\mathrm{ve}}_{ijkl}E_{klmn}+c^2+c^3\right)\frac{\mathrm{d}\varepsilon^z_{ij}}{\mathrm{d}t}$$
$$-B^*_{ijmn}\left(\frac{\mathrm{d}c^2}{\mathrm{d}t}-\frac{c^2}{c^3}\frac{\mathrm{d}c^3}{\mathrm{d}t}\right)\frac{\mathrm{d}\Delta^{z2}_{ij}}{\mathrm{d}t}+c^2Z_{mn}+c^3Z^*_{mn},$$

$$B^*_{ijmn}\underline{E}^{\mathrm{ve}}_{ijkl}\frac{\mathrm{d}\Delta^{z2}_{kl}}{\mathrm{d}t}-\varsigma^{z2}_{mn}+\underline{E}^{\mathrm{ve}}_{ijmn}Z_{ij}=0, \tag{4.29}$$

$$B^*_{ijmn}\underline{E}^{\mathrm{ve}}_{ijkl}\frac{\mathrm{d}\Delta^{z3}_{kl}}{\mathrm{d}t}-\varsigma^{z3}_{mn}+\underline{E}^{\mathrm{ve}}_{ijmn}Z^*_{ij}=0,$$

$$c^2 \frac{\mathrm{d}Z_{mn}}{\mathrm{d}t} = B^{\mathrm{T}}_{ijmn} \frac{\mathrm{d}\Delta^{z3}_{ij}}{\mathrm{d}t} + \sum_{\beta=1}^{r} D^*_{\beta mn}\Lambda_\beta - \sum_{\alpha=1}^{n} O^*_{\alpha mn}\left(\mu^1_\alpha - \mu^2_\alpha\right)$$
$$- \sum_{\alpha=1}^{n} Y^*_{\alpha mn}\left(\mu^3_\alpha - \mu^2_\alpha\right),$$

$$c^3 \frac{\mathrm{d}Z^*_{mn}}{\mathrm{d}t} = B_{mnij} \frac{\mathrm{d}\Delta^{z2}_{ij}}{\mathrm{d}t} + \sum_{\beta=1}^{r} D_{\beta mn}\Lambda_\beta - \sum_{\alpha=1}^{n} O_{\alpha mn}\left(\mu^1_\alpha - \mu^2_\alpha\right)$$
$$- \sum_{\alpha=1}^{n} Y_{\alpha mn}\left(\mu^3_\alpha - \mu^2_\alpha\right).$$

Finally, substituting ε^z_{ik} given by Eq. (4.14) and solving the resultant equations for σ_{11}, σ_{22} and σ_{12}, we obtain constitutive relations for the mechanochemically active biocomposite in terms of the deformations, curvature and twist of the middle surface of the shell. Applying σ_{ik} in Eqs. (3.25) and integrating it over the thickness of the shell, we find explicit relations for the in-plane forces $\overset{*}{T}_{ij}$ and moments $\overset{*}{M}_{ij}$. In general, the end formulas are very bulky and are not given here. In applications though, depending on the specific tissue, the formulas can be simplified to a certain degree and may even take an elegant form.

4.3 The biological factor

Fundamental mechanical functions of the gastrointestinal tract are closely related to electromechanical wave processes and the coordinated propagation of the waves of contraction–relaxation in the organs.

Consider smooth muscle syncytia to be an electrically excitable biological medium (Plonsey and Barr, 1984). Applying Ohm's law, we have

$$\overline{J}_{\mathrm{i}} = -\left(\hat{g}_{\mathrm{i}1}\frac{\partial\Psi_{\mathrm{i}}}{\partial x_1}\overline{e}_1 + \hat{g}_{\mathrm{i}2}\frac{\partial\Psi_{\mathrm{i}}}{\partial x_2}\overline{e}_2\right), \tag{4.30}$$

$$\overline{J}_{\mathrm{o}} = -\left(\hat{g}_{\mathrm{o}1}\frac{\partial\Psi_{\mathrm{o}}}{\partial x_1}\overline{e}_1 + \hat{g}_{\mathrm{o}2}\frac{\partial\Psi_{\mathrm{o}}}{\partial x_2}\overline{e}_2\right), \tag{4.31}$$

where $\overline{J}_{\mathrm{i}}$, and $\overline{J}_{\mathrm{o}}$ are the intracellular (i) and extracellular (o) currents, Ψ_i and Ψ_o are the scalar electrical potentials, $\hat{g}_{\mathrm{ij}}$ and $\hat{g}_{\mathrm{oj}}$ ($j = 1, 2$) are the conductivities, and $\overline{e}_1$ and $\overline{e}_2$ are the unit vectors in the directions of the α_1- and α_2-coordinate lines. The two cellular spaces are coupled through the transmembrane current $I_{\mathrm{m}1}$ and potential V_{m} as

$$I_{\mathrm{m}1} = -\mathrm{div}\overline{J}_{\mathrm{i}} = \mathrm{div}\overline{J}_{\mathrm{o}}, \tag{4.32}$$

$$V_{\mathrm{m}} = \Psi_{\mathrm{i}} - \Psi_{\mathrm{o}}. \tag{4.33}$$

On substituting Eqs. (4.30) and (4.31) into (4.32), we get

$$I_{\mathrm{m1}} = \hat{g}_{\mathrm{i1}} \frac{\partial^2 \Psi_{\mathrm{i}}}{\partial \alpha_1^2} \bar{e}_1 + \hat{g}_{\mathrm{i2}} \frac{\partial^2 \Psi_{\mathrm{i}}}{\partial \alpha_2^2} \bar{e}_2, \tag{4.34}$$

$$I_{\mathrm{m1}} = -\hat{g}_{\mathrm{o1}} \frac{\partial^2 \Psi_{\mathrm{o}}}{\partial \alpha_1^2} \bar{e}_1 + \hat{g}_{\mathrm{o2}} \frac{\partial^2 \Psi_{\mathrm{o}}}{\partial \alpha_2^2} \bar{e}_2. \tag{4.35}$$

On equating Eqs. (4.34) and (4.35), we find

$$(\hat{g}_{\mathrm{i1}} + \hat{g}_{\mathrm{o1}}) \frac{\partial^2 \Psi_{\mathrm{i}}}{\partial \alpha_1^2} + (\hat{g}_{\mathrm{i2}} + \hat{g}_{\mathrm{o2}}) \frac{\partial^2 \Psi_{\mathrm{i}}}{\partial \alpha_2^2} = \hat{g}_{\mathrm{o1}} \frac{\partial^2 V_{\mathrm{m}}}{\partial \alpha_1^2} + \hat{g}_{\mathrm{o2}} \frac{\partial^2 V_{\mathrm{m}}}{\partial \alpha_2^2}. \tag{4.36}$$

By solving Eq. (4.36) for Ψ_{i}, we obtain

$$\Psi_{\mathrm{i}} = \frac{1}{4\pi} \iint \left(\frac{\hat{g}_{\mathrm{o1}}}{\hat{g}_{\mathrm{i1}} + \hat{g}_{\mathrm{o1}}} \frac{\partial^2 V_{\mathrm{m}}}{\partial X'^2} + \frac{\hat{g}_{\mathrm{o2}}}{\hat{g}_{\mathrm{i2}} + \hat{g}_{\mathrm{o2}}} \frac{\partial^2 V_{\mathrm{m}}}{\partial Y'^2} \right) + \left[\log\left((X - X')^2 + (Y - Y')^2 \right) \right] \mathrm{d}X' \, \mathrm{d}Y',$$

where the following substitutions are used:

$$X = \alpha_1 / \sqrt{g_{\mathrm{i1}} + g_{\mathrm{o1}}}, \qquad Y = \alpha_2 / \sqrt{\hat{g}_{\mathrm{i2}} + \hat{g}_{\mathrm{o2}}}.$$

Here the integration variables are primed, and the unprimed variables indicate the space point (α_1', α_2') at which Ψ_{i} is evaluated. The reverse substitution of X and Y gives

$$\Psi_{\mathrm{i}} = \frac{1}{4\pi} \iint \left(\hat{g}_{\mathrm{o1}} \frac{\partial^2 V_{\mathrm{m}}}{\partial X'^2} + \hat{g}_{\mathrm{o2}} \frac{\partial^2 V_{\mathrm{m}}}{\partial Y'^2} \right) + \left[\log\left(\frac{(\alpha_1 - \alpha_1')^2}{\hat{g}_{\mathrm{i1}} + \hat{g}_{\mathrm{o1}}} + \frac{(\alpha_2 - \alpha_2')^2}{\hat{g}_{\mathrm{i2}} + \hat{g}_{\mathrm{o2}}} \right) \right] \frac{\mathrm{d}\alpha_1' \, \mathrm{d}\alpha_2'}{\sqrt{(\hat{g}_{\mathrm{i1}} + \hat{g}_{\mathrm{o1}})(\hat{g}_{\mathrm{i2}} + \hat{g}_{\mathrm{o2}})}}. \tag{4.37}$$

On introducing Eq. (4.37) into (4.34), after some algebra we obtain

$$I_{\mathrm{m1}} = \frac{\widetilde{\mu}_1 - \widetilde{\mu}_2}{2\pi G (1 + \widetilde{\mu}_1)(1 + \widetilde{\mu}_2)} \iint \left(\hat{g}_{\mathrm{o1}} \frac{\partial^2 V_{\mathrm{m}}}{\partial X'^2} + \hat{g}_{\mathrm{o2}} \frac{\partial^2 V_{\mathrm{m}}}{\partial Y'^2} \right) \times \left[\left(\frac{(\alpha_1 - \alpha_1')^2}{G_1} - \frac{(\alpha_2 - \alpha_2')^2}{G_2} \right) \Big/ \left(\frac{(\alpha_1 - \alpha_1')^2}{G_1} + \frac{(\alpha_2 - \alpha_2')^2}{G_2} \right)^2 \right] \mathrm{d}\alpha_1' \, \mathrm{d}\alpha_2', \tag{4.38}$$

where

$$G_1 = \hat{g}_{\mathrm{i1}} + \hat{g}_{\mathrm{o1}}, \qquad G_2 = \hat{g}_{\mathrm{i2}} + \hat{g}_{\mathrm{o2}},$$

$$G = \sqrt{G_1 G_2}, \qquad \widetilde{\mu}_1 = \hat{g}_{\mathrm{o1}} / \hat{g}_{\mathrm{i1}}, \qquad \widetilde{\mu}_2 = \hat{g}_{\mathrm{o2}} / \hat{g}_{\mathrm{i2}}.$$

On substituting Eq. (4.37) into (4.34), we find the contribution of an ε- neighbourhood of $(\alpha_1' = 0, \alpha_2' = 0)$ to I_{m1}. Using the transformations given by $X = \alpha_1/\sqrt{\hat{g}_{\mathrm{i1}}}$ and $Y = \alpha_2/\sqrt{\hat{g}_{\mathrm{i2}}}$, we find

$$I_{\mathrm{m2}} = \frac{\sqrt{\hat{g}_{\mathrm{i1}}\hat{g}_{\mathrm{i2}}}}{4\pi G}\left(\hat{g}_{\mathrm{o1}}\frac{\partial^2 V_{\mathrm{m}}}{\partial X'^2} + \hat{g}_{\mathrm{o2}}\frac{\partial^2 V_{\mathrm{m}}}{\partial Y'^2}\right)_{\alpha_1'=\alpha_2'=0} \times \int \operatorname{div}\operatorname{grad}\left[\log\left(\frac{X'^2}{G_1/\hat{g}_{\mathrm{i1}}} + \frac{Y'^2}{G_2/\hat{g}_{\mathrm{i2}}}\right)\right] \mathrm{d}\alpha_1'\,\mathrm{d}\alpha_2'. \tag{4.39}$$

On applying the divergence theorem and performing the gradient operation, the integral in Eq. (4.39) is converted to a line integral,

$$\int \frac{(2X'\hat{g}_{\mathrm{i1}}/G_1)\bar{e}_1 + (2Y'\hat{g}_{\mathrm{i2}}/G_2)\bar{e}_2}{(X'^2\hat{g}_{\mathrm{i1}}/G_1) + (Y'^2\hat{g}_{\mathrm{i2}}/G_2)} \cdot \bar{n}\,\mathrm{d}C', \tag{4.40}$$

where $\mathrm{d}C'$ is an element of the ε-contour. The result of integration yields

$$I_{\mathrm{m2}} = \left(\hat{g}_{\mathrm{o1}}\frac{\partial^2 V_{\mathrm{m}}}{\partial \alpha_1'^2} + \hat{g}_{\mathrm{o2}}\frac{\partial^2 V_{\mathrm{m}}}{\partial \alpha_2'^2}\right)\left(\frac{\hat{g}_{\mathrm{i2}}}{G_2} + \frac{2(\widetilde{\mu}_1 - \widetilde{\mu}_2)}{\pi(1+\widetilde{\mu}_1)(1+\widetilde{\mu}_2)}\tan^{-1}\sqrt{\frac{G_1}{G_2}}\right). \tag{4.41}$$

To simulate the excitation and propagation pattern in the anisotropic smooth muscle syncytium we employ the Hodgkin–Huxley formalism described by

$$C_{\mathrm{m}}\frac{\partial V_{\mathrm{m}}}{\partial t} = -(I_{\mathrm{m1}} + I_{\mathrm{m2}} + I_{\mathrm{ion}}),$$

where C_{m} is the membrane capacitance and I_{ion} is the total ion current through the membrane. On substituting expressions for I_{m1} and I_{m2} given by (4.38) and (4.41), we obtain

$$\begin{aligned} C_{\mathrm{m}}\frac{\partial V_{\mathrm{m}}}{\partial t} &= -\frac{\widetilde{\mu}_1 - \widetilde{\mu}_2}{2\pi G(1+\widetilde{\mu}_1)(1+\widetilde{\mu}_2)}\iint\left(\hat{g}_{\mathrm{o1}}\frac{\partial^2 V_{\mathrm{m}}}{\partial X'^2} + \hat{g}_{\mathrm{o2}}\frac{\partial^2 V_{\mathrm{m}}}{\partial Y'^2}\right) \\ &\times\left[\left(\frac{(\alpha_1-\alpha_1')^2}{G_1} - \frac{(\alpha_2-\alpha_2')^2}{G_2}\right)\Bigg/\left(\frac{(\alpha_1-\alpha_1')^2}{G_1} + \frac{(\alpha_2-\alpha_2')^2}{G_2}\right)^2\right]\mathrm{d}\alpha_1'\,\mathrm{d}\alpha_2' \\ &-\left(\hat{g}_{\mathrm{o1}}\frac{\partial^2 V_{\mathrm{m}}}{\partial \alpha_1'^2} + \hat{g}_{\mathrm{o2}}\frac{\partial^2 V_{\mathrm{m}}}{\partial \alpha_2^2}\right)\left(\frac{\hat{g}_{\mathrm{i2}}}{G_2} + \frac{2(\widetilde{\mu}_1 - \widetilde{\mu}_2)}{\pi(1+\widetilde{\mu}_1)(1+\widetilde{\mu}_2)}\tan^{-1}\sqrt{\frac{G_1}{G_2}}\right) - I_{\mathrm{ion}}, \end{aligned} \tag{4.42}$$

where I_{ion} is a function depending on the type and ion-channel properties of the biological tissue.

In the case of electrical isotropy, $\widetilde{\mu}_1 = \widetilde{\mu}_2 = \widetilde{\mu}$, the integral in Eq. (4.42) vanishes and we get

$$C_{\mathrm{m}}\frac{\partial V_{\mathrm{m}}}{\partial t} = -\frac{1}{1+\tilde{\mu}}\left(\hat{g}_{\mathrm{o}1}\frac{\partial^2 V_{\mathrm{m}}}{\partial \alpha_1^2} + \hat{g}_{\mathrm{o}2}\frac{\partial^2 V_{\mathrm{m}}}{\partial \alpha_2^2}\right) - I_{\mathrm{ion}}. \tag{4.43}$$

Finally, the constitutive relations of mechanochemically active electrogenic biological medium include Eqs. (4.6)–(4.8), (4.29), (4.42) and/or (4.43). The system is closed by providing the free energy, ion currents, initial and boundary conditions, and the function $Z_{ij} = Z_{ij}(V_{\mathrm{m}}, \mu_{\mathrm{i}}, \hat{g}_{ij}, \hat{g}_{oj})$. It is noteworthy that the closed system of equations describes the development of stresses in the absence of active strains and vice versa, a condition that is unique to all biological materials.

Models are generally evaluated in terms of the nature of parameters and constants involved as well as for their accurate and meaningful experimental determination. While phenomenological constitutive models are able to fit the experimental data with a high degree of accuracy, they are limited in that they do not give insight into the underlying cause of the particulars of mechanical behaviour. Fine molecular and structure-based models help one to avoid such ambiguities and are able to reveal the intricacies of functions of tissues. However, they are beyond the scope of this book.

Exercises

1. The foregut, midgut and hindgut are segments of the primitive gut. They give rise to the stomach and upper duodenum (foregut), lower duodenum, jejunum and ileum of the small intestine, caecum, appendix, ascending and the first two-thirds of the transverse colon (midgut), and the last third of the transverse colon, descending colon, rectum and upper part of the anal canal (hindgut). What are the morphofunctional similarities and differences of these different parts of the digestive tract?
2. The formulation of appropriate constitutive relations has always been of central importance in the biomechanics of living tissues. Formulate the basic steps in the formulation of a constitutive relation.
3. The inherent complexities of the microstructure and behaviour of biological tissues require theoretical frameworks that will guide the design and interpretation of new classes of experiments. Two conceptual mathematical frameworks – stochastic and deterministic – are currently adopted to synthesize and predict observations across multiple length- and time-scales. What are the advantages and disadvantages of each approach?
4. Derive a two-phase mechanochemically active model of a biocomposite, assuming that phase 1 is the connective tissue matrix and phase 2 is represented by active smooth muscle tissue.
5. Biological soft tissues exhibit both solid- and fluid-like behaviours, i.e. they possess characteristics of viscoelasticity. The standard Maxwell, Voigt and Kelvin, and Boltzmann models of linear viscoelasticity are not suitable to describe the complex behaviours of soft tissues. Discuss a quasi-linear viscoelasticity theory proposed by Y. C. Fung (1993), the single-integral finite-strain model of Johnson *et al.* (1996), the

generalized elastic-Maxwell model of Holzapfel *et al.* (2002) and the modified superposition model of Provenzano *et al.* (2002).

6. The continuum theory of mixtures assumes that (i) if all effects of diffusion are taken into account properly, the equations for the mean motion are the same as those governing the motion of a simple medium; and (ii) the increase in mass of one constituent must occur at the expense of a decrease in mass of another constituent in a closed thermodynamic system. Discuss the limitations of these fundamental assumptions.
7. Verify Eqs. (4.16) and (4.17).
8. The Onsager reciprocal relations express the equality of relations between flows and forces in thermodynamic systems that are away from the state of equilibrium, but where a notion of local thermodynamic equilibrium exists. Give examples of such relations in the wall of the gut.
9. Find expressions for B_{ijkl} and D_{nij} (Eqs. (4.22)–(4.24)) for a mechanically isotropic tissue.
10. Derive the final constitutive relations for a three-phase mechanochemically active biocomposite. (*Hint*: substitute ε_{ik}^{z} given by Eq. (4.14) into (4.29).)
11. A. L. Hodgkin and A. Huxley (1952) explained and formulated a mathematical model of the ionic mechanisms underlying the initiation and propagation of action potentials in a biologically excitable medium, namely the squid giant axon. The inward/outward ion currents were carried by Na^+, K^+ and Cl^- ions. Following the Hodgkin–Huxley formalism, what other current should be included in the model of abdominal viscera?

5

Boundary conditions

5.1 The geometry of the boundary

Consider the contour curve C on the boundary of the undeformed shell parameterized by arc length s,

$$\bar{r} = \bar{r}(s) = \bar{r}(\alpha_1(s), \alpha_2(s)). \tag{5.1}$$

Let $\{\bar{n}, \bar{\tau}, \bar{m}\}$ be the orthogonal basis of C (Fig. 5.1). Here $\bar{n}$ and $\bar{\tau}$ are unit vectors normal and tangent to C, respectively, and $\bar{m}$ is the vector normal to the middle surface S. The three unit vectors are linearly independent:

$$\bar{n} = \bar{\tau} \times \bar{m}, \qquad \bar{\tau} = \bar{m} \times \bar{n}, \qquad \bar{m} = \bar{n} \times \bar{\tau}. \tag{5.2}$$

On differentiating Eq. (5.1) with respect to s and using Eq. (1.13), the tangent vector $\bar{\tau}$ is found to be

$$\bar{\tau} = \frac{\mathrm{d}\bar{r}}{\mathrm{d}s} = \bar{r}_1 \frac{\mathrm{d}\alpha_1}{\mathrm{d}s} + \bar{r}_2 \frac{\mathrm{d}\alpha_2}{\mathrm{d}s} = \bar{e}_1\tau_1 + \bar{e}_2\tau_2. \tag{5.3}$$

Projections of $\bar{\tau}$ on unit vectors $\bar{e}_i \in S$ $(i = 1, 2)$ are given by

$$\tau_1 = A_1 \frac{\mathrm{d}\alpha_1}{\mathrm{d}s}, \qquad \tau_2 = A_2 \frac{\mathrm{d}\alpha_2}{\mathrm{d}s}. \tag{5.4}$$

Decomposing the normal vector $\bar{n}$ in the direction of $\bar{e}_i$ yields

$$\bar{n} = (\bar{e}_1\tau_1 + \bar{e}_2\tau_2) \times \bar{m} = \bar{e}_1\tau_2 - \bar{e}_2\tau_1, \qquad \bar{n} = \bar{e}_1 n_1 + \bar{e}_2 n_2. \tag{5.5}$$

From Eqs. (5.4) and (5.5), projections of $\bar{n}$ are found to be

$$n_1 = \tau_2 = A_2 \frac{\mathrm{d}\alpha_2}{\mathrm{d}s}, \qquad n_2 = -\tau_1 = -A_1 \frac{\mathrm{d}\alpha_1}{\mathrm{d}s}. \tag{5.6}$$

Since $\bar{n} \perp \bar{\tau}$, it follows that $\tau_1 n_1 + \tau_2 n_2 = 0$.

Let k_n and k_τ be normal curvatures in the direction of $\bar{n}$ and $\bar{\tau}$, respectively, and $k_{n\tau}$ be the twist of the contour line C:

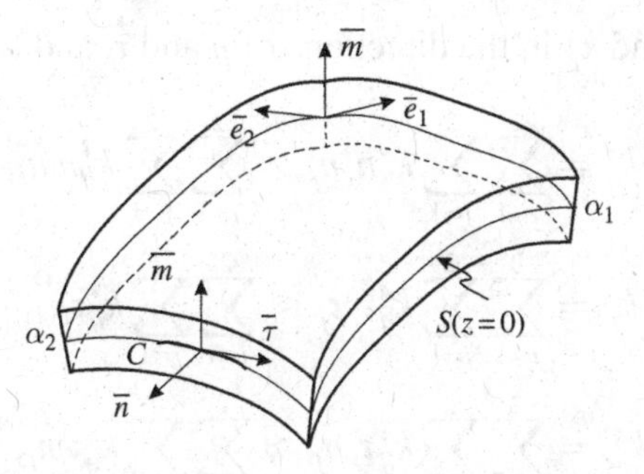

Fig. 5.1 An orthonormal basis $\{\bar{n}, \bar{\tau}, \bar{m}\}$ associated with the boundary.

$$k_n = \sum_{i=1}^{2}\sum_{j=1}^{2} k_{ij} n_i n_j, \qquad k_\tau = \sum_{i=1}^{2}\sum_{j=1}^{2} k_{ij}\tau_i\tau_j, \qquad k_{n\tau} = \sum_{i=1}^{2}\sum_{j=1}^{2} k_{ij}\tau_i n_j, \tag{5.7}$$

where k_{ij} satisfy Eqs. (1.21a) and (1.21b).

Let ds_z be the length of a line element on a contour curve C_z of the equidistant surface S_z ($S_z \parallel S$), and $\bar{\tau}^z$ and $\bar{n}^z$ be unit tangent vectors to C_z. Then

$$\bar{\tau}^z = \bar{e}_1^z \tau_1^z + \bar{e}_2^z \tau_2^z, \qquad \tau_1^z = H_1 \frac{d\alpha_1}{ds_z}, \qquad \tau_2^z = H_2 \frac{d\alpha_2}{ds_z}, \tag{5.8}$$

$$n_1^z = H_2 \frac{d\alpha_2}{ds_z}, \qquad n_2^z = H_1 \frac{d\alpha_1}{ds_z}, \tag{5.9}$$

where τ_i^z and n_i^z $(i=1,2)$ are the projections of $\bar{\tau}^z$ and $\bar{n}^z$ on vectors $\bar{e}_i^z \in S_z$. From Eqs. (5.6) and (5.9) for projections of $\bar{n}$ on tangents to the coordinate lines on S_z, we have

$$n_1^z = \frac{H_2}{A_2}\frac{ds}{ds_z}\, n_1, \qquad n_2^z = \frac{H_1}{A_1}\frac{ds}{ds_z}\, n_2. \tag{5.10}$$

Although Eqs. (5.2)–(5.10) are obtained in terms of the undeformed shell, they are also valid for the deformed configuration. Thus,

$$\overset{*}{\tau}_i = \overset{*}{A}_i \frac{d\alpha_i}{d\overset{*}{s}}, \qquad \overset{*}{n}_i = (-1)^{i+1} \overset{*}{A}_i \frac{d\alpha_i}{d\overset{*}{s}} \quad (i = 1, 2), \tag{5.11}$$

where $\overset{*}{\tau}_i, \overset{*}{n}_i$ and $d\overset{*}{s}$ are expressed in terms of the contour line $\overset{*}{C}$ and have the meanings described above. Assuming that deformations on the boundary are small, we have $\overset{*}{A}_i \approx A_i$ and $d\overset{*}{s} \approx ds$, so from Eq. (5.1) we have

$$\overset{*}{\tau}_i \approx \tau_i, \qquad \overset{*}{n}_i \approx n_i. \tag{5.12}$$

This implies that projections of $\overset{*}{\bar{\tau}}$ and $\overset{*}{\bar{n}}$ on the coordinate axes of the deformed middle surface $\overset{*}{S}$ equal projections of the same vectors on the undeformed middle surface S.

For curvatures k_n^* and k_τ^* in the directions of $\overset{*}{\bar{n}}$ and $\overset{*}{\bar{\tau}}$, and twist $k_{n\tau}^*$ of $\overset{*}{C}$, we have

$$k_n^* = \sum_{i=1}^{2}\sum_{j=1}^{2} k_{ij}^* n_i^* n_j^* \approx \sum_{i=1}^{2}\sum_{j=1}^{2} k_{ij}^* n_i n_j,$$
$$k_\tau^* = \sum_{i=1}^{2}\sum_{j=1}^{2} k_{ij}^* \tau_i^* \tau_j^* \approx \sum_{i=1}^{2}\sum_{j=1}^{2} k_{ij}^* \tau_i \tau_j, \tag{5.13}$$
$$k_{n\tau}^* = \sum_{i=1}^{2}\sum_{j=1}^{2} k_{ij}^* \tau_i^* n_j^* \approx \sum_{i=1}^{2}\sum_{j=1}^{2} k_{ij}^* \tau_i n_j,$$

where Eqs. (5.7) and (5.12) are used. Formulas for the deformed contour are similar to those given by Eq. (5.10) and have the form

$$\overset{*}{n}{}_1^z = \frac{\overset{*}{H}_2}{\overset{*}{A}_2}\frac{\mathrm{d}\overset{*}{s}}{\mathrm{d}\overset{*}{s}_z}\overset{*}{n}_1, \qquad \overset{*}{n}{}_2^z = \frac{\overset{*}{H}_1}{\overset{*}{A}_1}\frac{\mathrm{d}\overset{*}{s}}{\mathrm{d}\overset{*}{s}_z}\overset{*}{n}_2. \tag{5.14}$$

5.2 Stresses on the boundary

Let $\overset{*}{\bar{p}}{}_n^z$ be the normal stress vector acting upon a differential element of the boundary $\overset{*}{\Sigma}{}^z$ of the deformed shell located at a distance $\overset{*}{z} \approx z$ from the middle surface $\overset{*}{S}$ (Fig. 5.2),

$$\overset{*}{\bar{p}}{}_n^z = \bar{p}_1 \overset{*}{n}{}_1^z + \bar{p}_2 \overset{*}{n}{}_2^z + \bar{p}_3 \overset{*}{\bar{m}}. \tag{5.15}$$

Here $\overset{*}{n}{}_i^z$ are the projections of $\overset{*}{\bar{n}}{}^z$ on $\overset{*}{\bar{e}}{}_i^z \in \overset{*}{S}$, and $\bar{p}_i$ are the stress vectors acting upon the faces $\alpha_1 = \text{constant}$ and $\alpha_2 = \text{constant}$. Since $\overset{*}{\bar{n}}{}^z \perp \overset{*}{\bar{m}}$, the third component in Eq. (5.15) vanishes, $\bar{p}_3\,\overset{*}{\bar{m}} = 0$.

The surface area $\mathrm{d}\overset{*}{\Sigma}_z$ of a differential element on the edge is given by

$$\mathrm{d}\overset{*}{\Sigma}_z = \mathrm{d}\overset{*}{s}_z\,\mathrm{d}\overset{*}{z}. \tag{5.16}$$

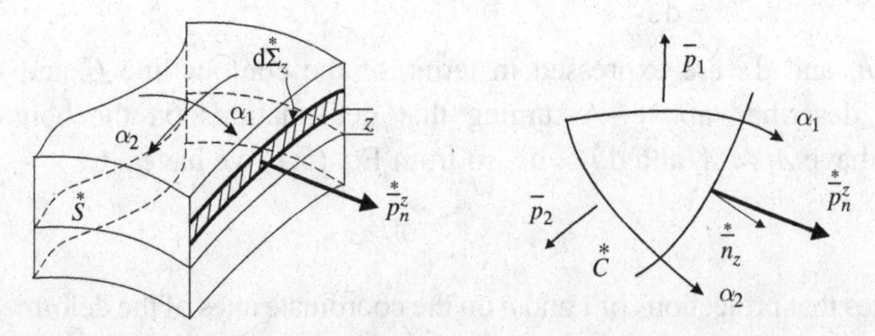

Fig. 5.2 Stresses on the boundary of a shell.

The resultant force $\overline{R}_{\overset{*}{n}}$ and moment $\overline{M}_{\overset{*}{n}}$ vectors per unit length of $\overset{*}{C}$ acting upon $\mathrm{d}\overset{*}{\Sigma}_z$ are given by

$$\overline{R}_{\overset{*}{n}} = \int_{z_1}^{z_2} \overset{*}{\overline{p}}{}_n^z \frac{\mathrm{d}\overset{*}{\Sigma}_z}{\mathrm{d}\overset{*}{s}}, \qquad \overline{M}_{\overset{*}{n}} = \int_{z_1}^{z_2} \left(\overset{*}{\overline{m}}\,\overset{*}{z} \times \overset{*}{\overline{p}}{}_n^z\right) \frac{\mathrm{d}\overset{*}{\Sigma}_z}{\mathrm{d}\overset{*}{s}}. \tag{5.17}$$

Using Eqs. (5.15) and (5.16) they can be written as

$$\overline{R}_{\overset{*}{n}} = \int_{z_1}^{z_2} \left(\overline{p}_1 \overset{*}{n}{}_1^z + \overline{p}_2 \overset{*}{n}{}_2^z\right) \frac{\mathrm{d}\overset{*}{s}_z\, \mathrm{d}\overset{*}{z}}{\mathrm{d}\overset{*}{s}},$$

$$\overline{M}_{\overset{*}{n}} = \int_{z_1}^{z_2} \left(\overset{*}{\overline{m}}\,\overset{*}{z} \times \left(\overline{p}_1\, \overset{*}{n}{}_1^z + \overline{p}_2\, \overset{*}{n}{}_2^z\right)\right) \frac{\mathrm{d}\overset{*}{s}_z\, \mathrm{d}\overset{*}{z}}{\mathrm{d}\overset{*}{s}}.$$

On substituting $\overset{*}{n}{}_i^z$ given by Eq. (5.14), and using approximations (5.12), $\overline{R}_{\overset{*}{n}}$ and $\overline{M}_{\overset{*}{n}}$ on the skewed faces of the boundary are found to be

$$\begin{aligned} \overline{R}_{\overset{*}{n}} &= \overline{R}_1 \overset{*}{n}_1 + \overline{R}_2 \overset{*}{n}_2 \approx \overline{R}_1 n_1 + \overline{R}_2 n_2, \\ \overline{M}_{\overset{*}{n}} &= \overline{M}_1 \overset{*}{n}_1 + \overline{M}_2 \overset{*}{n}_2 \approx \overline{M}_1 n_1 + \overline{M}_2 n_2. \end{aligned} \tag{5.18}$$

Decomposing $\overline{R}_{\overset{*}{n}}$ and $\overline{M}_{\overset{*}{n}}$ along the base $\{\overset{*}{\overline{n}}, \overset{*}{\overline{\tau}}, \overset{*}{\overline{m}}\}$ and substituting $\overline{R}_i$ and $\overline{M}_i$ given by Eqs. (3.22) and (3.23), we get

$$\begin{aligned} \overline{R}_{\overset{*}{n}} &= \sum_{i=1}^{2} \sum_{k=1}^{2} \overset{*}{\overline{e}}_1 \overset{*}{T}_{ik}\, \overset{*}{n}_i + \overset{*}{\overline{m}} \left(\overset{*}{N}_1\, \overset{*}{n}_1 + \overset{*}{N}_2\, \overset{*}{n}_2\right), \\ \overline{M}_{\overset{*}{n}} &= \overset{*}{\overline{e}}_2 \left(\overset{*}{M}_{11} \overset{*}{n}_1 + \overset{*}{M}_{21} \overset{*}{n}_2\right) - \overset{*}{\overline{e}}_1 \left(\overset{*}{M}_{12} \overset{*}{n}_1 + \overset{*}{M}_{22} \overset{*}{n}_2\right), \end{aligned} \tag{5.19}$$

where $\overset{*}{T}_{ik}$, $\overset{*}{N}_i$ and $\overset{*}{M}_{ik}$ are the in-plane forces and moments.

Let $T_{\overset{*}{n}}$, $T_{\overset{*}{n}\overset{*}{\tau}}$ and $\overset{*}{N}$ be the normal, tangent and lateral forces acting on $\overset{*}{\Sigma}$

$$T_{\overset{*}{n}} = \overline{R}_{\overset{*}{n}}\, \overset{*}{\overline{n}}, \qquad T_{\overset{*}{n}\overset{*}{\tau}} = \overline{R}_{\overset{*}{n}}\, \overset{*}{\overline{\tau}}, \qquad \overset{*}{N} = \overline{R}_{\overset{*}{n}}\, \overset{*}{\overline{m}}\,.$$

On use of $\overline{R}_{\overset{*}{n}}$ (5.19) and approximations (5.12), we find

$$\begin{aligned} T_{\overset{*}{n}} &= \sum_{i=1}^{2} \sum_{k=1}^{2} \overset{*}{T}_{ik}\, \overset{*}{n}_i\, \overset{*}{n}_k \approx \sum_{i=1}^{2} \sum_{k=1}^{2} \overset{*}{T}_{ik} n_i n_k, \\ T_{\overset{*}{n}\overset{*}{\tau}} &= \sum_{i=1}^{2} \sum_{k=1}^{2} \overset{*}{T}_{ik}\, \overset{*}{n}_i\, \overset{*}{\tau}_k \approx \sum_{i=1}^{2} \sum_{k=1}^{2} \overset{*}{T}_{ik} n_i \tau_k, \\ \overset{*}{N} &= \overset{*}{N}_1\, \overset{*}{n}_1 + \overset{*}{N}_2\, \overset{*}{n}_2 \approx \overset{*}{N}_1\, n_1 + \overset{*}{N}_2\, n_2. \end{aligned} \tag{5.20}$$

By projecting $\overline{M}_{\overset{*}{n}}$ on the tangent and normal planes to the boundary we get

$$\overset{*}{G} = \overline{M}_{\overset{*}{n}} \overset{*}{\bar{\tau}}, \qquad \overset{*}{H} = \overline{M}_{\overset{*}{n}} \bar{n},$$

where $\overset{*}{G}$ and $\overset{*}{H}$ are the bending and twisting moments. On substituting for $\overline{M}_{\overset{*}{n}}$ (Eq. (5.19)), we obtain

$$\overset{*}{G} = \overset{*}{\bar{\tau}}_2 \left(\overset{*}{M}_{11} \overset{*}{n}_1 + \overset{*}{M}_{21} \overset{*}{n}_2 \right) - \overset{*}{\bar{\tau}} \left(\overset{*}{M}_{21} \overset{*}{n}_1 + \overset{*}{M}_{22} \overset{*}{n}_2 \right),$$
$$\overset{*}{H} = \overset{*}{\bar{n}}_2 \left(\overset{*}{M}_{11} \overset{*}{n}_1 + \overset{*}{M}_{21} \overset{*}{n}_2 \right) - \overset{*}{\bar{n}}_1 \left(\overset{*}{M}_{21} \overset{*}{n}_1 + \overset{*}{M}_{22} \overset{*}{n}_2 \right).$$

Making use of Eq. (5.6), after some simple algebra, we find

$$\begin{aligned} \overset{*}{G} &= \sum_{i=1}^{2} \sum_{k=1}^{2} \overset{*}{M}_{ik} \overset{*}{n}_i \overset{*}{n}_k \approx \sum_{i=1}^{2} \sum_{k=1}^{2} \overset{*}{M}_{ik} n_i n_k, \\ \overset{*}{H} &= -\sum_{i=1}^{2} \sum_{k=1}^{2} \overset{*}{M}_{ik} \overset{*}{n}_i \overset{*}{\tau}_k \approx -\sum_{i=1}^{2} \sum_{k=1}^{2} \overset{*}{M}_{ik} n_i \tau_k. \end{aligned} \tag{5.21}$$

Decomposing the resultant force and moment vectors in the directions of $\{\overset{*}{\bar{n}}, \overset{*}{\bar{\tau}}, \overset{*}{\bar{m}}\}$, we get

$$\overline{R}_{\overset{*}{n}} = T_{\overset{*}{n}} \overset{*}{\bar{n}} + T_{\overset{*}{n}\overset{*}{\tau}} \overset{*}{\bar{\tau}} + \overset{*}{N} \overset{*}{\bar{m}}, \qquad \overline{M}_{\overset{*}{n}} = H_{\overset{*}{n}} \bar{n}_* + \overset{*}{G} \overset{*}{\bar{\tau}}. \tag{5.22}$$

The result of the above considerations is that the stressed state on the edge of the thin shell is determined in terms of five variables, namely $T_{\overset{*}{n}}$, $T_{\overset{*}{n}\overset{*}{\tau}}$, $\overset{*}{N}$, $\overset{*}{G}$ and $\overset{*}{H}$. However, as we shall show soon, the twisting moment $\overset{*}{H}$ can be replaced by a statically equivalent force $-\partial(\overset{*}{H}\overset{*}{\bar{m}})/\partial \overset{*}{s}$ per unit arc length of $\widehat{c_0 c_1}$ of the contour $\overset{*}{C}$. The substitution, according to the Saint-Venant principle, will have no effect on the stress state in the thin shell at distances sufficiently far away from the boundary.

Let $\overset{*}{H}$ be the twisting-moment vector acting at point $c_0 \in \overset{*}{C}$ (Fig. 5.3). Consider the vicinity of the point c_0. Approximating the arc $\widehat{c_0 c_1}$ by a straight line $\overline{c_0 c_1}$ of length $\mathrm{d}\overset{*}{s}$, the resultant vector of twisting moment $\overset{*}{H}\overset{*}{\bar{n}}\mathrm{d}\overset{*}{s}$ acting upon $\overline{c_0 c_1}$ can be substituted by a statically equivalent couple given by $\left(-\overset{*}{H}\overset{*}{\bar{m}}, +\overset{*}{H}\overset{*}{\bar{m}}\right)$,

$$\overset{*}{H}\overset{*}{\bar{n}}\,\mathrm{d}\overset{*}{s} \propto \left(-\overset{*}{H}\overset{*}{\bar{m}}, +\overset{*}{H}\overset{*}{\bar{m}}\right).$$

The vector $\overset{*}{\bar{m}}$ is orthogonal to $\overset{*}{S}$ and the vector $\overset{*}{\bar{n}}$ is orthogonal to $\overset{*}{\bar{m}}$ at c_0 and points towards the reader. The forces $\pm\overset{*}{H}\overset{*}{\bar{m}}$ are collinear with the vector $\overset{*}{\bar{m}}$ at c_{01} – the middle point of $\overline{c_0 c_1}$. Then the moment of the couple about point c_{01} is indeed equal to

$$\overset{*}{\bar{\tau}}\,\mathrm{d}\overset{*}{s} \times \overset{*}{H}\overset{*}{\bar{m}} = \left(\overset{*}{\bar{\tau}} \times \overset{*}{\bar{m}}\right) \overset{*}{H}\,\mathrm{d}\overset{*}{s} = \overset{*}{H}\overset{*}{\bar{n}}\,\mathrm{d}\overset{*}{s},$$

where $\overset{*}{\bar{\tau}}\,\mathrm{d}\overset{*}{s} \approx \overline{c_0 c_1}$.

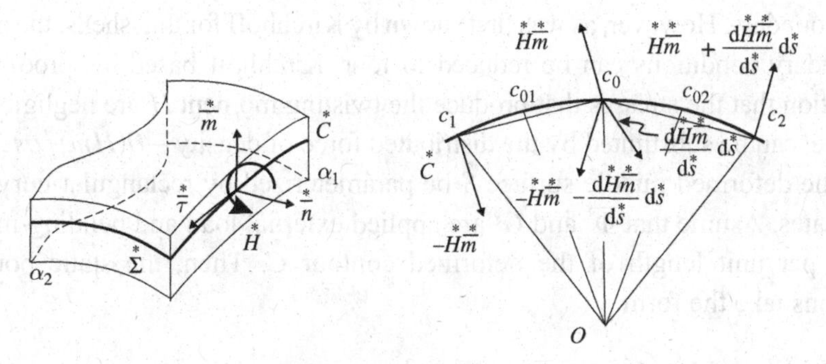

Fig. 5.3 Substitution of the twist moment H by a statically equivalent distributed force $\partial(\overset{*}{H}\overset{*}{\bar m}/\partial\overset{*}{s}$ acting on the boundary of a shell.

In just the same way, it can be shown that the torque exerted on the segment $\overline{c_0c_2}$ is statically equivalent to the couple applied at points c_0 and c_2. They are oriented along $\overset{*}{\bar m}$ at c_{02}, namely the middle point of $\overline{c_0c_2}$, and equal

$$-\left(\overset{*}{H}\overset{*}{\bar m} + \frac{\partial(\overset{*}{H}\overset{*}{\bar m})}{\partial \overset{*}{s}}\, \mathrm{d}\overset{*}{s}\right) \quad \text{and} \quad \left(\overset{*}{H}\,\overset{*}{\bar m} + \frac{\partial(\overset{*}{H}\,\overset{*}{\bar m})}{\partial \overset{*}{s}}\, \mathrm{d}\overset{*}{s}\right).$$

The geometric sum of forces applied at point c_0 is $-(\partial(\overset{*}{H}\,\overset{*}{\bar m})/\partial\overset{*}{s})\mathrm{d}\overset{*}{s}$. It follows that the twisting moment $\overset{*}{H}$ per unit length of the contour $\overset{*}{C}$ is indeed statically equivalent to the distributed force of density $-(\partial(\overset{*}{H}\overset{*}{\bar m})/\partial\overset{*}{s})\mathrm{d}\overset{*}{s}$.

The resultant force and moment vectors $\overline{R}_{\overset{*}{n}}$ and $\overline{M}_{\overset{*}{n}}$ acting upon $\overset{*}{\Sigma}$ are statically equivalent to the generalized force vector

$$\overline{\Phi} = \overline{R}_{\overset{*}{n}} - \frac{\partial(\overset{*}{H}\overset{*}{\bar m})}{\partial \overset{*}{s}}\, \mathrm{d}\overset{*}{s} \tag{5.23}$$

and the bending moment

$$\overset{*}{G} = \sum_{i=1}^{2}\sum_{k=1}^{2} \overset{*}{M}_{ik}\, \overset{*}{n}_i\, \overset{*}{n}_k \approx \sum_{i=1}^{2}\sum_{k=1}^{2} \overset{*}{M}_{ik} n_i n_k. \tag{5.24}$$

Here $\overset{*}{H}$ and $\overline{R}_{\overset{*}{n}}$ satisfy Eqs. (5.21) and (5.22), respectively.

In the above derivations we assumed that the boundary is non-singular and closed. If there are singularities, e.g. corner points along the edge of the boundary, then the force $-\partial(\overset{*}{H}\overset{*}{\bar m})/\partial\overset{*}{s}$ should be supplemented by forces $\left(\pm\overset{*}{H}\overset{*}{\bar m}\right)$ acting at the corners.

5.3 Static boundary conditions

Since stresses on the boundary $\overset{*}{\Sigma}$ of the shell are statically equivalent to three forces, $T_{\overset{*}{n}}$, $T_{\overset{*}{n}\overset{*}{\tau}}$ and $\overset{*}{N}$, and two moments, $\overset{*}{G}$ and $\overset{*}{H}$, five static conditions should be prescribed on the boundary. However, as was first shown by Kirchhoff for thin shells, the number of boundary conditions can be reduced to four. Kirchhoff based his proof on the assumption that the stresses that produce the twisting moment $\overset{*}{H}$ are negligible and, therefore, can be substituted by the distributed force of density $-\partial(\overset{*}{H}\overset{*}{\bar{m}})/\partial\overset{*}{s}$.

Let the deformed middle surface $\overset{*}{S}$ be parameterized by rectangular curvilinear coordinates. Asume that $\overline{\Phi}^s$ and $\overset{*}{G^s}$ are applied external load and bending-moment vectors per unit length of the deformed contour $\overset{*}{C}$. Then, the static boundary conditions take the form

$$\overline{\Phi}^s = \overline{R}_{\overset{*}{n}} - \frac{\partial(\overset{*}{H}\overset{*}{\bar{m}})}{\partial\overset{*}{s}}, \qquad \overset{*}{G^s} = \sum_{i=1}^{2}\sum_{k=1}^{2} \overset{*}{M}_{ik}\overset{*}{n}_i\overset{*}{n}_k \approx \sum_{i=1}^{2}\sum_{k=1}^{2} \overset{*}{M}_{ik} n_i n_k. \tag{5.25}$$

where $\overline{R}_{\overset{*}{n}}$ and $\overset{*}{H}$ satisfy Eqs. (5.21) and (5.22), respectively. The projections of $\overline{\Phi}^s$ on $\{\overset{*}{\bar{n}}, \overset{*}{\bar{\tau}}, \overset{*}{\bar{m}}\}$ are given by

$$\overset{*}{\bar{n}}\,\Phi^s_{\overset{*}{n}} + \overset{*}{\bar{\tau}}\,\Phi^s_{\overset{*}{\tau}} + \overset{*}{\bar{m}}\,\Phi^s_{\overset{*}{m}} = \overline{R}_{\overset{*}{n}} - \frac{\partial(\overset{*}{H}\overset{*}{\bar{m}})}{\partial\overset{*}{s}}.$$

Taking the scalar product of the above with $\overset{*}{\bar{n}}, \overset{*}{\bar{\tau}}$ and $\overset{*}{\bar{m}}$ yields

$$\Phi^s_{\overset{*}{n}} = T_{\overset{*}{n}} - \overset{*}{\bar{n}}\overset{*}{H}\frac{\partial\overset{*}{\bar{m}}}{\partial\overset{*}{s}}, \qquad \Phi^s_{\overset{*}{\tau}} = T_{\overset{*}{n}\overset{*}{\tau}} - \overset{*}{\bar{\tau}}\overset{*}{H}\frac{\partial\overset{*}{\bar{m}}}{\partial\overset{*}{s}}, \qquad \Phi^s_{\overset{*}{m}} = N_{\overset{*}{n}} - \frac{\partial\overset{*}{H}}{\partial\overset{*}{s}}.$$

Here use is made of $\overset{*}{\bar{n}}\overset{*}{\bar{m}} = \overset{*}{\bar{\tau}}\overset{*}{\bar{m}} = \overset{*}{\bar{m}}\,\partial\overset{*}{\bar{m}}/\partial\overset{*}{s} = 0$. Since

$$\overset{*}{\bar{m}}_i = \overset{*}{A}_i\left(\overset{*}{\bar{e}}_1\overset{*}{k}_{1i} + \overset{*}{\bar{e}}_2\overset{*}{k}_{2i}\right), \qquad \overset{*}{A}_i\frac{\partial\alpha_i}{\partial\overset{*}{s}} = \overset{*}{\tau}_i, \tag{5.26}$$

for the derivative $\partial\overset{*}{\bar{m}}/\partial\overset{*}{s}$ we have

$$\frac{\partial\overset{*}{\bar{m}}}{\partial\overset{*}{s}} = \overset{*}{\bar{m}}_1\frac{\partial\alpha_1}{\partial\overset{*}{s}} + \overset{*}{\bar{m}}_2\frac{\partial\alpha_2}{\partial\overset{*}{s}} = \sum_{i=1}^{2}\sum_{j=1}^{2}\overset{*}{A}_i\overset{*}{\bar{e}}_j\overset{*}{k}_{ij}\frac{\partial\alpha_i}{\partial\overset{*}{s}} = \sum_{i=1}^{2}\sum_{j=1}^{2}\overset{*}{\bar{e}}_j\overset{*}{k}_{ij}\overset{*}{\tau}_i.$$

The scalar products of $\partial\overset{*}{\bar{m}}/\partial\overset{*}{s}$ by $\overset{*}{\bar{n}}$ and $\overset{*}{\bar{\tau}}$ are found to be

$$\begin{aligned} \overset{*}{\bar{n}}\frac{\partial\overset{*}{\bar{m}}}{\partial\overset{*}{s}} &= \sum_{i=1}^{2}\sum_{j=1}^{2}\left(\overset{*}{\bar{n}}\overset{*}{\bar{e}}_j\right)\overset{*}{k}_{ij}\overset{*}{\tau}_i = \overset{*}{k}_{n\tau}, \\ \overset{*}{\bar{\tau}}\frac{\partial\overset{*}{\bar{m}}}{\partial\overset{*}{s}} &= \sum_{i=1}^{2}\sum_{j=1}^{2}\left(\overset{*}{\bar{\tau}}\overset{*}{\bar{e}}_j\right)\overset{*}{k}_{ij}\overset{*}{\tau}_i = \overset{*}{k}_{\tau}, \end{aligned} \tag{5.27}$$

where use is made of approximations given by Eqs. (5.12). Applying Eqs. (5.27) in (5.23), the boundary conditions are found to be

$$\Phi^s_{\overset{*}{n}} = T_{\overset{*}{n}} - \overset{*}{H}\overset{*}{k}_{n\tau}, \qquad \Phi^s_{\overset{*}{\tau}} = T_{\overset{*}{n}\overset{*}{\tau}} - \overset{*}{H}\overset{*}{k}_{\tau},$$

$$\Phi^s_{\overset{*}{m}} = N_{\overset{*}{n}} - \frac{\partial \overset{*}{H}}{\partial \overset{*}{s}} \approx N_{\overset{*}{n}} - \frac{\partial \overset{*}{H}}{\partial s}, \qquad \overset{*}{G^s} \approx \sum_{i=1}^{2}\sum_{k=1}^{2} \overset{*}{M}_{ik} n_i n_k. \tag{5.28}$$

In the above we assumed that $d\overset{*}{s} \approx ds$. Equations (5.28) contain nonlinearities that are introduced by $\overset{*}{k}_{n\tau}$ and $\overset{*}{k}_{\tau}$ and projections of the external load vector $\overline{\Phi}^s$. For example, if f is the angle between the positive orientation of the α_1 axis and the vector $\bar{n}$, then the projections of $\bar{n}$ and $\bar{\tau}$ on the undeformed contour of the shell are given by

$$\tau_1 = -n_2 = -\sin\varphi, \qquad \tau_2 = n_1 = \cos\varphi. \tag{5.29}$$

Given below are some commonly used boundary conditions.

1. Clamped edge:

$$u_n = u\tau = \omega = 0, \qquad \frac{\partial \omega}{\partial n} = 0, \tag{5.30}$$

where u_n is the projection of the displacement vector $\bar{v}$ on the vector $\bar{n}$ perpendicular to the contour $C \in \Sigma$, u_τ is the projection of $\bar{v}$ on the vector $\bar{\tau}$, and ω is the normal displacement (deflection)

$$u_n = \bar{v}\bar{n} = u_1 n_1 + u_2 n_2, \qquad u_n = \bar{v}\bar{\tau} = u_1\tau_1 + u_2\tau_2.$$

The last condition in (5.30) implies that the rotation about the vector $\bar{\tau}$ equals zero: $\bar{n}\,\overset{*}{\bar{m}} = 0$.

2. Simply supported edge:

$$u_n = u_\tau = \omega = 0, \qquad \overset{*}{G^s} = \sum_{i=1}^{2}\sum_{k=1}^{2} \overset{*}{M}_{ik}\,\overset{*}{n}_i\,\overset{*}{n}_k \approx \sum_{i=1}^{2}\sum_{k=1}^{2} \overset{*}{M}_{ik} n_i n_k = 0, \tag{5.31}$$

where $\overset{*}{G^s}$ is the bending moment.

3. Freely supported edge with a single degree of freedom in the normal direction:

$$u_n = u_\tau = 0, \qquad \overset{*}{G^s} = 0, \qquad N_{\overset{*}{n}} - \frac{\partial \overset{*}{H}}{\partial \overset{*}{s}} \approx N_{\overset{*}{n}} - \frac{\partial \overset{*}{H}}{\partial s} = 0. \tag{5.32}$$

4. Free edge:

$$T_{\overset{*}{n}} - \overset{*}{H}\overset{*}{k}_{n\tau} = 0, \qquad T_{\overset{*}{n}\overset{*}{\tau}} - \overset{*}{H}\overset{*}{k}_{\tau} = 0,$$

$$N_{\overset{*}{n}} - \frac{\partial \overset{*}{H}}{\partial \overset{*}{s}} \approx N_{\overset{*}{n}} - \frac{\partial \overset{*}{H}}{\partial s} = 0,$$

$$\sum_{i=1}^{2}\sum_{k=1}^{2} \overset{*}{M}_{ik}\,\overset{*}{n}_i\,\overset{*}{n}_k \approx \sum_{i=1}^{2}\sum_{k=1}^{2} \overset{*}{M}_{ik} n_i n_k = 0, \tag{5.33}$$

where the meanings of the parameters are as above.

5.4 Deformations of the edge

Consider deformations of a line element on $d\Sigma$. Assume that it is orthogonal to the undeformed middle surface S of the shell. The tangent, ε_n and ε_τ, bending, æ_n and æ_τ, and shear, $\varepsilon_{n\tau}$, deformations and twist $\text{æ}_{n\tau}$ of the edge of the shell are defined by

$$\varepsilon_n = \sum_{i=1}^{2}\sum_{k=1}^{2} \varepsilon_{ik} n_i n_k, \qquad \varepsilon_\tau = \sum_{i=1}^{2}\sum_{k=1}^{2} \varepsilon_{ik} \tau_i \tau_k, \qquad \varepsilon_{n\tau} = \sum_{i=1}^{2}\sum_{k=1}^{2} \varepsilon_{ik} \tau_i n_k, \tag{5.34}$$

$$\text{æ}_n = \sum_{i=1}^{2}\sum_{k=1}^{2} \text{æ}_{ik} n_i n_k, \qquad \text{æ}_\tau = \sum_{i=1}^{2}\sum_{k=1}^{2} \text{æ}_{ik} \tau_i \tau_k, \qquad \text{æ}_{n\tau} = \sum_{i=1}^{2}\sum_{k=1}^{2} \text{æ}_{ik} \tau_i n_k, \tag{5.35}$$

where n_1 and n_2 satisfy Eqs. (5.29).

Applying Eqs. (5.34) and expressions for the Lamé parameters,

$$\overset{*}{A}{}_i^2 = A_i^2(1 + 2\varepsilon_{ii}), \qquad \overset{*}{A}_1 \overset{*}{A}_2 \cos \overset{*}{\chi} = 2\varepsilon_{12} A_1 A_2, \qquad \tau_i \, ds = A_i \, d\alpha_i,$$

the length of a line element $d\overset{*}{s}$ on $\overset{*}{C}$ is given by

$$(d\overset{*}{s})^2 = \overset{*}{A}{}_1^2 \, d\alpha_1^2 + 2\overset{*}{A}_1 \overset{*}{A}_2 \cos \overset{*}{\chi} \, d\alpha_1 \, d\alpha_2 + \overset{*}{A}{}_2^2 \, d\alpha_2^2 = (1 + 2\varepsilon_\tau) ds^2. \tag{5.36}$$

Unit vectors $\overset{*}{\bar{\tau}}$ and $\bar{\tau}$ are given by

$$\overset{*}{\bar{\tau}} = \frac{d\overset{*}{\bar{r}}}{d\overset{*}{s}} = \frac{d\overset{*}{\bar{r}}}{ds} \frac{ds}{d\overset{*}{s}} = \left(\bar{\tau} + \frac{d\bar{v}}{ds}\right) \Big/ \sqrt{1 + 2\varepsilon_\tau}, \qquad \bar{\tau} = \frac{d\bar{r}}{ds}, \tag{5.37}$$

where use is made of Eqs. (5.35).

The vector $\overset{*}{\bar{m}}$ normal to the surface $\overset{*}{S}$ is found from

$$\overset{*}{\bar{m}} = \left(\overset{*}{\bar{r}}_1 \times \overset{*}{\bar{r}}_2\right) \Big/ \sqrt{\overset{*}{a}}, \tag{5.38}$$

where $\overset{*}{a}$ is the invariant of the first fundamental form given by

$$\overset{*}{a} = (A_1 A_2)^2 \mathbf{A}, \qquad \mathbf{A} = 1 + 2(\varepsilon_n + \varepsilon_\tau) + 4(\varepsilon_n \varepsilon_\tau - \varepsilon_{n\tau}^2). \tag{5.39}$$

The derivatives of the displacement vector in the direction of $\bar{n}$ and $\bar{\tau}$ are

$$\frac{d\bar{v}}{ds} = \frac{\bar{\tau}_1}{A_1} \frac{d\bar{v}}{d\alpha_1} + \frac{\bar{\tau}_2}{A_2} \frac{d\bar{v}}{d\alpha_2}, \qquad \frac{d\bar{v}}{ds_n} = \frac{n_1}{A_1} \frac{d\bar{v}}{d\alpha_1} + \frac{n_2}{A_2} \frac{d\bar{v}}{d\alpha_2}. \tag{5.40}$$

Here s and s_n are the lengths of line elements on C and C_n, such that $C_n \perp C$, and $\bar{\tau} = d\bar{r}/ds$ and $\bar{n} = d\bar{r}/ds_n$. Solving Eqs. (5.40) for $d\bar{v}/d\alpha_i$, we find

$$\frac{1}{A_i}\frac{\mathrm{d}\bar{v}}{\mathrm{d}\alpha_i} = \tau_i \frac{\mathrm{d}\bar{v}}{\mathrm{d}s} + n_i \frac{\mathrm{d}\bar{v}}{\mathrm{d}s_n}. \tag{5.41}$$

Since $\bar{e}_i = \bar{\tau}\tau_i + \bar{n}n_i$, from Eq. (3.41) for vectors $\overset{*}{\bar{r}}_i$ we obtain

$$\overset{*}{\bar{r}}_i = \overset{*}{\bar{r}} + \frac{\mathrm{d}\bar{v}}{\mathrm{d}\alpha_i} = A_i\left(\bar{e}_i + \frac{1}{A_i}\frac{\mathrm{d}\bar{v}}{\mathrm{d}s}\right) = A_i(\underline{a}\tau_i + \underline{b}n_i). \tag{5.42}$$

Here we have introduced the following notation:

$$\underline{a} = \frac{\mathrm{d}\overset{*}{\bar{r}}}{\mathrm{d}s} = \bar{\tau} + \frac{\mathrm{d}\bar{v}}{\mathrm{d}s}, \qquad \underline{b} = \frac{\mathrm{d}\overset{*}{\bar{r}}}{\mathrm{d}s_n} = \bar{n} + \frac{\mathrm{d}\bar{v}}{\mathrm{d}s_n}. \tag{5.43}$$

On substituting Eq. (5.42) into (5.38), we get

$$\overset{*}{\bar{m}} = (b \times a)/\sqrt{\mathbf{A}} = \left\{\bar{m} + \left(\bar{n} \times \frac{\mathrm{d}\bar{v}}{\mathrm{d}s}\right) + \left(\frac{\mathrm{d}\bar{v}}{\mathrm{d}s_n} \times \bar{\tau}\right) + \left(\frac{\mathrm{d}\bar{v}}{\mathrm{d}s_n} \times \frac{\mathrm{d}\bar{v}}{\mathrm{d}s}\right)\right\} \Big/ \sqrt{\mathbf{A}}. \tag{5.44}$$

On use of Eq. (5.42) and the equality $\bar{r}_i = A_i(\bar{\tau}\tau_i + \bar{n}n_i)$, from $2A_iA_k\varepsilon_{ik} = \overset{*}{\bar{r}}_i\overset{*}{\bar{r}}_k - \bar{r}_i\bar{r}_k$ (see Eqs. (1.67)) for deformation on the edge of the shell, we have

$$2\varepsilon_{ik} = (\underline{a}\tau_i + \underline{b}n_i)(\underline{a}\tau_k + \underline{b}n_k) - (\bar{\tau}\tau_i + \bar{n}n_i)(\bar{\tau}\tau_k + \bar{n}n_k). \tag{5.45}$$

Further, with the help of Eq. (5.45), from Eqs. (5.34) we obtain

$$\varepsilon_n = \bar{n}\,\frac{\mathrm{d}\bar{v}}{\mathrm{d}s_n} + \frac{1}{2}\left(\frac{\mathrm{d}\bar{v}}{\mathrm{d}s_n}\right)^2, \qquad \varepsilon_\tau = \bar{\tau}\,\frac{\mathrm{d}\bar{v}}{\mathrm{d}s_n} + \frac{1}{2}\left(\frac{\mathrm{d}\bar{v}}{\mathrm{d}s_n}\right)^2,$$
$$2\varepsilon_{n\tau} = \bar{n}\,\frac{\mathrm{d}\bar{v}}{\mathrm{d}s_n} + \bar{\tau}\,\frac{\mathrm{d}\bar{v}}{\mathrm{d}s_n} + \frac{\mathrm{d}\bar{v}}{\mathrm{d}s}\frac{\mathrm{d}\bar{v}}{\mathrm{d}s_n}, \tag{5.46}$$

where use is made of the facts $\bar{n}\perp\bar{\tau}$ and $n_2^1 + n_2^2 = \tau_2^1 + \tau_2^2 = 1$.

To express the vector $\overset{*}{\bar{n}}$ in terms of the displacements, we substitute $\overset{*}{\bar{\tau}}$ and $\overset{*}{\bar{m}}$ given by Eqs. (5.37) and (5.44) into the equality $\overset{*}{\bar{n}} = \overset{*}{\bar{\tau}} \times \overset{*}{\bar{m}}$. After some simple algebra we find

$$\overset{*}{\bar{n}}\sqrt{\mathbf{A}(1+2\varepsilon_\tau)} = (1+2\varepsilon_\tau)\underline{b} - 2\varepsilon_{n\tau}\cdot\underline{a}. \tag{5.47}$$

To decompose the right-hand sides of Eqs. (5.46) in the base $\{\bar{n}, \bar{\tau}, \bar{m}\}$, we proceed from formulas for derivatives of $\bar{n}, \bar{\tau}$ and $\bar{m}$ with respect to s and s_n, which are given by

$$\frac{\mathrm{d}\bar{n}}{\mathrm{d}s} = æ\bar{\tau} - \bar{m}k_{n\tau}, \qquad \frac{\mathrm{d}\bar{\tau}}{\mathrm{d}s} = -\bar{m}k_\tau - æ\bar{n}, \qquad \frac{\mathrm{d}\bar{m}}{\mathrm{d}s} = \bar{n}k_{n\tau} + \bar{\tau}k_\tau, \tag{5.48}$$

$$\frac{\mathrm{d}\bar{n}}{\mathrm{d}s_n} = -\bar{m}k_\tau - æ'\bar{\tau}, \qquad \frac{\mathrm{d}\bar{\tau}}{\mathrm{d}s_n} = æ'\bar{n} - \bar{m}k_{n\tau}, \qquad \frac{\mathrm{d}\bar{m}}{\mathrm{d}s_n} = \bar{\tau}k_{n\tau} + \bar{n}k_n. \tag{5.49}$$

Here k_n, k_τ and $k_{n\tau}$ satisfy Eqs. (5.7),

$$k_n = \bar{n}\frac{d\bar{\tau}}{ds_n} = -\bar{\tau}\frac{d\bar{n}}{ds_n} \qquad k_\tau = \bar{\tau}\frac{d\bar{m}}{ds_n} = -\bar{m}\frac{d\bar{\tau}}{ds_n} \qquad k_{n\tau} = \bar{n}\frac{d\bar{m}}{ds_n} = -\bar{m}\frac{d\bar{n}}{ds_n}, \tag{5.50}$$

and æ and æ′ are the geodesic curvatures of the contour lines C and C_n ($C_n \perp C$), which are described by

$$æ = \bar{\tau}\frac{d\bar{n}}{ds} = -\bar{n}\frac{d\bar{\tau}}{ds}, \qquad æ' = \bar{\tau}\frac{d\bar{n}}{ds_n} = -\bar{n}\frac{d\bar{\tau}}{ds_n}, \tag{5.51}$$

or in the expanded form

$$\begin{aligned} æ &= \frac{d\varphi}{ds} + \frac{\cos\varphi}{A_1A_2}\frac{\partial A_2}{\partial \alpha_1} + \frac{\sin\varphi}{A_1A_2}\frac{\partial A_1}{\partial \alpha_2}, \\ æ' &= -\frac{d\varphi}{ds_n} - \frac{\sin\varphi}{A_1A_2}\frac{\partial A_2}{\partial \alpha_1} + \frac{\cos\varphi}{A_1A_2}\frac{\partial A_1}{\partial \alpha_2}. \end{aligned} \tag{5.52}$$

Expanding the displacement vector $\bar{v}$ along the base $\{\bar{n}, \bar{\tau}, \bar{m}\}$,

$$\bar{v} = \bar{n}u_n + \bar{\tau}u_\tau + \bar{m}\omega, \tag{5.53}$$

and differentiating Eq. (5.53) with respect to s and s_n, we find

$$\frac{d\bar{v}}{ds} = \bar{n}e_{\tau n} + \bar{\tau}e_{\tau\tau} + \bar{m}\omega_\tau, \qquad \frac{d\bar{v}}{ds_n} = \bar{n}e_{nn} + \bar{\tau}e_{n\tau} + \bar{m}\omega_n. \tag{5.54}$$

Here $e_{nn}, e_{n\tau}, e_{n\tau}, e_{\tau n}, \omega_n$ and ω_τ are the rotation angles given by

$$\begin{aligned} e_{\tau n} &= \frac{du_n}{ds} - æu_\tau + \bar{\omega}k_{n\tau}, & e_{\tau\tau} &= \frac{du_\tau}{ds} + æu_n + \bar{\omega}k_\tau, \\ e_{nn} &= \frac{du_n}{ds_n} + æ'u_\tau + \bar{\omega}k_n, & e_{n\tau} &= \frac{du_\tau}{ds_n} - æ'u_n + \bar{\omega}k_{n\tau}, \end{aligned} \tag{5.55}$$

$$\omega_\tau = \frac{d\bar{\omega}}{ds} - k_{n\tau}u_n - k_\tau u_\tau, \qquad \omega_n = \frac{d\bar{\omega}}{ds_n} - k_n u_n - k_{n\tau}u_\tau. \tag{5.56}$$

In the above, use is made of Eqs. (5.48) and (5.49). On substituting Eqs. (5.54) into Eq. (5.44), we have

$$\overset{*}{m}\sqrt{\mathbf{A}} = \bar{n}S_n + \bar{\tau}S_\tau + \bar{m}S_3, \tag{5.57}$$

where

$$\begin{aligned} S_n &= \omega_\tau e_{n\tau} - \omega_n(1 + e_{\tau\tau}), \\ S_\tau &= \omega_n e_{\tau n} - \omega_\tau(1 + e_{nn}), \\ S_3 &= (1 + e_{\tau\tau})(1 + e_{nn}) - e_{\tau n}e_{n\tau}. \end{aligned} \tag{5.58}$$

Let $\overset{*}{k}_n, \overset{*}{k}_\tau$ and $\overset{*}{k}_{n\tau}$ be the curvatures and twist of the contour $\overset{*}{C}$ (see Eq. (5.13)):

$$\overset{*}{k}_n = \sum_{i=1}^{2}\sum_{j=1}^{2} \overset{*}{k}_{ij}\, n_i n_j = \underline{b}\,\frac{\mathrm{d}\overset{*}{\bar{m}}}{\mathrm{d}s_n} = -\overset{*}{\bar{m}}\,\frac{\mathrm{d}\underline{b}}{\mathrm{d}s_n},$$

$$\overset{*}{k}_\tau = \sum_{i=1}^{2}\sum_{j=1}^{2} \overset{*}{k}_{ij}\, \tau_i \tau_j = \underline{a}\,\frac{\mathrm{d}\overset{*}{\bar{m}}}{\mathrm{d}s} = -\overset{*}{\bar{m}}\,\frac{\mathrm{d}\underline{a}}{\mathrm{d}s}, \tag{5.59}$$

$$\overset{*}{k}_{n\tau} = \sum_{i=1}^{2}\sum_{j=1}^{2} \overset{*}{k}_{ij}\, \tau_i n_j = \underline{b}\,\frac{\mathrm{d}\overset{*}{\bar{m}}}{\mathrm{d}s} = -\overset{*}{\bar{m}}_i\,\frac{\mathrm{d}\underline{b}}{\mathrm{d}s}.$$

Using Eq. (5.59), the bending deformation of the boundary of the shell is found to be

$$æ_n = \overset{*}{k}_n - k_n = \bar{m}\,\frac{\mathrm{d}\bar{n}}{\mathrm{d}s_n} - \overset{*}{\bar{m}}\,\frac{\mathrm{d}\underline{b}}{\mathrm{d}s_n}, \qquad æ_\tau = \bar{m}\,\frac{\mathrm{d}\bar{\tau}}{\mathrm{d}s} - \overset{*}{\bar{m}}\,\frac{\mathrm{d}\underline{a}}{\mathrm{d}s}, \qquad æ_{n\tau} = \bar{m}\,\frac{\mathrm{d}\bar{n}}{\mathrm{d}s} - \overset{*}{\bar{m}}\,\frac{\mathrm{d}\underline{b}}{\mathrm{d}s}. \tag{5.60}$$

The curvatures of $\overset{*}{C}$ are calculated as

$$\overset{*}{k}_{\overset{*}{\tau}} = -\overset{*}{\bar{m}}\,\frac{\mathrm{d}\overset{*}{\bar{\tau}}}{\mathrm{d}\overset{*}{s}} = -\frac{\overset{*}{\bar{m}}}{\sqrt{(1+2\varepsilon_\tau)}}\,\frac{\mathrm{d}\overset{*}{\bar{\tau}}}{\mathrm{d}\overset{*}{s}},$$

$$\overset{*}{k}_{\overset{*}{n}\overset{*}{\tau}} = -\overset{*}{\bar{m}}\,\frac{\mathrm{d}\overset{*}{\bar{n}}}{\mathrm{d}\overset{*}{s}} = -\frac{\overset{*}{\bar{m}}}{\sqrt{(1+2\varepsilon_\tau)}}\,\frac{\mathrm{d}\overset{*}{\bar{n}}}{\mathrm{d}\overset{*}{s}}. \tag{5.61}$$

Substituting $\overset{*}{\bar{\tau}}$ and $\overset{*}{\bar{n}}$ given by Eqs. (5.37) and (5.47) into (5.61) and on use of the fact that $\overset{*}{\bar{m}}\underline{a} = \overset{*}{\bar{m}}\underline{b} = 0$, we obtain

$$\overset{*}{k}_{\overset{*}{\tau}} = -\frac{\overset{*}{k}_\tau}{\sqrt{(1+2\varepsilon_\tau)}} \qquad \overset{*}{k}_{\overset{*}{n}\overset{*}{\tau}} = -\frac{\overset{*}{k}_{n\tau} + 2\,\overset{*}{k}_{n\tau}\varepsilon_\tau - 2\,\overset{*}{k}_\tau \varepsilon_{n\tau}}{\sqrt{\mathbf{A}(1+2\varepsilon_\tau)}}. \tag{5.62}$$

Here $\overset{*}{k}_\tau = k_\tau + æ_\tau$ and $\overset{*}{k}_{n\tau} = k_{n\tau} + æ_{n\tau}$, and they satisfy Eqs. (5.59).

5.5 Gauss–Codazzi equations for the boundary

As a final point of our discussion, we derive the Gauss–Codazzi equations for the undeformed boundary of the thin shell. For the integral of a vector (scalar) function $f(\alpha_1\alpha_2)$ to exist, the following should hold:

$$\frac{\partial}{\partial s_n}\left(\frac{\partial f}{\partial s}\right) - \frac{\partial}{\partial s}\left(\frac{\partial f}{\partial s_n}\right) = æ'\frac{\partial f}{\partial s_n} - æ\frac{\partial f}{\partial s}. \tag{5.63}$$

On substituting the vector $\bar{m}(\bar{m}\perp S)$ for f and using Eqs. (5.48) and (5.49), we obtain

$$\frac{\partial}{\partial s_n}(\bar{n}k_{n\tau} + \bar{\tau}k_\tau) - \frac{\partial}{\partial s}(\bar{\tau}k_{n\tau} + \bar{n}k_n) = æ'(\bar{\tau}k_{n\tau} + \bar{n}k_n) - æ(\bar{n}k_{n\tau} + \bar{\tau}k_\tau). \tag{5.64}$$

Carrying on differentiation of (5.64) and equating the coefficients of $\bar{n}, \bar{\tau}$ and $\bar{m}$ to zero, the Codazzi formulas are found to be

$$\begin{aligned} &\frac{\partial k_{n\tau}}{\partial s_n} - \frac{\partial k_n}{\partial s} + æ'(k_\tau - k_n) + 2æk_{n\tau} = 0, \\ &\frac{\partial k_\tau}{\partial s_n} - \frac{\partial k_{n\tau}}{\partial s} + æ(k_\tau - k_n) - 2æ'k_{n\tau} = 0. \end{aligned} \tag{5.65}$$

Similarly, substituting $\bar{\tau}$ for f, we find

$$-\frac{\partial}{\partial s_n}(\bar{m}k_\tau + æ\bar{n}) - \frac{\partial}{\partial s}(æ'\bar{n} - \bar{m}k_{n\tau}) = æ'(\bar{n}æ' - \bar{m}k_{n\tau}) + æ(\bar{m}k_\tau + \bar{n}æ),$$

from which, after differentiation and setting to zero the coefficients of $\bar{n}$, we obtain the Gauss formula

$$\frac{\partial æ}{\partial s_n} - \frac{\partial æ'}{\partial s} + æ'^2 + æ^2 = k_{n\tau}^2 + k_n k_\tau. \tag{5.66}$$

Equating the coefficients of $\bar{m}$ to zero, we again obtain the Codazzi formulas. Formulas (5.63)–(5.66) can also be expressed in terms of the deformed boundary of the shell. This is left to the reader as an exercise.

Exercises

1. Throughout the book we assume that deformations on the boundary are small. Provide supportive physiological bases for this assumption.
2. Derive formulas for the resultant force $\overline{R}_{\overset{*}{n}}$ and moment $\overline{M}_{\overset{*}{n}}$ vectors per unit length of $\overset{*}{C}$ in the case of finite deformation of the boundary.
3. Verify Eqs. (5.21).
4. What types of boundary conditions are most appropriate in problems of biomechanics of the stomach?
5. Verify Eq. (5.41).
6. Verify Eqs. (5.46).
7. Verify the formulas for bending deformations given by Eqs. (5.60).
8. Derive the Gauss–Codazzi equations (5.63)–(5.66) in terms of the deformed boundary of the shell.

6

Soft shells

6.1 Deformation of soft shell

Members of a class of thin shells ($h/L \sim 10^{-5}$–10^{-2}, where h is the thickness and L is the characteristic dimension of a shell) that

(i) possess low resistance to stretching and zero-order flexural rigidity,
(ii) undergo finite deformations,
(iii) withstand only stretch but not compression forces,
(iv) have actual configurations defined by internal//external loads per unit surface area only and
(v) have stress–strain states that are fully described by in-plane membrane forces per unit length

are called soft shells. Because soft shells acquire multiple forms in the absence of loads, it is instructive to introduce into consideration the cut configuration $\overset{0}{S}$ of the shell, in addition to the undeformed S and deformed $\overset{*}{S}$ configurations (Fig. 6.1). It defines with accuracy the configuration of bending in the absence of loads.

Assume that the middle surface S of an undeformed soft shell coincides with its cut surface $\overset{0}{S}$ ($\overset{0}{S} \equiv S$). Let S be parameterized by curvilinear coordinates α_1 and α_2. A point $M(\alpha_1, \alpha_2) \in S$ is described by the position vector $\bar{r}(\alpha_1, \alpha_2)$. As a result of the action of external and or internal loads the shell will deform to attain a new configuration $\overset{*}{S}$. We assume that the deformation is such that any $M(\alpha_1, \alpha_2) \rightarrow \overset{*}{M}\left(\overset{*}{\alpha}_1, \overset{*}{\alpha}_2\right)$ and is a homeomorphism. Thus, the inverse transformation exists.

The deformation of linear elements along the α_1- and α_2-coordinate lines is described by stretch ratios λ_i ($i = 1, 2$) and elongations $e_{\alpha i}$ given by Eqs. (1.68) and (2.3), respectively,

$$e_{\alpha i} = \frac{d\overset{*}{s}_i - ds_i}{ds_i} = \lambda_i - 1 = \frac{\sqrt{\overset{*}{a}_{ii}}}{\sqrt{a_{ii}}} - 1. \tag{6.1}$$

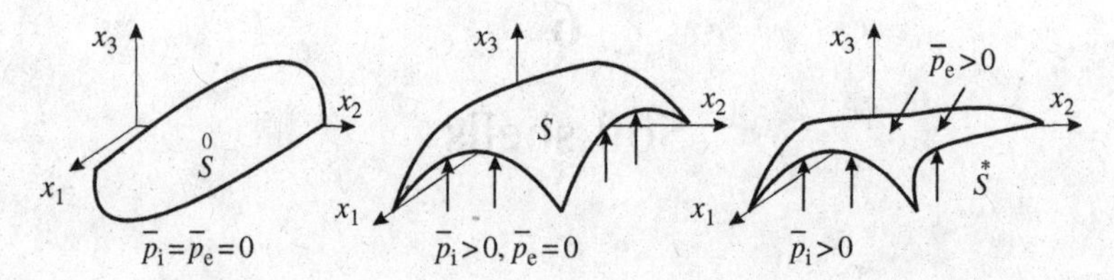

Fig. 6.1 Definition of the cut (ironed out), initial (undeformed) and actual (deformed) configurations of the soft shell.

Changes in the angle between coordinate lines and the surface area are described by Eqs. (1.69) and (2.4) or, in equivalent form,

$$\gamma = \overset{(0)}{\chi} - \overset{*}{\chi} = \overset{(0)}{\chi} - \cos^{-1}\left(\frac{a_{12}}{\sqrt{a_{11}a_{22}}}\right). \tag{6.2}$$

$$\delta s_\Delta = \frac{d\overset{*}{s}_\Delta}{ds_\Delta} = \frac{\sqrt{\overset{*}{a}}}{\sqrt{a}} = \frac{\sqrt{\overset{*}{a}_{11}\overset{*}{a}_{22}}\sin\overset{*}{\chi}}{\sqrt{a_{11}a_{22}}\sin\overset{(0)}{\chi}} = \lambda_1\lambda_2\frac{\sin\overset{*}{\chi}}{\sin\overset{(0)}{\chi}}. \tag{6.3}$$

In the above use is made of Eqs. (1.4), (1.12) and (2.3).

Vectors $\bar{r}_i$ and $\bar{\overset{*}{r}}_i$ tangent to coordinate lines on S and $\overset{*}{S}$ are defined by Eqs. (1.3) and (1.4). Making use of Eqs. (2.3) and (2.4), we have

$$\bar{r}_i = \sum_{k=1}^{2} C_i^k \bar{\overset{*}{r}}_k, \qquad \bar{\overset{*}{r}}_i = \sum_{k=1}^{2} \overset{*}{C}{}_i^k \bar{r}_k. \tag{6.4}$$

Hence, the unit vectors $\bar{e}_i \in S$ and $\bar{\overset{*}{e}}_i \in \overset{*}{S}$ are found to be

$$\bar{e}_i = \frac{\bar{r}_i}{|\bar{r}_i|} = \frac{\bar{r}_i}{\sqrt{a_{ii}}} = \sum_{k=1}^{2} C_i^k\left(\bar{\overset{*}{r}}_k\sqrt{\frac{a_{ii}}{\overset{*}{a}_{kk}}}\right) = \sum_{k=1}^{2} \hat{C}_i^k \bar{\overset{*}{e}}_k,$$
$$\bar{\overset{*}{e}}_i = \sum_{k=1}^{2} \hat{\overset{*}{C}}{}_i^k \bar{e}_k, \tag{6.5}$$

where the following notation is introduced:

$$\hat{C}_i^k = C_i^k\sqrt{\frac{\overset{*}{a}_{kk}}{a_{ii}}}, \qquad \hat{\overset{*}{C}}{}_i^k = \overset{*}{C}{}_i^k\sqrt{\frac{a_{kk}}{\overset{*}{a}_{ii}}}. \tag{6.6}$$

With the help of Eq. (6.6), the scalar and vector products of unit vectors $\bar{\overset{*}{e}}_i$ and $\bar{\overset{*}{e}}_k$ are found to be

$$
\begin{aligned}
\overset{*}{\bar{e}}_i \cdot \overset{*}{\bar{e}}_k &:= \cos \overset{*}{\chi}_{ik} = \sum_{j=1}^{2}\sum_{n=1}^{2} \bar{e}_j \cdot \bar{e}_n \overset{*}{\hat{C}}{}_i^j \overset{*}{\hat{C}}{}_k^n = \sum_{j=1}^{2}\sum_{n=1}^{2} \overset{*}{\hat{C}}{}_i^j \overset{*}{\hat{C}}{}_k^n \cos \overset{(0)}{\chi}_{jn}, \\
\overset{*}{\bar{e}}_i \times \overset{*}{\bar{e}}_k &:= \overset{*}{\bar{m}} \sin \overset{*}{\chi}_{ik} = \sum_{j=1}^{2}\sum_{n=1}^{2} \bar{e}_j \times \bar{e}_n \overset{*}{\hat{C}}{}_i^j \overset{*}{\hat{C}}{}_k^n = \sum_{j=1}^{2}\sum_{n=1}^{2} \overset{*}{\hat{C}}{}_i^j \overset{*}{\hat{C}}{}_k^n \overset{*}{\bar{m}} \sin \overset{(0)}{\chi}_{jn}.
\end{aligned}
\tag{6.7}
$$

In just the same way, proceeding from the scalar and vector multiplication of $\bar{e}_i$ by $\bar{e}_k$, it can be shown that

$$
\begin{aligned}
\cos \overset{(0)}{\chi}_{ik} &= \sum_{j=1}^{2}\sum_{n=1}^{2} \hat{C}_i^j \hat{C}_k^n \cos \overset{*}{\chi}_{jn}, \\
\sin \overset{(0)}{\chi}_{ik} &= \sum_{j=1}^{2}\sum_{n=1}^{2} \hat{C}_i^j \hat{C}_k^n \sin \overset{*}{\chi}_{jn}.
\end{aligned}
\tag{6.8}
$$

To calculate the coefficients C_i^k and $\overset{*}{\hat{C}}{}_i^k$, we proceed from geometrical considerations. Let the vectors $\bar{e}_i$ and $\overset{*}{\bar{e}}_i$ at point $M\left(\overset{*}{\alpha}_1, \overset{*}{\alpha}_2\right) \in \overset{*}{S}$ be oriented as shown in Fig. 6.2. Decomposing $\overset{*}{\bar{e}}_i$ in the directions of $\bar{e}_k$, we have

$$
\overset{*}{\hat{C}}{}_1^1 = \overline{MC}, \qquad \overset{*}{\hat{C}}{}_1^2 = \overline{CD}, \qquad \overset{*}{\hat{C}}{}_2^1 = -\overline{AB}, \qquad \overset{*}{\hat{C}}{}_2^2 = \overline{MB}. \tag{6.9}
$$

Solving for ΔMCD and ΔMBA, we find

$$
\begin{aligned}
&\overset{*}{\hat{C}}{}_1^1 = \sin(\overset{*}{\chi} - \overset{*}{\chi}_2)/\sin \overset{*}{\chi}, \qquad \overset{*}{\hat{C}}{}_1^2 = \sin \overset{*}{\chi}_2/\sin \overset{*}{\chi}, \\
&\overset{*}{\hat{C}}{}_2^1 = -\sin(\overset{*}{\chi}_1 + \overset{*}{\chi}_2 - \overset{*}{\chi})/\sin \overset{*}{\chi}, \qquad \overset{*}{\hat{C}}{}_2^2 = \sin(\overset{*}{\chi}_1 + \overset{*}{\chi}_2)/\sin \overset{*}{\chi}, \\
&\overset{*}{\hat{C}} = \det \overset{*}{\hat{C}}{}_i^k = \sin \overset{*}{\chi}_1/\sin \overset{*}{\chi}.
\end{aligned}
\tag{6.10}
$$

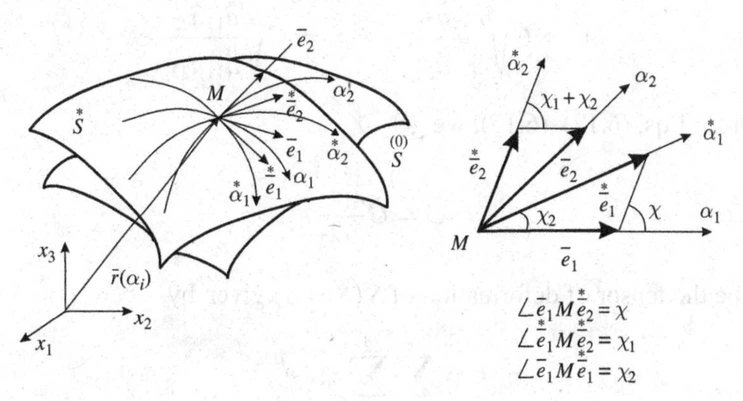

Fig. 6.2 Deformation of an element of the soft shell.

Similarly, expanding unit vectors $\bar{e}_i$ along $\overset{*}{\bar{e}}_k$, we obtain

$$\hat{C}_1^1 = \sin(\overset{0}{\chi}_1 + \overset{0}{\chi}_2)/\sin\overset{0}{\chi}_1, \qquad \hat{C}_1^2 = -\sin\overset{0}{\chi}_2/\sin\overset{0}{\chi}_1,$$
$$\hat{C}_2^1 = \sin(\overset{0}{\chi}_1 + \overset{0}{\chi}_2 - \overset{0}{\chi})/\sin\overset{0}{\chi}_1, \qquad \hat{C}_2^2 = \sin(\overset{0}{\chi} - \overset{0}{\chi}_2)/\sin\overset{0}{\chi}_1, \tag{6.11}$$
$$\hat{C} = \det \hat{C}_i^k = \sin\overset{0}{\chi}/\sin\overset{0}{\chi}_1$$

Note that the coefficients C_i^k and $\overset{*}{C}_i^k$ are functions of $\overset{0}{\chi}_i$ and $\overset{*}{\chi}_i$, whereas $\hat{C}_i^k$ and $\overset{*}{\hat{C}}_i^k$ depend on $\bar{r}_i$ and $\overset{*}{\bar{r}}_i$ and the actual configuration of a shell.

Let the cut configuration of a soft shell $\overset{0}{S}$ be different from the undeformed configuration S. We introduce the coefficients of transformation $\bar{r}_i \in \overset{0}{S} \to \overset{*}{S}$ by putting

$$\hat{C}_i^k = C_i^k \sqrt{\frac{\overset{*}{a}_{kk}}{\overset{0}{a}_{ii}}}, \qquad \overset{*}{\hat{\hat{C}}}{}_i^k = \overset{*}{C}_i^k \sqrt{\frac{\overset{0}{a}_{kk}}{\overset{*}{a}_{ii}}}, \tag{6.12}$$

where $\overset{0}{a}_{ii}$ and $\overset{*}{a}_{ii}$ are the components of the metric tensor $\mathbf{A}$ on $\overset{0}{S}$ and $\overset{*}{S}$, respectively. By eliminating C_i^k and $\overset{*}{C}_i^k$ from Eqs. (6.6), for the coefficients of the cut and deformed surfaces we obtain

$$\hat{C}_i^k = \hat{\hat{C}}_i^k \frac{\overset{*}{\lambda}_k}{\lambda_i}, \qquad \overset{*}{\hat{C}}{}_i^k = \overset{*}{\hat{\hat{C}}}{}_i^k \frac{\lambda_k}{\overset{*}{\lambda}_i}. \tag{6.13}$$

Similarly to Eqs. (6.14), we introduce the coefficients

$$\overset{*}{\hat{C}}{}_i^k = [\hat{C}_k^i]/\hat{C}, \qquad \overset{*}{\hat{\hat{C}}}{}_i^k = \left[\hat{\hat{C}}_k^i\right]/\hat{\hat{C}}, \tag{6.14}$$

where

$$\hat{C} = C\sqrt{\frac{\overset{*}{a}_{11}\overset{*}{a}_{22}}{\overset{0}{a}_{11}\overset{0}{a}_{22}}}, \qquad \hat{\hat{C}} = C\sqrt{\frac{\overset{*}{a}_{11}\overset{*}{a}_{22}}{\overset{0}{a}_{11}\overset{0}{a}_{22}}}. \tag{6.15}$$

Finally, from Eqs. (6.12)–(6.15), we get

$$\hat{C} = \hat{\hat{C}} \frac{\overset{*}{\lambda}_1\overset{*}{\lambda}_2}{\lambda_1\lambda_2}. \tag{6.16}$$

Let $\mathbf{E}$ be the tensor of deformation of S $(S = \overset{0}{S})$ given by

$$E = \sum_{i=1}^{2}\sum_{k=1}^{2} \varepsilon_{ik}\bar{r}^i\bar{r}^k, \tag{6.17}$$

where

$$\varepsilon_{ik} = \frac{\overset{*}{a}_{ik} - a_{ik}}{2}. \tag{6.18}$$

Substituting Eqs. (1.4) and (2.3) into (6.18) for ε_{ik}, we find

$$\varepsilon_{ik} = \frac{\left(\lambda_i \lambda_k \cos \overset{*}{\chi} - \cos \chi\right) \sqrt{a_{ii} a_{kk}}}{2}. \tag{6.19}$$

It is easy to show that the following relations hold:

$$\begin{aligned} \varepsilon_{ik} &= \sum_{j=1}^{2} \sum_{n=1}^{2} \overset{*}{\varepsilon}_{jn} C_i^j C_k^n, \qquad & \varepsilon^{ik} &= \sum_{j=1}^{2} \sum_{n=1}^{2} \overset{*}{\varepsilon}^{jn} C_j^i C_n^k, \\ \overset{*}{\varepsilon}_{ik} &= \sum_{j=1}^{2} \sum_{n=1}^{2} \varepsilon_{jn} \overset{*}{C}_i^j \overset{*}{C}_k^n, \qquad & \overset{*}{\varepsilon}^{ik} &= \sum_{j=1}^{2} \sum_{n=1}^{2} \varepsilon^{jn} C_j^i C_n^k. \end{aligned} \tag{6.20}$$

In the theory of soft thin shells, stretch ratios and membrane forces per unit length of a differential element are preferred to traditional deformations and stresses per unit cross-sectional area of the shell. Thus, dividing (6.19) by the surface area $\sqrt{a_{ii} a_{kk}}$ of an element, we get

$$\tilde{\varepsilon}_{ik} := \frac{\varepsilon_{ik}}{\sqrt{a_{ii} a_{kk}}} = \frac{\left(\lambda_i \lambda_k \cos \overset{*}{\chi} - \cos \chi\right)}{2}, \tag{6.21}$$

where $\tilde{\varepsilon}_{ik}$ are called the physical components of $\mathbf{E}$. Using Eqs. (6.20), for $\overset{*}{\tilde{\varepsilon}}_{ik}$ in terms of the deformed configuration we obtain

$$\overset{*}{\tilde{\varepsilon}}_{ik} = \sum_{j=1}^{2} \sum_{n=1}^{2} \tilde{\varepsilon}_{jn} \overset{*}{C}_i^j \overset{*}{C}_k^n \frac{\sqrt{a_{jj} a_{nn}}}{\sqrt{\overset{*}{a}_{jj} \overset{*}{a}_{nn}}}, \tag{6.22}$$

where the coefficients $\overset{*}{C}_i^j$ satisfy Eqs. (2.3) and (2.4).

Making use of Eq. (6.21) in (2.3) and (6.2) for λ_i and γ, we find

$$\begin{aligned} \lambda_i &= 1 + \varepsilon_i = \sqrt{1 + 2\tilde{\varepsilon}_{ii}}, \\ \gamma &= \overset{(0)}{\chi} - \cos^{-1}\left(\frac{2\tilde{\varepsilon}_{12} + \cos \overset{(0)}{\chi}}{\sqrt{(1 + 2\tilde{\varepsilon}_{11})}\sqrt{(1 + 2\tilde{\varepsilon}_{22})}}\right). \end{aligned} \tag{6.23}$$

Substituting $\hat{\hat{C}}_i^k$ and $\overset{*}{\hat{\hat{C}}}{}_i^k$ given by Eqs. (6.12) into (6.22), we have

$$\overset{*}{\tilde{\varepsilon}}_{ik} = \sum_{j=1}^{2} \sum_{n=1}^{2} \tilde{\varepsilon}_{jn} \overset{*}{\hat{\hat{C}}}{}_i^j \overset{*}{\hat{\hat{C}}}{}_k^n, \qquad \tilde{\varepsilon}_{ik} = \sum_{j=1}^{2} \sum_{n=1}^{2} \tilde{\varepsilon}_{jn}^{m} \hat{\hat{C}}_i^j \hat{\hat{C}}_k^n. \tag{6.30}$$

Finally, the formulas for $\overset{*}{\tilde{\varepsilon}}_{ik}$ in terms of the $\overset{0}{S}$ configuration of the soft shell take the forms

$$\begin{aligned}
\overset{*}{\tilde{\varepsilon}}_{11} &= \left[\tilde{\varepsilon}_{11}\sin^2(\overset{0}{\chi}-\overset{0}{\chi}_2) + \tilde{\varepsilon}_{22}\sin^2\overset{0}{\chi}_2 + 2\tilde{\varepsilon}_{12}\sin(\overset{0}{\chi}-\overset{0}{\chi}_2)\sin\overset{0}{\chi}_2\right]\Big/\sin^2\overset{0}{\chi},\\
\overset{*}{\tilde{\varepsilon}}_{12} &= \Big[-\tilde{\varepsilon}_{11}\sin(\overset{0}{\chi}-\overset{0}{\chi}_2)\sin(\overset{0}{\chi}_1+\overset{0}{\chi}_2-\overset{0}{\chi}) + \tilde{\varepsilon}_{22}\sin\overset{0}{\chi}_2\sin(\overset{0}{\chi}_1+\overset{0}{\chi}_2)\\
&\quad + \tilde{\varepsilon}_{12}\left(\cos(\overset{0}{\chi}_1+2\overset{0}{\chi}_2-\overset{0}{\chi}) - \cos\overset{0}{\chi}\cos\overset{0}{\chi}_1\right)\Big]\Big/\sin^2\overset{0}{\chi}, \qquad (6.31)\\
\overset{*}{\tilde{\varepsilon}}_{22} &= \Big[\tilde{\varepsilon}_{11}\sin^2(\overset{0}{\chi}_1+\overset{0}{\chi}_2-\overset{0}{\chi}) + \tilde{\varepsilon}_{22}\sin^2(\overset{0}{\chi}_1+\overset{0}{\chi}_2)\\
&\quad - 2\tilde{\varepsilon}_{12}\sin(\overset{0}{\chi}_1+\overset{0}{\chi}_2-\overset{0}{\chi})\sin(\overset{0}{\chi}_1+\overset{0}{\chi}_2)\Big]\Big/\sin^2\overset{0}{\chi}.
\end{aligned}$$

With the help of Eq. (6.21) in (6.30) the physical components can also be expressed in terms of stretch ratios and shear angles as

$$\overset{*}{\lambda}_i\overset{*}{\lambda}_k\cos\overset{*}{\chi}_{ik} - \cos\overset{0}{\chi}_{ik} = \sum_{j=1}^{2}\sum_{n=1}^{2}(\lambda_j\lambda_n\cos\chi_{jn} - \cos\overset{0}{\chi}_{jn})\overset{*}{\hat{\hat{C}}}{}^j_i\overset{*}{\hat{\hat{C}}}{}^n_k. \qquad (6.32)$$

Further, on use of Eqs. (6.7) and (6.10), Eq. (6.32) in terms of the $\overset{0}{S}$ configuration takes the form

$$\overset{*}{\lambda}_i\overset{*}{\lambda}_k\cos\overset{*}{\chi}_{ik} = \sum_{j=1}^{2}\sum_{n=1}^{2}\lambda_j\lambda_n\hat{\hat{C}}{}^j_i\hat{\hat{C}}{}^n_k\cos\overset{0}{\chi}_{jn}, \qquad (6.33)$$

or, in expanded form,

$$\begin{aligned}
\overset{*}{\lambda}_1 &= \left[\lambda_1^2\sin^2(\overset{0}{\chi}-\overset{0}{\chi}_2) + \lambda_2^2\sin^2\overset{0}{\chi}_2 + 2\lambda_1\lambda_2\sin(\overset{0}{\chi}-\gamma)\sin(\overset{0}{\chi}-\overset{0}{\chi}_2)\sin\overset{0}{\chi}_2\right]^{1/2}\Big/\sin^2\overset{0}{\chi},\\
\overset{*}{\gamma} &= \overset{0}{\chi}_1 - \cos^{-1}\Big[\Big(-\lambda_1^2\sin(\overset{0}{\chi}-\overset{0}{\chi}_2)\sin(\overset{0}{\chi}_1+\overset{0}{\chi}_2-\overset{0}{\chi})\\
&\quad + \lambda_2^2\sin\overset{0}{\chi}_2\sin(\overset{0}{\chi}_1+\overset{0}{\chi}_2) + \lambda_1\lambda_2\Big(\cos(\hat{\delta}+2\overset{0}{\chi}_2-\overset{0}{\chi})\\
&\quad - \cos\overset{0}{\chi}\cos\overset{0}{\chi}_1\Big)\cos(\overset{0}{\chi}-\gamma)\Big)(\overset{*}{\lambda}_1\overset{*}{\lambda}_2\sin^2\overset{0}{\chi})^{-1}\Big],\\
\overset{*}{\lambda}_2 &= \Big[\lambda_1^2\sin^2(\overset{0}{\chi}_1+\overset{0}{\chi}_2-\overset{0}{\chi}) + \lambda_2^2\sin^2(\overset{0}{\chi}_1+\overset{0}{\chi}_2)\\
&\quad - 2\lambda_1\lambda_2\cos(\overset{0}{\chi}-\Delta\overset{*}{\chi})\sin(\overset{0}{\chi}_1+\overset{0}{\chi}_2-\overset{0}{\chi})\sin(\overset{0}{\chi}_1+\overset{0}{\chi}_2)\Big]^{1/2}\Big/\sin^2\overset{0}{\chi}.
\end{aligned}$$

(6.34)

Formulas (6.33) and (6.34) are preferred in practical applications, particularly when dealing with finite deformations of shells.

6.2 Principal deformations

At any point $\overset{*}{M} \in \overset{*}{S}$, there exist two mutually orthogonal directions that remain orthogonal during deformation and along which the components of **E** attain their maximum and minimum values. They are called the principal directions.

To find the orientation of the principal axes, we proceed as follows. Let $\overset{(0)}{\varphi}$ and $\overset{*}{\varphi}$ be the angles of the direction away from the base vectors $\bar{e}_1 \in \overset{(0)}{S}$ and $\overset{*}{\bar{e}}_1 \in \overset{*}{S}$, respectively. We assume that the cut and undeformed configurations are indistinguishable, $\overset{0}{S} = S$. Then, setting $\overset{0}{\chi}_2 = \overset{0}{\varphi}$ in the first equation of (6.31), we have

$$\overset{*}{\tilde{\varepsilon}}_{11} \sin^2 \overset{0}{\chi} = \tilde{\varepsilon}_{11} \sin^2 (\overset{0}{\chi} - \overset{0}{\varphi}) + 2\tilde{\varepsilon}_{12} \sin(\overset{0}{\chi} - \overset{0}{\varphi}) \sin \overset{0}{\varphi} + \tilde{\varepsilon}_{22} \sin^2 \overset{0}{\varphi}. \tag{6.35}$$

After simple rearrangements it can be written in the form

$$\varepsilon = a_0 + b_0 \cos\left(2\overset{0}{\varphi}\right) + c_0 \sin\left(2\overset{0}{\varphi}\right), \tag{6.36}$$

where

$$\begin{aligned}
a_0 &= \left[\frac{1}{2}(\tilde{\varepsilon}_{11} + \tilde{\varepsilon}_{22}) - \tilde{\varepsilon}_{12} \cos \overset{0}{\chi}\right] \Big/ \sin^2 \overset{0}{\chi}, \\
b_0 &= \left[\frac{1}{2}(\tilde{\varepsilon}_{11} - \tilde{\varepsilon}_{22}) + \left(\tilde{\varepsilon}_{12} - \tilde{\varepsilon}_{11} \cos \overset{0}{\chi}\right) \Big/ \cos \overset{0}{\chi}\right] \Big/ \sin^2 \overset{0}{\chi}, \\
c_0 &= \left(\tilde{\varepsilon}_{12} - \tilde{\varepsilon}_{11} \cos \overset{0}{\chi}\right) \Big/ \sin \overset{0}{\chi}.
\end{aligned} \tag{6.37}$$

On differentiating Eq. (6.36) with respect to $\overset{0}{\varphi}$ and equating the resultant equation to zero, for the principal axes on the surface $\overset{(0)}{S}$, we find ($b_0 \neq 0$)

$$\tan\left(2\overset{0}{\varphi}\right) = \frac{c_0}{b_0} = \frac{2\left(\tilde{\varepsilon}_{11} \cos \overset{0}{\chi} - \tilde{\varepsilon}_{12}\right) \sin \overset{0}{\chi}}{\tilde{\varepsilon}_{11} \cos\left(2\overset{0}{\chi}\right) - 2\tilde{\varepsilon}_{12} \cos \overset{0}{\chi} + \tilde{\varepsilon}_{22}}. \tag{6.38}$$

By substituting Eq. (6.38) into (6.36), we obtain the principal physical components ε_1 and ε_2 of **E**:

$$\begin{aligned}
\varepsilon_{1,2} = a_0^2 \pm \sqrt{b_0^2 + c_0^2} &= \frac{(\tilde{\varepsilon}_{11} + \tilde{\varepsilon}_{22}) - 2\tilde{\varepsilon}_{12} \cos \overset{0}{\chi}}{2 \sin^2 \overset{0}{\chi}} \\
&\pm \frac{1}{\sin^2 \overset{0}{\chi}} \sqrt{\frac{(\tilde{\varepsilon}_{11} - \tilde{\varepsilon}_{22})^2}{4} + \tilde{\varepsilon}_{12}^2 + \tilde{\varepsilon}_{11}\tilde{\varepsilon}_{22} \cos^2 \overset{0}{\chi} - \tilde{\varepsilon}_{12}(\tilde{\varepsilon}_{11} + \tilde{\varepsilon}_{22}) \cos \overset{0}{\chi}}.
\end{aligned} \tag{6.39}$$

Henceforth, we assume that $\max \varepsilon_1$ is achieved in the direction of the principal axis defined by the angle $\overset{0}{\varphi}_1 = \overset{0}{\varphi}$, and $\min \varepsilon_2$ along the axis defined by the angle $\overset{0}{\varphi}_2 = \overset{0}{\varphi} + \pi/2$. Since for the principal directions $\overset{0}{\chi}_1 \equiv \pi/2$, from the second Eq. (6.31), we find

$$\overset{*}{\tilde{\varepsilon}}_{12} = -b_0 \sin\left(2\overset{0}{\varphi}\right) + c_0 \cos\left(2\overset{0}{\varphi}\right). \tag{6.40}$$

Dividing both sides of Eq. (6.40) by $c_0 \cos\left(2\overset{0}{\varphi}\right)$ and using Eq. (6.38), we find $\overset{*}{\tilde{\varepsilon}}_{12} = 0$. Thus, there do indeed exist two mutually orthogonal directions for all $M(\alpha_i) \in \overset{(0)}{S}$ that remain orthogonal throughout deformation. A similar result, i.e. $\overset{*}{\gamma} = 0$, can be obtained from Eq. (6.9) by setting $\overset{0}{\chi} = \pi/2$.

By substituting Eq. (6.21) into (6.38) and (6.39) for the orientation of the principal axes on $\overset{0}{S}$ and the principal stretch ratios, we obtain

$$\tan\left(2\overset{0}{\varphi}_1\right) = \frac{2\left[\lambda_1\lambda_2 \cos(\overset{0}{\chi} - \overset{*}{\gamma}) - \lambda_1^2 \cos\overset{0}{\chi}\right] \sin\overset{0}{\chi}}{\lambda_1^2 - \lambda_2^2 + 2\left[\lambda_1\lambda_2 \cos(\overset{0}{\chi} - \overset{*}{\gamma}) - \lambda_1^2 \cos\overset{0}{\chi}\right] \cos\overset{0}{\chi}},$$
$$\overset{0}{\varphi}_2 = \overset{0}{\varphi}_1 + \pi/2, \tag{6.41}$$

$$\begin{aligned}\Lambda_{1,2}^2 = \Bigg[&(\lambda_1^2 + \lambda_2^2)/2 - \lambda_1\lambda_2 \cos(\overset{0}{\chi} - \overset{*}{\gamma})\cos\overset{0}{\chi}.\\ &\pm \left(\left(\lambda_1^2 + \lambda_2^2\right)^2/4 + \lambda_1^2\lambda_2^2 \cos(2\overset{0}{\chi} - \overset{*}{\gamma})\cos\overset{*}{\gamma}\right.\\ &\left.- \lambda_1\lambda_2(\lambda_1^2 + \lambda_2^2)\cos(\overset{0}{\chi} - \overset{*}{\gamma}) \cos\overset{0}{\chi}\right)^{1/2}\Bigg]^{1/2} \Bigg/ \sin\overset{0}{\chi}.\end{aligned} \tag{6.42}$$

To find the orientation of the principal axes on the deformed surface $\overset{*}{S}$, consider a triangular element on S bounded by the two principal axes and the α_1-coordinate line (Fig. 6.3). Geometrical analysis leads to the following obvious equalities:

$$\begin{aligned}&\cos\overset{*}{\varphi}_1 = \frac{d\overset{*}{s}_1}{ds_1} = \frac{\Lambda_1}{\lambda_1}\cos\overset{(0)}{\varphi}_1, \qquad \sin\overset{*}{\varphi}_1 = \frac{d\overset{*}{s}_2}{ds_1} = \frac{\Lambda_2}{\lambda_1}\sin\overset{(0)}{\varphi}_1,\\ &\tan\overset{*}{\varphi}_1 = \frac{d\overset{*}{s}_2}{d\overset{*}{s}_1} = \frac{\Lambda_2}{\Lambda_1}\tan\overset{(0)}{\varphi}_1.\end{aligned} \tag{6.43}$$

On use of Eqs. (6.41) and (6.42) from the above, we find the angles for the principal axes $\overset{*}{\varphi}_1$ and $\overset{*}{\varphi}_2 = \overset{*}{\varphi}_1 + \pi/2$.

Finally, substituting Eqs. (6.3) and (6.21) into expressions for the first and second invariants of the tensor of deformation $\mathbf{E}$ defined by

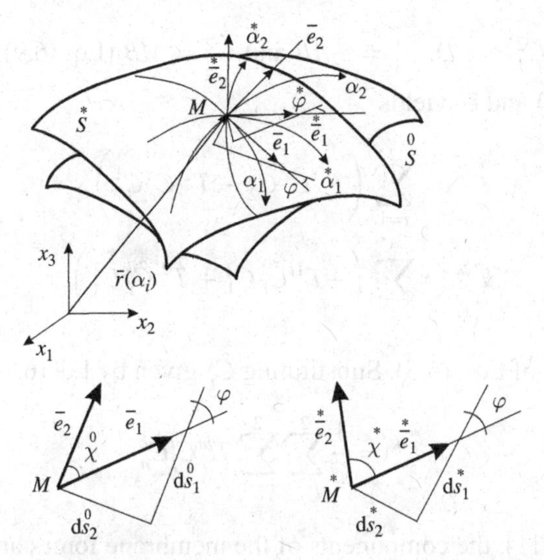

Fig. 6.3 Principal deformations.

$$
\begin{aligned}
I^{(\mathbf{E})}{}_1 &= \varepsilon_1 + \varepsilon_2 = \overset{*}{\tilde{\varepsilon}}_{11} + \overset{*}{\tilde{\varepsilon}}_{22} = \left[\tilde{\varepsilon}_{11} - 2\tilde{\varepsilon}_{12}\cos\overset{0}{\chi} + \tilde{\varepsilon}_{22}\right] \Big/ \sin^2\overset{0}{\chi}, \\
I^{(\mathbf{E})}{}_2 &= \varepsilon_1 \varepsilon_2 = \overset{*}{\tilde{\varepsilon}}_{11}\overset{*}{\tilde{\varepsilon}}_{22} - (\overset{*}{\tilde{\varepsilon}}_{12})^2 = \left[\tilde{\varepsilon}_{11}\tilde{\varepsilon}_{22} - (\tilde{\varepsilon}_{12})^2\right] \Big/ \sin^2\overset{0}{\chi}.
\end{aligned}
\tag{6.44}
$$

for the principal stretch ratios and the shear angle, we get

$$
\begin{aligned}
\Lambda^2{}_1 + \Lambda^2{}_2 &= (\overset{*}{\lambda}_1)^2 + (\overset{*}{\lambda}_2)^2 = \left(\lambda_1^2 + \lambda_2^2 - 2\lambda_1\lambda_2\cos(\overset{0}{\chi} - \gamma)\cos\overset{0}{\chi}\right) \Big/ \sin^2\overset{0}{\chi}, \\
\Lambda_1 \Lambda_2 &= \sqrt{1 + 2I^{(\mathbf{E})}{}_1 + 4I^{(\mathbf{E})}{}_2} = \overset{*}{\lambda}_1\overset{*}{\lambda}_2\cos\gamma = \lambda_1\lambda_2\sin(\overset{0}{\chi} - \gamma) / \sin\overset{0}{\chi}.
\end{aligned}
\tag{6.45}
$$

The last equation is also used to calculate the change in surface area of S.

6.3 Membrane forces

The stress state of a differential element of the soft shell is described entirely by in-plane tangent T_{ii} (T^{ii}) and shear T_{ik} (T^{ik}) $(i \neq k)$ forces per unit length of the element. To study the equilibrium of the shell, we proceed from consideration of triangular elements ΔMAB and ΔMCD on $\overset{*}{S}$ (Fig. 6.4). Analysis of force distribution in the elements yields

$$
\begin{aligned}
-\overline{MA}\sum_{k=}^{2}\overset{*}{T}{}^{1k}\overset{*}{\bar{e}}_k + \overline{MB}\sum_{i=1}^{2} T^{1i}\,\bar{e}_i + \overline{AB}\sum_{i=1}^{2} T^{2i}\,\bar{e}_i &= 0, \\
\overline{MD}\sum_{k=1}^{2}\overset{*}{T}{}^{2k}\overset{*}{\bar{e}}_k + \overline{CD}\sum_{i=1}^{2} T^{1i}\,\bar{e}_i - \overline{MC}\sum_{i=1}^{2} T^{2i}\,\bar{e}_i &= 0,
\end{aligned}
\tag{6.46}
$$

where $\overset{*}{\hat{C}}{}_1^1 = \overline{MC}$, $\overset{*}{\hat{C}}{}_1^2 = \overline{CD}$, $\overset{*}{\hat{C}}{}_2^1 = -\overline{AB}$ and $\overset{*}{\hat{C}}{}_2^2 = \overline{MB}$ (Eq. (6.9)). The scalar product of Eqs. (6.46) and $\overset{*}{\bar{e}}{}^k$ yields

$$\overset{*}{T}{}^{1k} = \sum_{i=1}^{2}\left(T^{1i}\hat{C}_i^k\overset{*}{\hat{C}}{}_2^2 - T^{2i}\hat{C}_i^k\overset{*}{\hat{C}}{}_2^1\right),$$

$$\overset{*}{T}{}^{2k} = \sum_{i=1}^{2}\left(-T^{1i}\hat{C}_i^k\overset{*}{\hat{C}}{}_1^2 + T^{2i}\hat{C}_i^k\overset{*}{\hat{C}}{}_1^1\right),$$

where use is made of Eqs. (6.5). Substituting $\overset{*}{\hat{C}}{}_i^k$ given by Eq. (6.14) for $\hat{C}_i^k$, we find

$$\overset{*}{T}{}^{ik} = \frac{1}{C}\sum_{j=1}^{2}\sum_{n=1}^{2} T^{jn}\hat{C}_j^i\hat{C}_n^k. \tag{6.47}$$

On use of Eqs. (6.11), the components of the membrane forces are found to be

$$\begin{aligned}
\overset{*}{T}{}^{11} &= \Big\{T^{11}\sin^2(\overset{0}{\chi}_1 + \overset{0}{\chi}_2) + T^{22}\sin^2(\overset{0}{\chi}_1 + \overset{0}{\chi}_2 - \overset{0}{\chi}) \\
&\qquad + 2T^{12}\sin(\overset{0}{\chi}_1 + \overset{0}{\chi}_2 - \overset{0}{\chi})\sin(\overset{0}{\chi}_1 + \overset{0}{\chi}_2)\Big\}\Big/\Big(\sin\overset{0}{\chi}\sin\overset{0}{\chi}_1\Big), \\
\overset{*}{T}{}^{12} &= \Big\{-T^{11}\sin\overset{0}{\chi}_2\sin(\overset{0}{\chi}_1 + \overset{0}{\chi}_2) + T^{22}\sin(\overset{0}{\chi}_1 + \overset{0}{\chi}_2 - \overset{0}{\chi})\sin(\overset{0}{\chi} - \overset{0}{\chi}_2) \\
&\qquad + T^{12}\Big[\cos(\overset{0}{\chi}_1 + 2\overset{0}{\chi}_2 - \overset{0}{\chi}) - \cos\overset{0}{\chi}\cos\overset{0}{\chi}_1\Big]\Big\}\Big/\Big(\sin\overset{0}{\chi}\sin\overset{0}{\chi}_1\Big), \\
\overset{*}{T}{}^{22} &= \Big\{T^{11}\sin^2\overset{0}{\chi}_2 - 2T^{12}\sin(\overset{0}{\chi} - \overset{0}{\chi}_2)\sin\overset{0}{\chi}_2 \\
&\qquad + T^{22}\sin^2(\overset{0}{\chi} - \overset{0}{\chi}_2)\Big\}\Big/\Big(\sin\overset{0}{\chi}\sin\overset{0}{\chi}_1\Big).
\end{aligned} \tag{6.48}$$

Introducing the tensor of membrane forces $\mathbf{T}$,

$$\mathbf{T} = \sum_{j=1}^{2}\sum_{n=1}^{2} T^{ik}\overline{r_i}\overline{r_k} = \frac{1}{\sin\overset{*}{\chi}}\sum_{j=1}^{2}\sum_{n=1}^{2}\tilde{T}^{ik}\bar{e}_i\bar{e}_k, \tag{6.49}$$

where $\tilde{T}^{ik}$ are the physical components of $\mathbf{T}$, and using Eq. (6.5), $\tilde{T}^{ik}$ can be expressed in terms of $\overset{*}{\tilde{T}}{}^{ik}$ as

$$\mathbf{T} = \frac{1}{\sin\overset{*}{\chi}}\sum_{j=1}^{2}\sum_{n=1}^{2}\tilde{T}^{ik}\bar{e}_i\bar{e}_k = \frac{1}{\sin\overset{*}{\chi}_1}\sum_{j=1}^{2}\sum_{n=1}^{2}\overset{*}{\tilde{T}}{}^{jn}\,\overset{*}{\bar{e}}_i\,\overset{*}{\bar{e}}_k = \frac{1}{\sin\overset{*}{\chi}_1}\sum_{j=1}^{2}\sum_{n=1}^{2}\overset{*}{\tilde{T}}{}^{jn}\,\bar{e}_i\bar{e}_k\overset{*}{\hat{C}}{}_j^i\overset{*}{\hat{C}}{}_n^k.$$

Further, making use of Eq. (6.11), we obtain

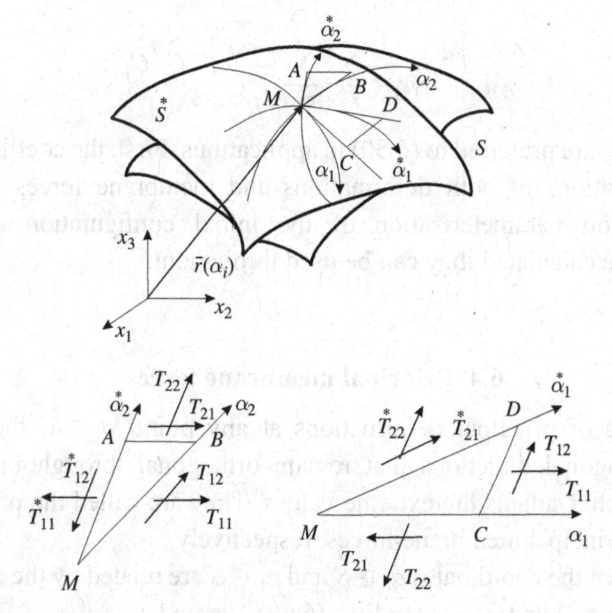

Fig. 6.4 Membrane forces in the soft shell.

$$\mathbf{T} = \frac{1}{\overset{*}{C}} \sum_{j=1}^{2} \sum_{n=1}^{2} \tilde{\overset{*}{T}}{}^{jn} \hat{\overset{*}{C}}{}_j^i \hat{\overset{*}{C}}{}_n^k. \tag{6.50}$$

Substituting $\hat{\overset{*}{C}}{}_j^i$ given by Eqs. (6.10), we get

$$\begin{aligned}
\tilde{T}^{11} &= \Big\{ \tilde{\overset{*}{T}}{}^{11} \sin^2(\overset{*}{\chi} - \overset{*}{\chi}_2) + \tilde{\overset{*}{T}}{}^{22} \sin^2(\overset{*}{\chi}_1 + \overset{*}{\chi}_2 - \overset{*}{\chi}) \\
&\quad - 2\tilde{\overset{*}{T}}{}^{12} \sin(\overset{*}{\chi}_1 + \overset{*}{\chi}_2 - \overset{*}{\chi}) \sin(\overset{*}{\chi} - \overset{*}{\chi}_2) \Big\} \Big/ \Big(\sin\overset{*}{\chi} \sin\overset{*}{\chi}_1 \Big), \\
\tilde{T}^{12} &= \Big\{ \tilde{\overset{*}{T}}{}^{11} \sin\overset{*}{\chi}_2 \sin(\overset{*}{\chi} - \overset{*}{\chi}_2) + \tilde{\overset{*}{T}}{}^{22} \sin(\overset{*}{\chi}_1 + \overset{*}{\chi}_2 - \overset{*}{\chi}) \sin(\overset{*}{\chi}_1 + \overset{*}{\chi}_2) \\
&\quad + \tilde{\overset{*}{T}}{}^{12} \Big[\cos(\overset{*}{\chi}_1 + 2\overset{*}{\chi}_2 - \chi) - \cos\overset{*}{\chi} \cos\overset{*}{\chi}_1 \Big] \Big\} \Big/ \Big(\sin\overset{*}{\chi} \sin\overset{*}{\chi}_1 \Big), \\
\tilde{T}^{22} &= \Big\{ \tilde{\overset{*}{T}}{}^{11} \sin\overset{*}{\chi}_2 - 2\tilde{\overset{*}{T}}{}^{12} \sin(\overset{*}{\chi}_1 + \overset{*}{\chi}_2) \sin\overset{*}{\chi}_2 + \tilde{\overset{*}{T}}{}^{22} \sin^2(\overset{*}{\chi}_1 + \overset{*}{\chi}_2) \Big\} \Big/ \Big(\sin\overset{*}{\chi} \sin\overset{*}{\chi}_1 \Big).
\end{aligned} \tag{6.51}$$

Using Eqs (6.13) after simple rearrangements, Eq. (6.50) takes the form

$$\frac{\lambda_1\lambda_2}{\lambda_i\lambda_k}\tilde{T}^{ik} = \frac{1}{\hat{\hat{C}}^*}\sum_{j=1}^{2}\sum_{n=1}^{2}\frac{\overset{*}{\lambda}_1\overset{*}{\lambda}_2}{\overset{*}{\lambda}_j\overset{*}{\lambda}_n}\overset{*}{\tilde{T}}^{jn}\hat{\hat{C}}_j^{*j}\hat{\hat{C}}_n^{*k}. \tag{6.52}$$

Formulas (6.52) are preferred to (6.50) in applications. First, the coefficients $\hat{\hat{C}}_n^{*k}$ are used in calculations of both deformations and membrane forces. Second, $\hat{\hat{C}}_n^{*k}$ depend only on parameterization of the initial configuration of the shell. Therefore, once calculated they can be used throughout.

6.4 Principal membrane forces

As in the case of principal deformations at any point $\overset{*}{M} \in \overset{*}{S}$, there exist two mutually orthogonal directions that remain orthogonal throughout deformation and along which **T** attains the extreme values. They are called the principal directions and the principal membrane forces, respectively.

Assuming that the coordinates $\overset{*}{\alpha}_i \in \overset{*}{S}$ and $\alpha_i \in S$ are related by the angle $\overset{*}{\psi}$, then, on setting $\overset{*}{\chi}_1 = \pi/2$ and $\overset{*}{\chi}_2 = \psi$ in Eqs. (6.48), we find

$$\begin{aligned}
\overset{*}{T}^{11} &= \left\{T^{11}\cos^2\psi^* + T^{22}\cos^2(\overset{*}{\chi}-\overset{*}{\psi}) + 2T^{12}\cos\overset{*}{\psi}\cos(\overset{*}{\chi}-\overset{*}{\psi})\right\}\Big/\sin\overset{*}{\chi},\\
\overset{*}{T}^{12} &= \Big\{-T^{11}\cos\overset{*}{\psi}\sin\overset{*}{\psi} + T^{12}\sin(\overset{*}{\chi}-2\overset{*}{\psi})\\
&\qquad + T^{22}\cos(\overset{*}{\chi}-\overset{*}{\psi})\sin(\overset{*}{\chi}-\overset{*}{\psi})\Big\}\Big/\sin\overset{*}{\chi},\\
\overset{*}{T}^{22} &= \left\{T^{11}\sin^2\overset{*}{\psi} - 2T^{12}\sin(\overset{*}{\chi}-\overset{*}{\psi})\sin\overset{*}{\psi} + T^{22}\sin^2(\overset{*}{\chi}-\overset{*}{\psi})\right\}\Big/\sin\overset{*}{\chi}.
\end{aligned} \tag{6.53}$$

Equations (6.53) can be written in the form

$$\begin{aligned}
\overset{*}{T}^{11} &= a_1 + b_1\cos\left(2\overset{*}{\psi}\right) + c_1\sin\left(2\overset{*}{\psi}\right),\\
\overset{*}{T}^{12} &= -b_1\sin\left(2\overset{*}{\psi}\right) + c_1\cos\left(2\overset{*}{\psi}\right),\\
\overset{*}{T}^{22} &= a_1 - b_1\cos\left(2\overset{*}{\psi}\right) - c_1\sin\left(2\overset{*}{\psi}\right),
\end{aligned} \tag{6.54}$$

where the following notation is introduced:

$$\begin{aligned}
a_1 &= \left[\frac{1}{2}(T^{11}+T^{22}) - T^{12}\cos\overset{*}{\chi}\right]\Big/\sin\overset{*}{\chi},\\
b_1 &= \left[\frac{1}{2}(T^{11}+T^{22}) + \left(T^{12}+T^{22}\cos\overset{*}{\chi}\right)\cos\overset{*}{\chi}\right]\Big/\sin\overset{*}{\chi},\\
c_1 &= T^{12} + T^{22}\cos\overset{*}{\chi}.
\end{aligned} \tag{6.55}$$

Differentiating $\overset{*}{T}^{ii}$ with respect to $\overset{*}{\psi}$ and equating the result to zero, we obtain

$$\tan\left(2\overset{*}{\psi}\right) = \frac{c_1}{b_1} = \frac{2\left(T^{12} + T^{22}\cos\overset{*}{\chi}\right)\sin\overset{*}{\chi}}{T^{11} + 2T^{12}\cos\overset{*}{\chi} + T^{22}\cos\left(2\overset{*}{\chi}\right)}. \tag{6.56}$$

Solving the above for $\overset{*}{\psi}$ for the directional angles of the principal axes, we get

$$\tan\left(2\,\overset{*}{\psi}_1\right) = \frac{2\left(T^{12} + T^{22}\cos\overset{*}{\chi}\right)\sin\overset{*}{\chi}}{T^{11} + 2T^{12}\cos\overset{*}{\chi} + T^{22}\cos\left(2\overset{*}{\chi}\right)},$$

$$\overset{*}{\psi}_2 = \overset{*}{\psi}_1 + \pi/2. \tag{6.57}$$

On substituting Eq. (6.55) into (6.54), we have

$$\overset{*}{T}^{11} = a_1 + \sqrt{b_1^2 + c_1^2}, \qquad \overset{*}{T}^{12} = 0, \qquad \overset{*}{T}^{22} = a_1 - \sqrt{b_1^2 + c_1^2}.$$

From Eq. (6.54) the principal membrane forces T_1 and T_2 are found to be

$$T_{1,2} = a_1^2 \pm \sqrt{b_1^2 + c_1^2} = \frac{1}{\sin\overset{*}{\chi}}\left\{\frac{T^{11} + T^{22}}{2} + T^{12}\cos\overset{*}{\chi}\right.$$
$$\left.\pm\sqrt{\frac{1}{4}(T^{11} - T^{22})^2 + (T^{12})^2 + T^{12}(T^{11} + T^{22})\cos\overset{*}{\chi} + T^{11}T^{22}\cos^2\overset{*}{\chi}}\right\}. \tag{6.58}$$

Thus, at each point of the surface of the soft shell there are two mutually orthogonal directions that remain orthogonal throughout deformation. Henceforth, we assume that $T_1 \geq T_2$, i.e. the maximum stress is in the direction of the principal axis defined by the angle $\overset{*}{\psi}_1$, and the minimum is in that defined by the angle $\overset{*}{\psi}_2$.

Analogously to the invariants of the tensor of deformation described by Eqs. (6.44), we introduce the first and second invariants of **T**

$$I^{(\mathrm{T})}{}_1 = T_1 + T_2 = \overset{*}{T}^{11} + \overset{*}{T}^{22} = \left(T^{11} + T^{22} + 2T^{12}\cos\overset{*}{\chi}\right) / \sin\overset{*}{\chi},$$
$$I^{(\mathrm{T})}{}_2 = T_1T_2 = \overset{*}{T}^{11}\overset{*}{T}^{22} - (\overset{*}{T}^{12})^2 = T^{11}T^{22} - (T^{12})^2. \tag{6.59}$$

6.5 Corollaries of the fundamental assumptions

The fundamental assumptions stated at the beginning of the chapter have several corollaries specific to thin soft shells.

1. The zero-flexural-rigidity state is natural and unique to thin soft shells, in contrast to thin elastic shells with finite bending rigidity.
2. Soft shells do not resist compression forces and thus $T_1 \geq 0$, $T_2 \geq 0$ and $I^{(\mathrm{T})}{}_1 \geq 0, I^{(\mathrm{T})}{}_2 \geq 0$.
3. Shear membrane forces are significantly smaller than stretch forces, $T_{12} \approx 10^{-3} \max T_{ii}$.
4. Areas of the soft shell where $\Lambda_1 \leq 1$ and $\Lambda_2 \leq 1$ attain multiple configurations and are treated as the zero-stressed areas.
5. Stress states of the soft shell are classified as (i) biaxial, if $T_1 > 0$, $T_2 > 0$ and $(I^{(\mathrm{T})}{}_1 > 0, I^{(\mathrm{T})}{}_2 > 0)$, (ii) uniaxial, if either $T_1 = 0$ and $T_2 > 0$ or $T_1 > 0$ and $T_2 = 0$ $(I^{(\mathrm{T})}{}_1 > 0, I^{(\mathrm{T})}{}_2 = 0)$ and (iii) unstressed, if $T_1 = 0$ and $T_2 = 0$ $(I^{(\mathrm{T})}{}_1 = I^{(\mathrm{T})}{}_2 = 0)$.
6. Constitutive relations for the uniaxial stress–strain state (Fig. 6.5) are functions of either Λ_1 or Λ_2 and empirical mechanical constants c_m given by

$$
\begin{aligned}
T_1 &= f_1(\Lambda_1, c_1, \ldots, c_m, Z_{ij}) \quad \text{for } \Lambda_1 \geq 1, \Lambda_2 < 1, \\
T_2 &= f_2(\Lambda_2, c_1, \ldots, c_m, Z_{ij}) \quad \text{for } \Lambda_1 \leq 1, \Lambda_2 \geq 1.
\end{aligned}
\tag{6.60}
$$

7. Constitutive relations for the in-plane biaxial state (Fig. 6.5), $\Lambda_1 \geq 1, \Lambda_2 > 1$ ($T_1 > 0$, $T_2 > 0$), have the form

$$
\begin{aligned}
T_1 &= F_1(\Lambda_1, \Lambda_2, \varphi, c_1, \ldots, c_m, Z_{ij}), \\
T_2 &= F_2(\Lambda_1, \Lambda_2, \varphi, c_1, \ldots, c_m, Z_{ij}), \\
\psi &= \psi(\Lambda_1, \Lambda_2, \varphi, c_1, \ldots, c_m, Z_{ij}).
\end{aligned}
\tag{6.61}
$$

In general $f_n(\ldots) \neq F_n(\ldots)$, however, $f_n(\ldots)$ can be defined uniquely if $F_n(\ldots)$ is known.

Constitutive relations for biological tissues are either derived analytically or obtained experimentally. Continuum models typically allow greater computational efficiency and are easily integrated into multicomponent mathematical models. However, the identification of homogenized parameters and constants of the models can be a formidable challenge. Therefore, it is common practice to use approximations of experimental results from uniaxial, biaxial and shear tests conducted on isolated tissue samples. Constitutive relations for soft biological tissues, e.g. the skin, the

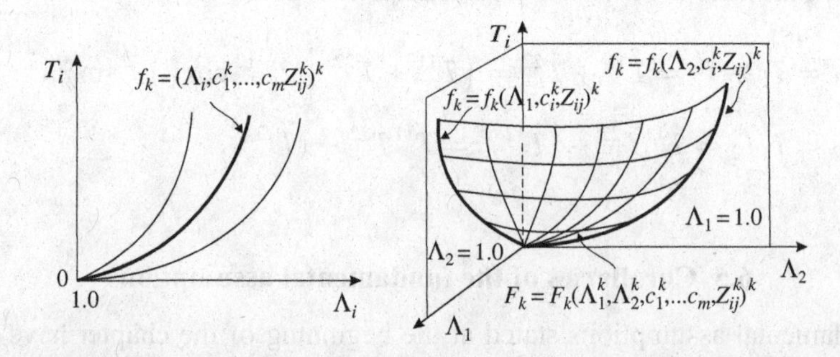

Fig. 6.5 Uniaxial (A) and biaxial (B) constitutive relations for soft biological tissues.

stomach and the gallbladder, are usually obtained along structurally preferred directions that are defined by the orientation of reinforced smooth muscle, collagen and elastin fibres, thus facilitating their use in calculations.

If constitutive relations are obtained in directions different from the actual parameterization of the shell, the task then is to learn how to calculate membrane forces in the principal directions. Consider two typical situations.

Case 1. Constitutive relations are given by Eq. (6.61). Then,

(i) from Eqs. (6.41) and (6.42), we calculate the principal deformations Λ_1, Λ_2 and the angle $\overset{*}{\varphi}_1$,
(ii) using Eqs. (6.61), we compute the principal membrane forces T_1 and T_2 and the angle ψ;
(iii) finally, setting $\overset{*}{\chi}_1 = \pi/2, \overset{*}{\chi}_2 = \psi, \overset{*}{T}{}^{11} = T_1, \overset{*}{T}{}^{22} = T_2, \overset{*}{T}{}^{12} = 0$ in Eqs. (6.51), we find

$$\begin{aligned} \tilde{T}^{11} &= \{T_1 \sin^2(\chi - \psi) + T_2 \cos^2(\chi - \psi)\}/\sin\chi, \\ \tilde{T}^{12} &= \{T_1 \sin\psi \sin(\chi - \psi) - T_2 \cos\psi \cos(\chi - \psi)\}/\sin\chi, \\ \tilde{T}^{22} &= \{T_1 \sin^2\psi + T_2 \cos^2\psi\}/\sin\chi. \end{aligned} \tag{6.62}$$

Case 2. Constitutive relations are formulated for the orientation of reinforced fibres, with superscript r,

$$\begin{aligned} T^{\mathrm{r}}{}_1 &= F^{\mathrm{r}}{}_1(\lambda^{\mathrm{r}}{}_1, \lambda^{\mathrm{r}}{}_2, \gamma^{\mathrm{r}}, c_1, \ldots, c_m, Z_{ij}), \\ T^{\mathrm{r}}{}_2 &= F^{\mathrm{r}}{}_2(\lambda^{\mathrm{r}}{}_1, \lambda^{\mathrm{r}}{}_2, \gamma^{\mathrm{r}}, c_1, \ldots, c_m, Z_{ij}), \\ S^{\mathrm{r}} &= S^{\mathrm{r}}(\lambda^{\mathrm{r}}{}_1, \lambda^{\mathrm{r}}{}_2, \gamma^{\mathrm{r}}, c_1, \ldots, c_m, Z_{ij}). \end{aligned} \tag{6.63}$$

Let $\overset{*}{\alpha}_i \in \overset{*}{S}$ be an auxiliary orthogonal coordinate system oriented with respect to a set of reinforced fibres at angle $\overset{*}{\psi}$. Then,

(i) setting $\chi_1 = \pi/2, \chi_2 = \psi$ and $\overset{*}{\chi}_1 = \pi/2 - \gamma^{\mathrm{r}}$ in Eq. (6.34), where $\lambda_1^{\mathrm{r}} := \overset{*}{\lambda}_1$ and $\lambda_2^{\mathrm{r}} := \overset{*}{\lambda}_2$, for the stretch ratios and the shear angle γ^{r}, we have

$$\begin{aligned} \lambda_1^{\mathrm{r}} &= \Big(\lambda_1^2 \sin^2(\overset{0}{\chi} - \overset{0}{\psi}) + \lambda_2^2 \sin^2\overset{0}{\psi} \\ &\quad + 2\lambda_1\lambda_2 \cos\overset{0}{\chi} \sin(\overset{0}{\chi} - \overset{0}{\psi}) \sin\overset{0}{\psi}\Big)^{1/2} \Big/ \sin\overset{0}{\chi}, \\ \gamma^{\mathrm{r}} &= \sin^{-1}\left(\frac{1}{\lambda_1\lambda_2 \sin^2\chi^0}\left(-\frac{1}{2}\lambda_1^2 \sin\left[2(\overset{0}{\chi} - \overset{0}{\psi})\right]\right.\right. \\ &\quad \left.\left. + \frac{1}{2}\lambda_2^2 \sin^2\overset{0}{\psi} + \lambda_1\lambda_2 \cos\overset{0}{\chi} \sin(\overset{0}{\chi} - \overset{0}{\psi})\right)\right) \\ \lambda_2^{\mathrm{r}} &= \Big(\lambda_1^2 \cos^2(\overset{0}{\chi} - \overset{0}{\psi}) + \lambda_2^2 \cos^2\overset{0}{\psi} \\ &\quad - 2\lambda_1\lambda_2 \cos(\overset{0}{\chi} - \overset{0}{\psi}) \cos\overset{0}{\chi} \cos\overset{0}{\psi}\Big)^{1/2} \Big/ \sin\overset{0}{\chi}; \end{aligned} \tag{6.64}$$

(ii) using Eqs. (6.63) we find $T_1^{\mathrm{r}}, T_2^{\mathrm{r}}$ and S^{r};

(iii) the angle $\overset{*}{\psi}$ is found from Eq. (6.34) by putting $\overset{*}{\chi}_1 = \overset{*}{\psi}, \overset{*}{\chi}_2 = 0, \overset{*}{\lambda}_1 = \lambda_1$ and $\overset{*}{\lambda}_2 = \lambda_1^{\mathrm{r}}$, with

$$\overset{*}{\psi} = \cos^{-1}\left(\frac{1}{\lambda_1^{\mathrm{r}} \sin\overset{0}{\chi}}\left(\lambda_1 \sin(\overset{0}{\chi} - \overset{0}{\psi}) + \lambda_2 \cos\overset{0}{\chi} \sin\overset{0}{\psi}\right)\right); \tag{6.65}$$

(iv) finally, setting $\overset{*}{\chi}_1 = \pi/2 - \gamma^{\mathrm{r}}, \overset{*}{\chi}_2 = \overset{*}{\psi}$, $T_1^{\mathrm{r}} := \tilde{\overset{*}{T}}{}^{11}$, $T_2^{\mathrm{r}} := \tilde{\overset{*}{T}}{}^{22}$ and $S^{\mathrm{r}} := \tilde{\overset{*}{T}}{}^{12}$ in Eqs. (6.51), we obtain

$$\begin{aligned}
\tilde{T}^{11} &= \Big\{T_1^{\mathrm{r}} \sin^2(\overset{*}{\chi} - \overset{*}{\psi}) + T_2^{\mathrm{r}} \cos^2(\overset{*}{\chi} - \overset{*}{\psi} + \gamma^{\mathrm{r}}) \\
&\quad - 2S^{\mathrm{r}} \cos(\overset{*}{\chi} - \overset{*}{\psi} + \gamma^{\mathrm{r}}) \sin(\overset{*}{\chi} - \overset{*}{\psi})\Big\} \Big/ \left(\sin\overset{*}{\chi} \cos\gamma^{\mathrm{r}}\right), \\
\tilde{T}^{12} &= \Big\{T_1^{\mathrm{r}} \sin\mu \sin(\overset{*}{\chi} - \overset{*}{\psi}) - T_2^{\mathrm{r}} \cos(\overset{*}{\psi} + \gamma^{\mathrm{r}}) \cos(\overset{*}{\chi} - \overset{*}{\psi} + \gamma^{\mathrm{r}}) \\
&\quad + S^{\mathrm{r}}\left[\sin(\overset{*}{\chi} - 2\overset{*}{\psi} + \gamma^{\mathrm{r}}) - \cos\overset{*}{\chi} \sin\gamma^{\mathrm{r}}\right]\Big\} \Big/ \left(\sin\overset{*}{\chi} \cos\gamma^{\mathrm{r}}\right), \\
\tilde{T}^{22} &= \Big\{T_1^{\mathrm{r}} \sin^2\overset{*}{\psi} - 2S^{\mathrm{r}} \sin(\overset{*}{\psi} - \gamma^{\mathrm{r}}) \sin\overset{*}{\psi} + T_2^{\mathrm{r}} \cos^2(\overset{*}{\psi} - \gamma^{\mathrm{r}})\Big\} \Big/ \left(\sin\overset{*}{\chi} \cos\gamma^{\mathrm{r}}\right).
\end{aligned} \tag{6.66}$$

Formulas (6.66) can be written in more concise form if we introduce generalized forces defined by $N^{ik} = T^{ik}\lambda_k/\lambda_i$,

$$\overset{*}{N}{}^{11} = T_1^{\mathrm{r}} \frac{\lambda_2^{\mathrm{r}}}{\lambda_1^{\mathrm{r}}}, \qquad \overset{*}{N}{}^{22} = T_2^{\mathrm{r}} \frac{\lambda_1^{\mathrm{r}}}{\lambda_2^{\mathrm{r}}}, \qquad \overset{*}{N}{}^{12} = S^{\mathrm{r}}.$$

Then, Eq. (6.52) takes the form

$$N^{ik} = \frac{1}{\hat{\overset{*}{C}}} \sum_{j=1}^{2} \sum_{n=1}^{2} \overset{*}{N}{}^{jn} \hat{\overset{*}{C}}{}_j^{i} \hat{\overset{*}{C}}{}_n^{k}. \tag{6.67}$$

On substituting $\hat{\overset{*}{C}}{}_j^{i}$ given by Eqs. (6.12) and (6.14), we find

$$\begin{aligned}
N^{11} &= \Big\{\overset{*}{N}{}^{11} \sin^2(\overset{0}{\chi} - \overset{0}{\chi}_2) + \overset{*}{N}{}^{22} \sin^2(\overset{0}{\chi}_1 + \overset{0}{\chi}_2 - \overset{0}{\chi}) \\
&\quad - 2\overset{*}{N}{}^{12} \sin(\overset{0}{\chi}_1 + \overset{0}{\chi}_2 - \overset{0}{\chi}) \sin(\overset{0}{\chi} - \overset{0}{\chi}_2)\Big\} \Big/ \left(\sin\overset{0}{\chi} \sin\overset{0}{\chi}_1\right), \\
N^{12} &= \Big\{\overset{*}{N}{}^{11} \sin\overset{0}{\chi}_2 \sin(\overset{0}{\chi} - \overset{0}{\chi}_1) + \overset{*}{N}{}^{22} \sin(\overset{0}{\chi}_1 + \overset{0}{\chi}_2 - \overset{0}{\chi}) \sin(\overset{0}{\chi}_1 + \overset{0}{\chi}_2) \\
&\quad + \overset{*}{N}{}^{12}\left[\cos(\overset{0}{\chi}_1 + 2\overset{0}{\chi}_2 - \overset{0}{\chi}) - \cos\overset{0}{\chi} \cos\overset{0}{\chi}_1\right]\Big\} \Big/ \left(\sin\overset{0}{\chi} \sin\overset{0}{\chi}_1\right), \\
N^{22} &= \Big\{\overset{*}{N}{}^{11} \sin^2\overset{0}{\chi}_2 - 2\overset{*}{N}{}^{12} \sin(\overset{0}{\chi}_1 + \overset{0}{\chi}_2) \sin\overset{0}{\chi}_2 \\
&\quad + \overset{*}{N}{}^{22} \sin^2(\overset{0}{\chi}_1 + \overset{0}{\chi}_2)\Big\} \Big/ \left(\sin\overset{0}{\chi} \sin\overset{0}{\chi}_1\right).
\end{aligned} \tag{6.68}$$

Putting $\overset{0}{\chi}_1 = \pi/2$ and $\overset{0}{\chi}_{2_0} = \overset{0}{\psi}$ in Eqs. (6.68), for the membrane forces in terms of the undeformed surface S ($\overset{0}{S} = S$), we have

$$\tilde{T}^{11} = \frac{\lambda_1}{\lambda_2}\left(T_1^{\mathrm{r}}\frac{\lambda_2^{\mathrm{r}}}{\lambda_1^{\mathrm{r}}}\sin^2(\overset{0}{\chi} - \overset{0}{\psi}) + T_2^{\mathrm{r}}\frac{\lambda_1^{\mathrm{r}}}{\lambda_2^{\mathrm{r}}}\cos^2(\overset{0}{\chi} - \overset{0}{\psi}) - 2S^{\mathrm{r}}\sin\left[2(\overset{0}{\chi} - \overset{0}{\psi})\right]\right) \Big/ \sin\overset{0}{\chi},$$

$$\tilde{T}^{12} = \left(T_1^{\mathrm{r}}\frac{\lambda_2^{\mathrm{r}}}{\lambda_1^{r}}\sin^2(\overset{0}{\chi} - \overset{0}{\psi})\sin\overset{0}{\psi} - T_2^{\mathrm{r}}\frac{\lambda_1^{\mathrm{r}}}{\lambda_2^{\mathrm{r}}}\cos^2(\overset{0}{\chi} - \overset{0}{\psi})\cos\overset{0}{\psi}\right.$$
$$\left. + S^{\mathrm{r}}\sin(\overset{0}{\chi} - 2\overset{0}{\psi})\right) \Big/ \sin\overset{0}{\chi},$$

$$\tilde{T}^{22} = \frac{\lambda_2}{\lambda_1}\left(T_1^{\mathrm{r}}\frac{\lambda_2^{\mathrm{r}}}{\lambda_1^{\mathrm{r}}}\sin^2\overset{0}{\psi} + T_2^{\mathrm{r}}\frac{\lambda_1^{\mathrm{r}}}{\lambda_2^{\mathrm{r}}}\cos^2\overset{0}{\psi} + S^{\mathrm{r}}\sin\left(2\overset{0}{\psi}\right)\right) \Big/ \sin\overset{0}{\chi}. \tag{6.69}$$

Formulas (6.69) depend only on the parameterization of S and the axes of anisotropy. Thus, they are less computationally demanding compared to Eqs. (6.65) and (6.66).

6.6 Nets

Members of a special class of soft shell, in which discrete reinforced fibres are the main structural and weight-bearing elements, are called nets. Depending on the engineering design and practical needs, the fibres may remain discrete or be embedded in the connective matrix. Although the nets have distinct discrete structure, they are modelled as a solid continuum. Since the nets have very low resistance to shear forces, $T_{12} = 0$ ($S^{\mathrm{r}} \equiv 0$) and the resultant formulas obtained in the previous paragraphs are valid for modelling nets.

Consider a net with the cell structure of a parallelogram. Let the sides of the cell be formed by two distinct families of reinforced fibres (Fig. 6.6). Their mechanical properties are described by

$$\begin{aligned} T_1^{\mathrm{r}} &= F_1^{\mathrm{r}}(\lambda_1^{\mathrm{r}}, \lambda_2^{\mathrm{r}}, \gamma^{\mathrm{r}}, c_1, \ldots, c_m, Z_{ij}), \\ T_2^{\mathrm{r}} &= F_2^{\mathrm{r}}(\lambda_1^{\mathrm{r}}, \lambda_2^{\mathrm{r}}, \gamma^{\mathrm{r}}, c_1, \ldots, c_m, Z_{ij}), \end{aligned} \tag{6.70}$$

where the meanings of the parameters and constants are as discussed above.

Let the undeformed ($S \equiv \overset{0}{S}$) configuration of the net be parameterized by $\overset{*}{\alpha}_1$- and $\overset{*}{\alpha}_2$-coordinates oriented along the reinforced fibres. For the force distribution in the net we have to

(i) find the stretch ratios $\lambda_1^{\mathrm{r}} = \overset{*}{\lambda}_1$ and $\lambda_2^{\mathrm{r}} = \overset{*}{\lambda}_2$, using Eqs. (6.34);
(ii) by substituting $\lambda_1^{\mathrm{r}} = \lambda_1^*$ and $\lambda_2^{\mathrm{r}} = \lambda_2^*$ in Eq. (6.70), calculate T_1^{r} and T_2^{r};
(iii) making use of Eqs. (6.66) or (6.69), find for the membrane forces, in terms of the S configuration,

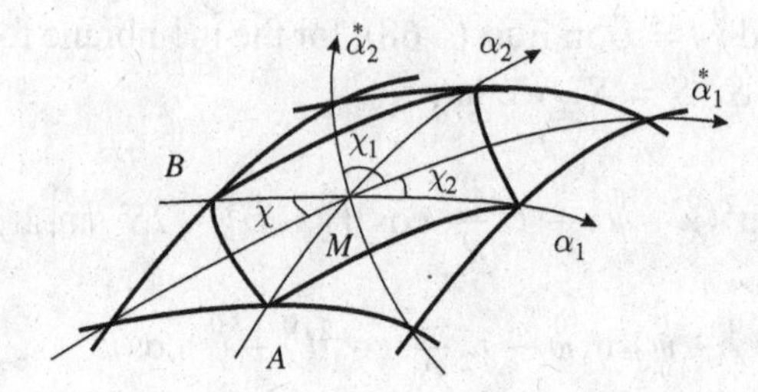

Fig. 6.6 A structural element of the net formed by two distinct types of reinforced fibres.

$$
\begin{aligned}
\tilde{T}^{11} &= \frac{\lambda_1}{\lambda_2}\left(T_1^{\mathrm{r}}\frac{\lambda_2^{\mathrm{r}}}{\lambda_1^{\mathrm{r}}}\sin^2(\overset{0}{\chi}-\overset{0}{\psi}) + T_2^{\mathrm{r}}\frac{\lambda_1^{\mathrm{r}}}{\lambda_2^{\mathrm{r}}}\cos^2(\overset{0}{\chi}-\overset{0}{\psi})\right)\Big/\sin\overset{0}{\chi},\\
\tilde{T}^{12} &= \left(T_1^{\mathrm{r}}\frac{\lambda_2^{\mathrm{r}}}{\lambda_1^{\mathrm{r}}}\sin^2(\overset{0}{\chi}-\overset{0}{\psi})\sin\overset{0}{\psi} - T_2^{\mathrm{r}}\frac{\lambda_1^{\mathrm{r}}}{\lambda_2^{\mathrm{r}}}\cos^2(\overset{0}{\chi}-\overset{0}{\psi})\cos\overset{0}{\psi}\right)\Big/\sin\overset{0}{\chi},\\
\tilde{T}^{22} &= \frac{\lambda_2}{\lambda_1}\left(T_1^{\mathrm{r}}\frac{\lambda_2^{\mathrm{r}}}{\lambda_1^{\mathrm{r}}}\sin^2\overset{0}{\psi} + T_2^{\mathrm{r}}\frac{\lambda_1^{\mathrm{r}}}{\lambda_2^{\mathrm{r}}}\cos^2\overset{0}{\psi}\right)\Big/\sin\overset{0}{\chi}.
\end{aligned}
\tag{6.71}
$$

The principal membrane forces and their directions are found from Eqs. (6.56) and (6.57) by putting $T^{12} = S^{\mathrm{r}} = 0, T^{11} = T_1^{\mathrm{r}}$ and $T^{22} = T_2^{\mathrm{r}}$,

$$
\begin{aligned}
T_{1,2} &= \frac{(T_1^{\mathrm{r}} + T_2^{\mathrm{r}}) \pm \sqrt{(T_1^{\mathrm{r}} - T_2^{\mathrm{r}})^2 + 4T_1^{\mathrm{r}}T_2^{\mathrm{r}}\cos^2\overset{*}{\chi}}}{2\sin\overset{*}{\chi}},\\
\tan\left(2\overset{*}{\psi}_1\right) &= \frac{T_2^{\mathrm{r}}\sin\left(2\overset{*}{\chi}\right)}{T_1^{\mathrm{r}} + T_2^{\mathrm{r}}\cos\left(2\overset{*}{\chi}\right)},\\
\overset{*}{\psi}_2 &= \overset{*}{\psi}_1 + \pi/2.
\end{aligned}
\tag{6.72}
$$

In particular,

(i) if $\overset{*}{\chi} = \pi/2$ then $\overset{*}{\psi}_1 = 0, T_1 = T_1^{\mathrm{r}}$ and $T_2 = T_2^{\mathrm{r}}$;
(ii) if $T_1^{\mathrm{r}} = 0$ then $\overset{*}{\psi}_1 = 0, T_1 = T_1^{\mathrm{r}}/\sin\overset{*}{\chi}$ and $T_2 = 0$;
(iii) if $T_2^{\mathrm{r}} = 0$ then $\overset{*}{\psi}_1 = \overset{*}{\chi}, T_1 = T_2^{\mathrm{r}}/\sin\overset{*}{\chi}$ and $T_2 = 0$.

Corollary 5 of the fundamental assumptions for the nets is given by

$$
\begin{aligned}
I^{(\mathbf{T})}{}_1 &= T_1 + T_2 = \frac{1}{\sin\overset{*}{\chi}}(T_1^{\mathrm{r}} + T_2^{\mathrm{r}}) \geq 0,\\
I^{(\mathbf{T})}{}_2 &= T_1 T_2 = T_1^{\mathrm{r}}T_2^{\mathrm{r}} \geq 0.
\end{aligned}
\tag{6.73}
$$

6.7 Equations of motion in general curvilinear coordinates

Let $\Delta\overset{(*)}{\sigma}$ and $\Delta\overset{(*)}{m}$ be the surface area and mass of a differential element of the soft shell in undeformed and deformed configurations. The position of a point $\overset{*}{M} \in \overset{*}{S}$ at any moment of time t is given by the vector $\bar{r}(\alpha_1, \alpha_2, t)$. The densities of the material in undeformed, ρ, and deformed, $\overset{*}{\rho}$, states are defined by

$$\rho = \lim_{\Delta\sigma\to 0} \frac{\Delta m}{\Delta\sigma} = \frac{\mathrm{d}m}{\mathrm{d}\sigma} \quad \text{and} \quad \overset{*}{\rho} = \lim_{\Delta\overset{*}{\sigma}\to 0} \frac{\Delta\overset{*}{m}}{\Delta\overset{*}{\sigma}} = \frac{\mathrm{d}\overset{*}{m}}{\mathrm{d}\overset{*}{\sigma}}, \tag{6.74}$$

where

$$\mathrm{d}\sigma = \sqrt{a}\,\mathrm{d}\alpha_1\,\mathrm{d}\alpha_2, \qquad \mathrm{d}\overset{*}{\sigma} = \sqrt{\overset{*}{a}}\,\mathrm{d}\alpha_1\,\mathrm{d}\alpha_2.$$

Applying the law of conservation of the mass to Eq. (6.74), we find

$$\mathrm{d}m = \mathrm{d}\overset{*}{m} = \overset{*}{\rho}\ \mathrm{d}\overset{*}{\sigma} = \overset{*}{\rho}\sqrt{\overset{*}{a}}\,\mathrm{d}\alpha_1\,\mathrm{d}\alpha_2 = \rho\sqrt{\overset{*}{a}}\,\mathrm{d}\alpha_1\,\mathrm{d}\alpha_2.$$

It follows that

$$\overset{*}{\rho} = \rho\sqrt{a/\overset{*}{a}}. \tag{6.75}$$

Let $\bar{p}_s(\alpha_1, \alpha_2, t)$ be the resultant of the external, $\bar{p}_{(+)}(\alpha_1, \alpha_2, t)$, and internal, $\bar{p}_{(-)}(\alpha_1, \alpha_2, t)$, forces distributed over the outer and inner surfaces of the shell,

$$\bar{p}_s(\alpha_1, \alpha_2, t) = \bar{p}_{(+)}(\alpha_1, \alpha_2, t) + \bar{p}_{(-)}(\alpha_1, \alpha_2, t).$$

The density of the resultant force per unit area of a deformed element $\bar{p}_s$ is defined by

$$\bar{p}(\alpha_1, \alpha_2, t) = \lim_{\Delta\sigma^*\to 0} \frac{\bar{p}_s}{\Delta\overset{*}{\sigma}}. \tag{6.76}$$

Similarly, we introduce the density of the mass force $\bar{F}(\alpha_1, \alpha_2, t)$ by

$$\bar{f}(\alpha_1, \alpha_2, t) = \lim_{\Delta m\to 0} \frac{\bar{F}}{\Delta m} = \frac{\mathrm{d}\bar{F}}{\mathrm{d}m} = \frac{1}{\rho}\frac{\mathrm{d}\bar{F}}{\mathrm{d}\sigma}. \tag{6.77}$$

The resultant stress vectors $\bar{R}_i$ acting upon the differential element are found to be

$$\begin{aligned}
\bar{R}_1 &= -(T^{11}\bar{e}_1 + T^{12}\bar{e}_2)\sqrt{\overset{*}{a}_{22}}\mathrm{d}\alpha_2, \qquad \bar{R}_2 = -(T^{21}\bar{e}_1 + T^{22}\bar{e}_2)\sqrt{\overset{*}{a}_{11}}\mathrm{d}\alpha_1,\\
-\left(\bar{R}_1 + \frac{\partial\bar{R}_1}{\partial\alpha_1}\mathrm{d}\alpha_1\right) &= -(T^{11}\bar{e}_1 + T^{12}\bar{e}_2)\sqrt{\overset{*}{a}_{22}}\mathrm{d}\alpha_2\\
&\quad - \frac{\partial}{\partial\alpha_1}(T^{11}\bar{e}_1 + T^{12}\bar{e}_2)\sqrt{\overset{*}{a}_{22}}\mathrm{d}\alpha_1\,\mathrm{d}\alpha_2,\\
-\left(\bar{R}_2 + \frac{\partial\bar{R}_2}{\partial\alpha_2}\mathrm{d}\alpha_2\right) &= -(T^{21}\bar{e}_1 + T^{22}\bar{e}_2)\sqrt{\overset{*}{a}_{11}}\mathrm{d}\alpha_1\\
&\quad - \frac{\partial}{\partial\alpha_2}(T^{21}\bar{e}_1 + T^{22}\bar{e}_2)\sqrt{\overset{*}{a}_{11}}\mathrm{d}\alpha_1\,\mathrm{d}\alpha_2.
\end{aligned} \tag{6.78}$$

Applying the law of conservation of momentum to Eqs. (6.76)–(6.78), for the equation of motion of the soft shell we get

$$\overset{*}{\rho}\frac{\mathrm{d}^2\bar{r}(\alpha_1,\alpha_2,t)}{\mathrm{d}t^2} = -\frac{\partial\overline{R}_1}{\partial\alpha_1}\mathrm{d}\alpha_1 - \frac{\partial\overline{R}_2}{\partial\alpha_2}\mathrm{d}\alpha_2 + \bar{p} + \bar{f}\overset{*}{\rho}, \tag{6.79}$$

where $\mathrm{d}^2\bar{r}/\mathrm{d}t^2$ is the acceleration. On substituting $\overline{R}_i$ and $\overset{*}{\rho}$ given by Eqs. (6.75) and (6.78) into (6.79), we get

$$\begin{aligned}\rho\sqrt{a}\frac{\mathrm{d}^2\bar{r}}{\mathrm{d}t^2} = \frac{\partial}{\partial\alpha_1}\left[\left(T^{11}\bar{e}_1 + T^{12}\bar{e}_2\right)\sqrt{\overset{*}{a}_{22}}\right] + \frac{\partial}{\partial\alpha_2}\left[\left(T^{21}\bar{e}_1 + T^{22}\bar{e}_2\right)\sqrt{\overset{*}{a}_{11}}\right]\\ + \bar{p}\sqrt{\overset{*}{a}} + \bar{f}\rho\sqrt{a}.\end{aligned} \tag{6.80}$$

Let $\overline{G}_i, \overline{M}_p$ and $\overline{M}_f$ be the resultant moment vectors acting on the element of the shell defined by

$$\begin{aligned}&\overline{G}_1 = \bar{r}\times\overline{R}_1, \qquad \overline{G}_2 = \bar{r}\times\overline{R}_2,\\ &-\left(\overline{G}_1 + \frac{\partial\overline{G}_1}{\partial\alpha_1}\mathrm{d}\alpha_1\right), \qquad -\left(\overline{G}_2 + \frac{\partial\overline{G}_2}{\partial\alpha_2}\mathrm{d}\alpha_2\right) \quad (\alpha_i + \mathrm{d}\alpha_i = \text{constant}),\\ &\overline{M}_p = (\bar{r}\times\bar{p})\sqrt{\overset{*}{a}}\mathrm{d}\alpha_1\,\mathrm{d}\alpha_2, \qquad \overline{M}_f = (\bar{r}\times\bar{f})\rho\sqrt{\overset{*}{a}}\mathrm{d}\alpha_1\,\mathrm{d}\alpha_2.\end{aligned} \tag{6.81}$$

Assuming the shell is in equilibrium, the sum of the moments vanishes. Hence

$$-\frac{\partial\overline{G}_1}{\partial\alpha_1}\mathrm{d}\alpha_1 - \frac{\partial\overline{G}_2}{\partial\alpha_2}\mathrm{d}\alpha_2 + \overline{M}_p + \overline{M}_q = 0. \tag{6.82}$$

Substituting $\overline{G}_i, \overline{M}_p$ and $\overline{M}_f$ in (6.82), we obtain

$$\begin{aligned}\left[-\left(\bar{r}\times\frac{\partial\overline{R}_1}{\partial\alpha_1}\mathrm{d}\alpha_1\right) - \left(\bar{r}\times\frac{\partial\overline{R}_2}{\partial\alpha_2}\mathrm{d}\alpha_2\right) + (\bar{r}\times\bar{p}) + (\bar{r}\times\bar{f})\right]\\ - \left(\bar{r}\times\overline{R}_1\right)\mathrm{d}\alpha_1 - \left(\bar{r}\times\overline{R}_2\right)\mathrm{d}\alpha_2 = 0.\end{aligned}$$

Further, on use of Eq. (6.78), we find

$$\begin{aligned}&\left[\bar{e}_1\times\left(T^{11}\bar{e}_1 + T^{12}\bar{e}_2\right) + \bar{e}_2\times\left(T^{21}\bar{e}_1 + T^{22}\bar{e}_2\right)\right]\sqrt{\overset{*}{a}_{11}}\sqrt{\overset{*}{a}_{22}}\mathrm{d}\alpha_1\,\mathrm{d}\alpha_2\\ &\qquad - \bar{r}\times\underline{\left[-\frac{\partial\overline{R}_1}{\partial\alpha_1}\mathrm{d}\alpha_1 - \frac{\partial\overline{R}_2}{\partial\alpha_2}\mathrm{d}\alpha_2 + \bar{p} + \bar{f}\right]} = 0.\end{aligned} \tag{6.83}$$

Since the underlined term equals zero, we have

$$(\bar{e}_1\times\bar{e}_2)T^{12} + (\bar{e}_2\times\bar{e}_1)T^{21} = 0. \tag{6.84}$$

It follows immediately from the above that $T^{12} = T^{21}$.

Remarks

1. If a soft shell is parameterized along the principal axes then $T^{11} = T_1, T^{22} = T_2, T^{12} = 0$ and $\overset{*}{a}_{12} = a_{12} = 0$, and the equation of motion (6.80) takes the simplest form

$$\rho\sqrt{a}\frac{\mathrm{d}^2\bar{r}}{\mathrm{d}t^2} = \frac{\partial}{\partial\alpha_1}\left[T_1\sqrt{\overset{*}{a}_{22}}\bar{e}_1\right] + \frac{\partial}{\partial\alpha_2}\left[T_2\sqrt{\overset{*}{a}_{11}}\bar{e}_2\right] + \bar{p}\sqrt{\overset{*}{a}} + \bar{f}\rho\sqrt{a}. \qquad (6.85)$$

2. During the dynamic process of deformation different parts of the soft shell may experience different stress–strain states. The biaxial stress state occurs when ${I^{(\mathrm{T})}}_1 = T_1 + T_2 > 0$ and ${I^{(\mathrm{T})}}_2 = T_1T_2 > 0$, the uniaxial state develops in areas where ${I^{(\mathrm{T})}}_1 > 0$ and ${I^{(\mathrm{T})}}_2 = 0$, and the zero stress state takes place anywhere in the shell where ${I^{(\mathrm{T})}}_1 = {I^{(\mathrm{T})}}_2 = 0$. The uniaxially stressed area ($T_2 = 0$) will develop wrinkles oriented along the direction of action of the positive principal membrane force T_1. The equation of motion for the wrinkled area becomes

$$\rho\sqrt{a}\frac{\mathrm{d}^2\bar{r}}{\mathrm{d}t^2} = \frac{\partial}{\partial\alpha_1}\left[T_1\sqrt{\overset{*}{a}_{22}}\bar{e}_1\right] + \bar{p}\sqrt{\overset{*}{a}} + \bar{f}\rho\sqrt{a}. \qquad (6.86)$$

To preserve smoothness and continuity of the surface $\overset{*}{S}$, the uniaxially stressed area is substituted by an ironed surface made out of an array of closely packed reinforced fibres. This approach allows one to use the equations of motion (6.79) throughout the deformed surface $\overset{*}{S}$.

The governing system of equations of dynamics of the soft shell includes the equations of motion (6.80) and (6.85), constitutive relations (6.60), (6.61) or (6.70), initial and boundary conditions, and the conditions given by Corollary 5.

6.8 Governing equations in orthogonal Cartesian coordinates

Let a soft shell be associated with an orthogonal Cartesian coordinate system x_1, x_2, x_3:

$$\begin{aligned} x_1 &= x_1(\alpha_1, \alpha_2, t), \\ x_2 &= x_2(\alpha_1, \alpha_2, t), \\ x_3 &= x_3(\alpha_1, \alpha_2, t). \end{aligned} \qquad (6.87)$$

The position vector of point $M(\alpha_1, \alpha_2) \in S$ and its derivatives are given by

$$\bar{r} = \bar{i}_1x_1 + \bar{i}_2x_2 + \bar{i}_3x_3 = \sum_{k=1}^{3} x_k\bar{i}_k, \qquad (6.88)$$

$$\bar{r}_i = \frac{\partial\bar{r}}{\partial\alpha_i} = \sum_{k=1}^{3}\frac{\partial x_k}{\partial\alpha_i}\bar{i}_k = \sum_{k=1}^{3}\bar{r}_{ik}\bar{i}_k \quad (i = 1, 2), \qquad (6.89)$$

where $\bar{r}_{ik} = \partial x_k / \partial \alpha_i$ is the projection of the ith basis vector on the x_1, x_2, x_3 axes. Decomposing the unit vectors $\bar{e}_i = \bar{r}_i / |\bar{r}_i|$ along the base $\{\bar{i}_1, \bar{i}_2, \bar{i}_3\}$, we get

$$\bar{e}_i = \sum_{k=1}^{3} l_{ik}\bar{i}_k = \frac{\bar{r}_i}{\sqrt{\overset{*}{a}_{ii}}} = \frac{\bar{r}_{ik}}{\sqrt{\overset{*}{a}_{ii}}}\bar{i}_k = l_{ik}\bar{i}_k, \tag{6.90}$$

where l_{ik} are the direction cosines defined by

$$l_{ik} := \cos(\bar{e}_i, \bar{i}_k) = \bar{r}_{ik} \Big/ \sqrt{\overset{*}{a}_{ii}}. \tag{6.91}$$

The vector $\overline{m}$ normal to $\bar{e}_i$ ($\bar{m} \perp \bar{e}_i$) is given by

$$\overline{m} = (\bar{e}_1 \times \bar{e}_2)\frac{\sqrt{\overset{*}{a}_{11}\overset{*}{a}_{22}}}{\sqrt{\overset{*}{a}}}. \tag{6.92}$$

With the help of Eqs. (6.90) and (6.92), the direction cosines $l_{3k} = \cos(\bar{m}, \bar{i}_k)$ are found to be

$$\begin{aligned} l_{31} &= (l_{12}l_{23} - l_{13}l_{22})\sqrt{\overset{*}{a}_{11}\overset{*}{a}_{22}} \Big/ \sqrt{\overset{*}{a}}, \\ l_{32} &= (l_{13}l_{21} - l_{11}l_{23})\sqrt{\overset{*}{a}_{11}\overset{*}{a}_{22}} \Big/ \sqrt{\overset{*}{a}}, \\ l_{33} &= (l_{11}l_{22} - l_{12}l_{21})\sqrt{\overset{*}{a}_{11}\overset{*}{a}_{22}} \Big/ \sqrt{\overset{*}{a}}. \end{aligned} \tag{6.93}$$

The scalar products $\bar{e}_i\bar{e}_k$ yield

$$\begin{aligned} \bar{e}_1\bar{e}_2 &= l_{11}l_{21} + l_{12}l_{22} + l_{13}l_{23} = \cos\overset{*}{\chi} = \overset{*}{a}_{12} \Big/ \sqrt{\overset{*}{a}_{11}\overset{*}{a}_{22}}, \\ \bar{e}_i\bar{e}_2 &= l_{i1}l_{31} + l_{i2}l_{32} + l_{i3}l_{33} = 0, \\ \bar{e}_k\bar{e}_k &= l_{k1}^2 + l_{k2}^2 + l_{k3}^2 = 1. \end{aligned} \tag{6.94}$$

Expanding $\bar{p}$ and $\bar{f}$ in the directions of $\bar{e}_i$ and $\bar{i}_i$, respectively, we obtain

$$\bar{p} = \bar{e}_1 p_1 + \bar{e}_2 p_2 + \bar{m} p_3, \tag{6.95}$$

$$\bar{f} = \bar{i}_1 f_1 + \bar{i}_2 f_2 + \bar{i}_3 f_3. \tag{6.96}$$

On substituting Eqs. (6.91), (6.93), (6.95) and (6.96) into (6.84), the equation of motion of the soft shell takes the form

$$\rho\sqrt{a}\frac{d^2x_1}{dt^2} = \frac{\partial}{\partial\alpha_1}\left[(T^{11}l_{11} + T^{12}l_{21})\sqrt{\overset{*}{a}_{22}}\right] + \frac{\partial}{\partial\alpha_2}\left[(T^{12}l_{11} + T^{22}l_{21})\sqrt{\overset{*}{a}_{11}}\right]$$
$$+ (p_1l_{11} + p_2l_{21})\sqrt{\overset{*}{a}} + p_3(l_{21}l_{23} - l_{13}l_{22})\sqrt{\overset{*}{a}_{11}\overset{*}{a}_{22}} + \rho f_1\sqrt{a},$$
$$\rho\sqrt{a}\frac{d^2x_2}{dt^2} = \frac{\partial}{\partial\alpha_1}\left[(T^{11}l_{12} + T^{12}l_{22})\sqrt{\overset{*}{a}_{22}}\right] + \frac{\partial}{\partial\alpha_2}\left[(T^{12}l_{12} + T^{22}l_{22})\sqrt{\overset{*}{a}_{11}}\right]$$
$$+ (p_1l_{12} + p_2l_{22})\sqrt{\overset{*}{a}} + p_3(l_{13}l_{21} - l_{11}l_{23})\sqrt{\overset{*}{a}_{11}\overset{*}{a}_{22}} + \rho f_2\sqrt{a},$$
$$\rho\sqrt{a}\frac{d^2x_3}{dt^2} = \frac{\partial}{\partial\alpha_1}\left[(T^{11}l_{13} + T^{12}l_{23})\sqrt{\overset{*}{a}_{22}}\right] + \frac{\partial}{\partial\alpha_2}\left[(T^{12}l_{13} + T^{22}l_{23})\sqrt{\overset{*}{a}_{11}}\right]$$
$$+ (p_1l_{13} + p_2l_{23})\sqrt{\overset{*}{a}} + p_3(l_{11}l_{22} - l_{12}l_{21})\sqrt{\overset{*}{a}_{11}\overset{*}{a}_{22}} + \rho f_3\sqrt{a}. \tag{6.97}$$

Here d^2x_i/dt^2 ($i = 1, 2, 3$) are the components of the vector of acceleration. Equations (6.97) should be complemented by constitutive relations (6.60), (6.61) or (6.70), initial and boundary conditions, and the conditions given by Corollary 5.

6.9 Governing equations in cylindrical coordinates

Let a soft shell be associated with a cylindrical coordinate system $\{r, \varphi, z\}$:

$$\begin{aligned} r &= r(\alpha_1, \alpha_2, t), \\ \varphi &= \varphi(\alpha_1, \alpha_2, t), \\ z &= z(\alpha_1, \alpha_2, t). \end{aligned} \tag{6.98}$$

It is related to the orthogonal Cartesian coordinates $\{x_1, x_2, x_3\}$ as

$$x_1 = r\cos\varphi, \qquad x_2 = r\sin\varphi, \qquad x_3 = z.$$

The position vector $\bar{r}$ of point $M(r, \varphi, z) \in S$ is given by

$$\bar{r} = r\bar{k}_1 + z\bar{k}3, \tag{6.99}$$

where

$$\bar{k}_1 = \bar{i}_1\cos\varphi + \bar{i}_2\sin\varphi, \qquad \bar{k}_2 = -\bar{i}_1\sin\varphi + \bar{i}_2\cos\varphi. \tag{6.100}$$

Differentiating Eq. (6.99) with respect to α_i with the help of (6.100), we find

$$\bar{r}_i = \frac{\partial \bar{r}}{\partial \alpha_i} = \frac{\partial r}{\partial \alpha_i}\bar{k}_1 + r\frac{\partial \bar{k}_1}{\partial \alpha_i} + \frac{\partial z}{\partial \alpha_i}\bar{k}_3 = \frac{\partial r}{\partial \alpha_i}\bar{k}_1 + r\frac{\partial \varphi}{\partial \alpha_i}\bar{k}_2 + \frac{\partial z}{\partial \alpha_i}\bar{k}_3, \tag{6.101}$$

where

$$\begin{aligned} \frac{\partial \bar{k}_1}{\partial \alpha_i} &= \frac{\partial}{\partial \alpha_i}\left(\bar{i}_1 \cos\varphi + \bar{i}_2 \sin\varphi\right) \\ &= \frac{\partial \varphi}{\partial \alpha_i}\left(-\bar{i}_1 \sin\varphi + \bar{i}_2 \cos\varphi\right) = \frac{\partial \varphi}{\partial \alpha_i}\bar{k}_2. \end{aligned} \tag{6.102}$$

Projections of $\bar{r}_i$ in the directions of the r, φ and z axes are given by

$$\bar{r}_{ir} = \frac{\partial r}{\partial \alpha_i}, \qquad \bar{r}_{i\varphi} = \frac{\partial \varphi}{\partial \alpha_i}, \qquad \bar{r}_{iz} = \frac{\partial z}{\partial \alpha_i}.$$

Hence,

$$\bar{r}_i = \bar{r}_{ik}\bar{k}_1 + \bar{r}_{i\varphi}\bar{k}_2 + \bar{r}_{iz}\bar{k}_3. \tag{6.103}$$

Decomposing $\bar{e}_i$ along the base $\{\bar{k}_1, \bar{k}_2, \bar{k}_3\}$, we find

$$\bar{e}_i = l_{ir}\bar{k}_1 + l_{i\varphi}\bar{k}_2 + l_{iz}\bar{k}_3, \tag{6.104}$$

where the direction cosines l_{ij} $(i=1, 2, 3; j=r, \varphi, z)$ are given by

$$\begin{aligned} \bar{e}_1\bar{e}_2 &= l_{1r}l_{2r} + l_{1\varphi}l_{2\varphi} + l_{1z}l_{2z} = \cos\overset{*}{\chi} = \overset{*}{a}_{12} \Big/ \sqrt{\overset{*}{a}_{11}\overset{*}{a}_{22}}, \\ \bar{e}_i\bar{e}_3 &= l_{ir}l_{3r} + l_{i\varphi}l_{3\varphi} + l_{iz}l_{3z} = 0, \\ \bar{e}_n\bar{e}_n &= l_{n1}^2 + l_{n2}^2 + l_{n3}^2 = 1 \qquad (n = 1, 2, 3). \end{aligned} \tag{6.105}$$

Expanding the resultant of external, $\bar{p}$, and mass, $\bar{f}$, forces in the directions of unit vectors $\bar{e}_i$, we get

$$\bar{p} = \bar{e}_1 p_1 + \bar{e}_2 p_2 + \bar{e}_3 p_3, \tag{6.106}$$

$$\bar{f} = \bar{k}_1 f_r + \bar{k}_2 f_\varphi + \bar{k}_3 f_z. \tag{6.107}$$

The vector of acceleration $\bar{a}\left(a_r, a_\varphi, a_z\right)$ in cylindrical coordinates is given by

$$a_r = -\frac{\mathrm{d}^2 r}{\mathrm{d}t^2} - r\left(\frac{\mathrm{d}\varphi}{\mathrm{d}t}\right)^2, \qquad a_\varphi = -r\frac{\mathrm{d}^2\varphi}{\mathrm{d}t^2} + 2r\frac{\mathrm{d}r}{\mathrm{d}t}\frac{\mathrm{d}\varphi}{\mathrm{d}t}, \qquad a_z = -\frac{\mathrm{d}^2 z}{\mathrm{d}t^2}. \tag{6.108}$$

Substituting Eqs. (6.104) and (6.106)–(6.108) into (6.84), the equations of motion of the soft shell in cylindrical coordinates take the form

$$\rho\sqrt{a}\left(\frac{\mathrm{d}^2 r}{\mathrm{d}t^2} - r\left(\frac{\mathrm{d}\varphi}{\mathrm{d}t}\right)^2\right) = \frac{\partial}{\partial\alpha_1}\left[\left(T^{1r}l_{1r} + T^{2r}l_{2r}\right)\sqrt{\overset{*}{a}_{22}}\right]$$
$$+\frac{\partial}{\partial\alpha_2}\left[\left(T^{1r}l_{1r} + T^{2r}l_{2r}\right)\sqrt{\overset{*}{a}_{11}}\right]$$
$$+p_3\left(l_{2\varphi}l_{2z} - l_{1z}l_{2\varphi}\right) + \left(p_1 l_{1r} + p_2 l_{2r}\right)\sqrt{\overset{*}{a}}$$
$$+\sqrt{\overset{*}{a}_{11}\overset{*}{a}_{22}} + \rho f_r\sqrt{a},$$
$$\rho\sqrt{a}\left(r\frac{\mathrm{d}^2\varphi}{\mathrm{d}t^2} + 2r\frac{\mathrm{d}r}{\mathrm{d}t}\frac{\mathrm{d}\varphi}{\mathrm{d}t}\right) = \frac{\partial}{\partial\alpha_1}\left[\left(T^{1\varphi}l_{1\varphi} + T^{1\varphi}l_{2\varphi}\right)\sqrt{\overset{*}{a}_{22}}\right]$$
$$+\frac{\partial}{\partial\alpha_2}\left[\left(T^{1\varphi}l_{1\varphi} + T^{2\varphi}l_{2\varphi}\right)\sqrt{\overset{*}{a}_{11}}\right]$$
$$+p_3\left(l_{1z}l_{2r} - l_{1r}l_{2z}\right)$$
$$+\left(p_1 l_{1\varphi} + p_2 l_{2\varphi}\right)\sqrt{\overset{*}{a}}\sqrt{\overset{*}{a}_{11}\overset{*}{a}_{22}} + \rho f_\varphi\sqrt{a},$$
$$\rho\sqrt{a}\frac{\mathrm{d}^2 z}{\mathrm{d}t^2} = \frac{\partial}{\partial\alpha_1}\left[\left(T^{1z}l_{1z} + T^{2z}l_{2z}\right)\sqrt{\overset{*}{a}_{22}}\right]$$
$$+\frac{\partial}{\partial\alpha_2}\left[\left(T^{1z}l_{1z} + T^{2z}l_{2z}\right)\sqrt{\overset{*}{a}_{11}}\right] + \left(p_1 l_{1z} + p_2 l_{2z}\right)\sqrt{\overset{*}{a}}$$
$$+p_3\left(l_{1r}l_{2\varphi} - l_{1\varphi}l_{2r}\right)\sqrt{\overset{*}{a}_{11}\overset{*}{a}_{22}} + \rho f_z\sqrt{a}. \tag{6.109}$$

Finally, to close a mathematical problem Eqs. (6.109) should be complemented by constitutive relations (6.60), (6.61) or (6.70), initial and boundary conditions, and the conditions given by Corollary 5.

Exercises

1. Give examples of the cut, undeformed and deformed configurations of a soft biological shell, namely the small intestine. Refer to the organ's anatomy and its accommodation in the abdominal cavity.
2. Verify Eqs. (6.10) and (6.11).
3. Confirm the relations given by Eqs. (6.20).
4. Verify Eqs. (6.31) and (6.32).
5. Verify Eqs. (6.39).
6. Verify Eqs. (6.48).
7. Confirm Eqs. (6.52).

8. Verify Eq. (6.57).
9. Show that the function $f_n(\ldots)$ (Eq. 6.60)) can be defined uniquely if $F_n(\ldots)$ (Eq. (6.61)) is known.
10. Show that $T_{1,2} = \frac{1}{2}I^{(\mathbf{T})}{}_1 \pm \sqrt{\frac{1}{4}I^{(\mathbf{T})2}_1 - I^{(\mathbf{T})2}_2}$.
11. Formulate a closed mathematical problem for a soft cylindrical shell, namely a functional unit of the small intestine.

7

Biomechanics of the stomach

Knowing is not enough. We must apply.

J.W. von Goethe

7.1 Anatomical and physiological background

The stomach is located in the left upper part of the abdomen immediately below the diaphragm. The shape of the organ is greatly modified by changes within itself and in the surrounding viscera such that no one form can be described as typical. The chief configurations are determined by the amount of the stomach contents, the stage of the digestive process, the degree of development of the gastric musculature and the condition of the adjacent loops of the small and large intestines. The stomach is more or less concave on its right side, convex on its left. The concave border is called the lesser curvature; the convex border, the greater curvature. The region that connects the lower oesophagus with the upper part of the stomach is called the cardia. The uppermost adjacent part to it is the fundus. The fundus adapts to the varying volume of ingested food and it frequently contains a gas bubble, especially after a meal. The largest part of the stomach is known simply as the body. The antrum, the lowermost part of the stomach, is usually funnel-shaped, with its narrow end connecting with the pyloric region. The latter empties into the duodenum – the upper division of the small intestine. The pyloric portion of the stomach tends to curve to the right and slightly upwards and backwards and thus gives the stomach its J-shaped appearance (Fig. 7.1).

The stomach functions as (i) an expansile reservoir that allows the rapid consumption of large meals; the process is facilitated by receptive relaxation of the proximal stomach in response to food and is called gastric accommodation; (ii) a digestive and absorptive organ that breaks down large protein and carbohydrate molecules and thus facilitates their absorption in the stomach and the small intestine; (iii) a part of the endocrine and immune system; it secretes hormones and neurotransmitters,

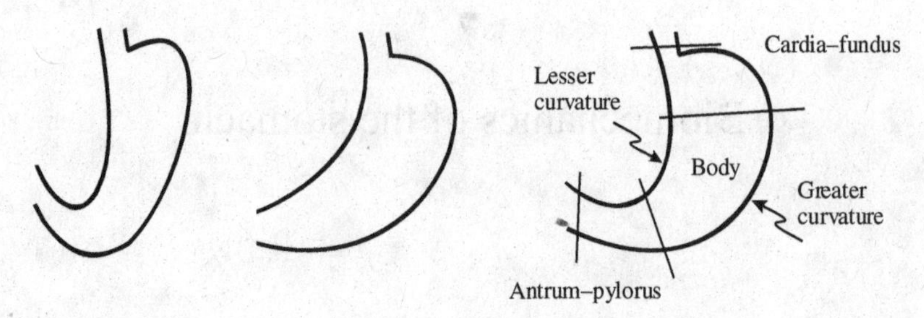

Fig. 7.1 Variations in anatomical shape of the human stomach.

e.g. gastrin, histamine, endorphins, somatostatin, serotonin and intrinsic factor; and (iv) a biomechanical system that grinds, mixes, forms and periodically discharges the preformed chyme into the duodenum as the physical and chemical condition of the mixture is rendered suitable for the next phase of digestion.

The effectiveness and diversity of physiological responses of the stomach to internal and external stimuli depend on the inherent activity of histomorphological elements, namely smooth muscle cells, neurons and interstitial cells of Cajal, and their topographical organization in gastric tissue (Suzuki, 2000; Hirst and Suzuki, 2006; Koh *et al.*, 2003; Carniero *et al.*, 1999). Smooth muscle cells are embedded into a network of collagenous and elastin fibres and are coupled via gap junctions into three distinct syncytia (muscle layers). The external longitudinal muscle layer continues from the oesophagus into the duodenum. The middle uniform circular layer is the strongest and completely covers the stomach. The circular fibres are best developed in the antrum and pylorus. At the pyloric end, the circular muscle layer greatly thickens to form the pyloric sphincter. The innermost oblique muscular layer is limited chiefly to the cardiac region.

Elastin and collagen fibres are structural proteins built in a three-dimensional supporting network. Elastin may be stretched to 250% of the unloaded configuration while collagen is relatively inextensible reinforced protein, and the main load-carrying element. Collagen fibres are undulated in the undeformed state and become stiff when straightened under the action of external applied loads. The overall strength of soft tissues is strongly correlated with their content (Cowin, 2000). The uttermost tunica serosa is formed of densely packed collagen and elastin fibres that coat the entire organ and thus provide its final shape.

The submucous coat and mucous membrane of the stomach consist of loose epithelial and glandular cells. Their main function is digestion and immune response. Their role in the biodynamics of the organ, i.e. force–stretch-ratio development, propulsion etc., is negligible.

Motor propulsive activity in the stomach originates in the upper part of the body of the organ. Three types of mechanical waves are observed. The first two types are small isolated contraction waves and peristaltic waves that slowly move from the point of origin down towards the pyloric sphincter. These types of contractions produce slight or deep indentations in the wall and serve as mixing, crushing and pumping mechanisms for the gastric contents. The third type of wave is non-propagating in nature and is a result of the tonic simultaneous contraction of all muscle layers that are normally superimposed on small and peristaltic contractions (Alvarez and Zimmermann, 1928).

The wide variety of rhythmical mechanical movements ranging from rapid phasic to slow sustained contractions is a result of the electromechanical coupling phenomenon. Although the mechanical component is highly variable, the electrical activity underlying it often shows remarkable similarities. Results from electrophysiological and molecular cloning studies suggest that L- and T-type Ca^{2+} channels, Ca^{2+}-activated K^{+}, selective K^{+}, Na^{+}, leak Cl^{-}, and non-selective cation channels are responsible for the electrical activity of smooth muscle syncytia represented by slow waves and spikes (Muraki *et al.*, 1991; Ou *et al.*, 2003; Lyford *et al.*, 2002; Lyford and Farrugia, 2003). Slow waves are low-amplitude 20–40-mV membrane potential oscillations occurring at a low frequency, 0.03–0.13 Hz. Their shape and dynamics differ from layer to layer: slow waves recorded from the circular smooth-muscle layer have a triangular shape whereas those from the longitudinal layer exhibit a square shape. Slow waves progress at a low velocity of 2.5–3.5 mm/s in the oro-anal direction and at a higher velocity of 13.9–14.7 mm/s in the circumferential direction. Spikes are high-amplitude 35–45-mV and low-duration action potentials that occur on the crests of slow waves. The electrical properties of gastric smooth muscle are region-dependent. For example, the smooth muscles of the fundus are electrically quiescent, bundles of muscle from the corpus generate slow waves, while those from the pyloric region produce slow waves and spikes on crests of slow waves. Antrum muscle fibres generate slow waves with an initial spike-like component.

Slow-wave and spiking activities in the smooth muscle of the gastrointestinal tract comprise the fundamental physiological phenomenon called the migrating myoelectrical complex. It is observed during the fasting state and has four distinct phases: a quiescent phase, 1, when only slow waves are recorded; a transitional phase, 2, when irregular spikes and smooth-muscle contractions are detected; a phase of regular prolonged spiking activity, 3, which is characterized by intensive contractions; and a further transitional phase, 4, which is similar to phase 2 in appearance. The rate of propagation of the myoelectric complex is approximately 5–10 mm/s. After feeding, the pattern of myoelectrical activity changes to a random-firing–propulsive type with no quiescent periods. It lasts for 6–8 hours until the stomach has emptied its content.

The role of pacemakers – initiators of myoelectrical activity – concerns the interstitial cells of Cajal (ICC). Each individual ICC is spontaneously active and generates square-shaped high-amplitude potentials of 50–100 mV. Their discharge always precedes electrical events in gastric muscles. The ion channels responsible for electrical activity comprise L- and T-type Ca^{2+} channels, large-conductance Ca^{2+}-activated K^+ channels, voltage-dependent K^+ channels, including inward rectifier K^+ channels, Na^+ channels, large-conductance Cl^- channels, and non-selective cation channels expressing transient receptor potential. The cells are located within the myenteric nervous plexus (ICC_{MY}) and smooth muscle syncytia (ICC_{IM}) at the distal end of the corpus and spread over the antrum and pylorus (Dickens *et al.*, 2001; Hennig *et al.*, 2004; Hirst *et al.*, 2006). Their density is highest near the greater curvature and falls towards the lesser curvature of the stomach. The ICC form multiple extensive networks that are coupled to the circular and longitudinal smooth muscle syncytia via gap junctions (ICC_{IM}) and to the neurons of the intrinsic nervous plexi via electrochemical synapses (ICC_{MY}). Thus, the three types of cells form an array of interconnected networks and electrical signals originated in one cell can spread to surrounding cells and tissues. Although they are intrinsically active, ICC per se cannot sustain and distribute pacemaker activity. Slow waves decay within a few millimetres in tissues lacking ICC_{MY}. Intact networks as described above are crucial for generation and non-decremental propagation of slow waves over long distances through the gastrointestinal organs.

In the intact stomach, slow waves commonly originate in the corpus and propagate along the antrum to the pylorus. However, the pacemaker site is not constant but changes its location within the organ. Thus, low-frequency irregular rhythmical slow-wave activity is always present in the antral and pyloric regions, which are classified as secondary and tertiary pacemaker sites. When all these regions are coupled together, activity in them occurs at the higher frequency entrained by corporal ICC_{IM} that play the dominant pacemaker role in the stomach.

Contractions in the smooth muscle are initiated by phosphorylation of the myosin light chain via the activation of Ca^{2+} calmodulin-dependent myosin-light-chain kinase. The key player in the process is free cytoplasmic Ca^{2+}. In the resting state, the concentration of intracellular $Ca^{2+} \approx 0.1$ μM, which is insufficient to trigger the chain of reactions leading to mechanical contraction. Upon stimulation, the concentration of free intracellular calcium rises, mainly as a result of influx and partly following the mobilization of internal stores.

There are at least four different pathways for calcium influx. The Ca^{2+} entering the cell through resting influx or leak mechanisms is taken up by storage sites, presumably the sacroplasmic reticulum, and does not directly increase the cytoplasmic free $[Ca^{2+}]$. Calcium entering the cell through L-type Ca^{2+} channels is partly accumulated in the storage sites while the rest reaches the cytoplasm and activates contractile filaments. The Ca^{2+} entering through T-type Ca^{2+} channels does not fill the store because the storage site cannot accumulate or retain calcium during bursting. Thus, all the calcium entering through voltage-dependent Ca^{2+} channels is available to activate contractile filaments directly. An additional source of the internal Ca^{2+} rise is the calcium-induced release of Ca^{2+} from the sarcoplasmic reticulum, which can be initiated by Ca^{2+} influx, caffeine and/or activation of the second messenger system, e.g. inositol triphosphate.

Disruptions in the chain of regulatory mechanisms result in disorganized movements of the stomach and its contents. The conditions are clinically described as functional dyspepsia and include gastroparesis, antral hypomotility, gastric dysrhythmia, tachygastria and abnormal tone, among others. For example, delayed emptying of the stomach may be a result of (i) disorganized contractions of the antro-pyloric part, (ii) partial or complete obstruction of the pyloric sphincter and/or (iii) a long-standing illness, e.g. Parkinson's disease, diabetes or amyloidosis. The pathophysiology of these processes is poorly understood and thus their treatment remains a medical challenge. A study of gastric function from a biomechanical perspective may offer unique insights and will provide otherwise-inaccessible information on intrinsic physiological events. Such an approach will have enormous implications for our understanding of the mechanisms of certain diseases, for improving diagnostic accuracy and for the planning of therapeutic interventions.

7.2 Constitutive relations for the tissue

The mechanical properties of the wall of the stomach are highly specific and depend on the topographical site and species, in addition to being greatly influenced by food, environmental factors and age. *In vitro* uniaxial tension tests, conducted along two structurally anisotropic directions on specimens collected from different regions of the organ, convincingly demonstrate that the tissue of the wall has nonlinear, viscoelastic properties. Since the experiments were performed on segments removed from the host, it was assumed that the muscle fibres were fully relaxed and the mechanical contribution was attributed to mechanochemically inert components of smooth muscle cells together with elastin and collagen fibres. To describe it quantitatively, the total force in the tissue can be decoupled into the passive and active components,

$$T_{\mathrm{c,l}} = T^{\mathrm{p}}_{\mathrm{c,l}}(\lambda_{\mathrm{c}}, \lambda_{\mathrm{l}}, c_1, \ldots, c_9) + T^{\mathrm{a}}_{\mathrm{c,l}}(\lambda_{\mathrm{c}}, \lambda_{\mathrm{l}}, Z^{(*)}_{mn}, [\mathrm{Ca}^{2+}], c_{10}, \ldots, c_{14}), \tag{7.1}$$

where $c_1, \ldots, c_{14}$ are empirically estimated material constants of the tissue, $Z^{(*)}_{mn}$ is the 'biofactor' (see Chapter 4), $[\mathrm{Ca}^{2+}]$ is the concentration of intracellular calcium and subscripts l and c denote the longitudinal and circumferential layers.

Assuming homogeneity of the stress and strain field and incompressibility of the tissue, the force and stretch ratios were calculated. Analysis of the force–stretch-ratio curves ($T^{\mathrm{p}}_{\mathrm{c,l}} - \lambda_{\mathrm{c,l}}$) showed a characteristic 'triphasic' response with a nonlinear transition between the low and high elastic states (Fig. 7.2). The tissue was highly

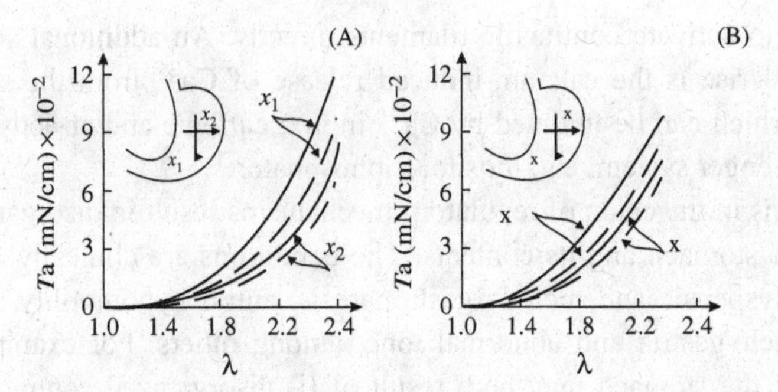

Fig. 7.2 Uniaxial constitutive relations for the human stomach. Solid lines refer to the anterior and dashed lines to the posterior wall of the organ: (A) age group 20–29 years; (B) age group 40–49 years.

compliant at low levels of stretching, $1.0 < \lambda_{c,l} \leq 1.2$, with the force values varying in the range $T^{p}_{c,l} \sim 0$–38 mN/cm, followed by a highly nonlinear transitory state, $1.2 < \lambda_{c,l} \leq 1.6$ and $T^{p}_{c,l} \sim 38$–377 mN/cm. For higher levels of stretching $1.6 < \lambda_{c,l} \leq 2.4$ specimens demonstrated pure linear elastic behaviour. In both directions the ultimate stretch ratios are similar $\lambda_{max,l} = \lambda_{max,c} = 2.4 \pm 0.1$ (Miftakhov, 1983a).

In general the wall of the stomach is stronger and stiffer longitudinally than circumferentially. Thus, the maximum loads the tissue specimens can withstand are $T^{p}_{max,l} = 1691 \pm 340$ mN/cm, and $T^{p}_{max,c} = 1175 \pm 181$ mN/cm. Comparison of experimental curves obtained from various regions of the organ confirmed the property of transverse curvilinear anisotropy. It is noteworthy that, while insignificant differences between the loading and unloading curves were present due to 'biological hysteresis', the force–stretch-ratio responses were independent of the stretching rate.

The uniaxial force–stretch-ratio approximation of data in the preferred axes of structural anisotropy yields

$$T^{p}_{(c,l)} = \begin{cases} 0, & \lambda_{(c,l)} \leq 1, \\ c_1 \left[\exp(c_2(\lambda_{(c,l)} - 1)) - 1\right], & \lambda_{(c,l)} > 1. \end{cases} \tag{7.2}$$

where the meanings of the parameters and constants are as described above.

The biaxial tests conducted on square-shaped tissue specimens collected from three different regions of the organ (i.e. the fundus, the body and the antro-pyloric regions) allowed us to deduce full in-plane mechanical properties of the wall of the stomach (Fig. 7.3) (Miftakhov, 1983b; Miftakhov, 1985). The edges of the specimens were aligned parallel and perpendicular to the orientation of the longitudinal and circular smooth muscle fibres. The experimental protocol applied in order to obtain quasi-static force–stretch-ratio curves used constant $\lambda_l : \lambda_c$ stretch ratios. The tissue under biaxial loading exhibits a complex response including nonlinear pseudo-elasticity, transverse anisotropy and finite deformability, and no dependence on the stretch rate.

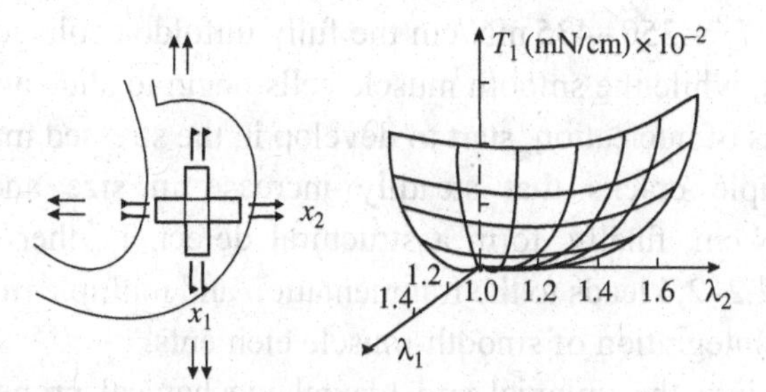

Fig. 7.3 A representative biaxial constitutive relation for the human stomach.

The $T^{\mathrm{p}}_{\mathrm{c,l}}(\lambda_{\mathrm{c}}, \lambda_{\mathrm{l}})$ curves show that, as the stretch ratio λ_{c} increases gradually from 1 to 1.4, the extensibility along λ_{l} decreases from 1.9 to 1.5. There is a concomitant increase in the stiffness of the biomaterial. For $\lambda_{\mathrm{c,l}} > 1.4$ and $1 \leq \lambda_{\mathrm{l,c}} \leq 1.6$ the force–stretch-ratio curves displayed linear relations. The maximum force the tissue bears during the biaxial tests is $T_{\mathrm{max\ c,l}}\ (\lambda_{\mathrm{c}}, \lambda_{\mathrm{l}}) = 2200 \pm 180$ mN/cm and $\lambda_{\mathrm{max\ c,l}} = 1.5–2.3$, which depends on the ratio $\lambda_{\mathrm{l}}:\lambda_{\mathrm{c}}$. Experiments showed that the shear force applied to the tissue was significantly less, $10^{-2}T^{\mathrm{p}}_{\mathrm{max\ c,l}}$, than the stretch force.

The in-plane passive $T^{\mathrm{p}}_{\mathrm{c,l}}$ forces under biaxial loading are calculated as

$$T^{\mathrm{p}}_{\mathrm{c,l}} = \frac{\partial(\rho W)}{\partial(\lambda_{\mathrm{c,l}} - 1)}. \tag{7.3}$$

Here W is the strain energy density function of the connective tissue network and for passive muscle tissue

$$\rho W = \frac{1}{2}\Big[c_3(\lambda_l - 1)^2 + 2c_4(\lambda_l - 1)(\lambda_c - 1) + c_4(\lambda_c - 1)^2 + c_6 \exp\Big(c_7(\lambda_l - 1)^2 + c_8(\lambda_c - 1)^2 + 2c_9(\lambda_l - 1)(\lambda_c - 1)\Big)\Big]; \tag{7.4}$$

where ρ is the density of the undeformed tissue.

Microscopic analysis of the dynamics of crack nucleation and crack growth within the wall of the stomach revealed that first small randomly oriented cracks occur along the cell contacts in the mucosal and submucosal layers at $1.1 < \lambda_{\mathrm{l}} \leq 1.3$ and $T^{\mathrm{p}}_{\mathrm{l}} \sim 20–30$ mN/cm (Fig. 7.4) (Miftakhov, 1981). They align perpendicularly to the axis of the applied force. Uncurling and reorientation of collagen and elastin fibres occur in the submucosal layer. Smooth muscle cells behave passively.

Collagen and elastin fibres become fully straightened at $\lambda_{\mathrm{l}} = 1.4–1.5$ and $T^{\mathrm{p}}_{\mathrm{l}} = 120–135$ mN/cm. There is a disruption in the dense packaging of the fibrillary matrix with the development of multiple small fractures of elastin fibres. Significant damage appears in the mucosa at $\lambda_{\mathrm{l}} = 1.4$ with the development of a fracture growing outside the thickness of the wall. The total rupture of the mucosal layer occurs at $\lambda_{\mathrm{l}} = 1.5$, $T^{\mathrm{p}}_{\mathrm{l}} \sim 130$ mN/cm.

At $\lambda_1 \sim 1.5$ and $T_1^p = 350$–435 mN/cm the fully unfolded collagen fibres bear the main stretch load, while the smooth muscle cells begin to slide against each other. Small pores, areas of nucleation, start to develop in the stressed muscle layer. They grow into multiple cracks that steadily increase in size and, at $\lambda_1 \sim 1.8$–2, $T_1^p = 650$–700 mN/cm, finally, form a structural defect. Further extension to the levels of $\lambda_{max,1} \sim 2.2$–2.4 leads to the fragmentation and multiple rupture of collagen fibres and the disintegration of smooth-muscle elements.

Investigations into the uniaxial and biaxial mechanical properties of actively contracting tissue remain a challenging area in biomechanics. Up to the time at

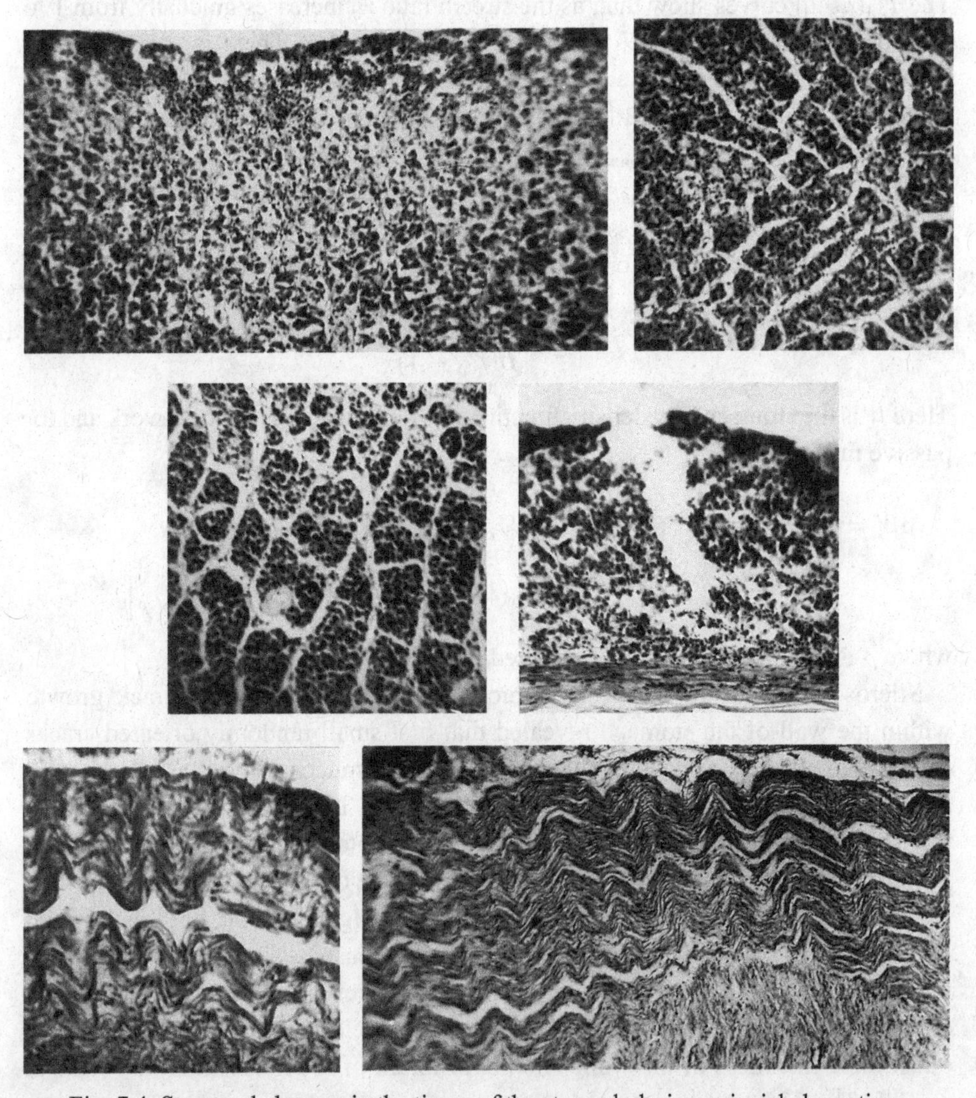

Fig. 7.4 Structural changes in the tissue of the stomach during uniaxial elongation.

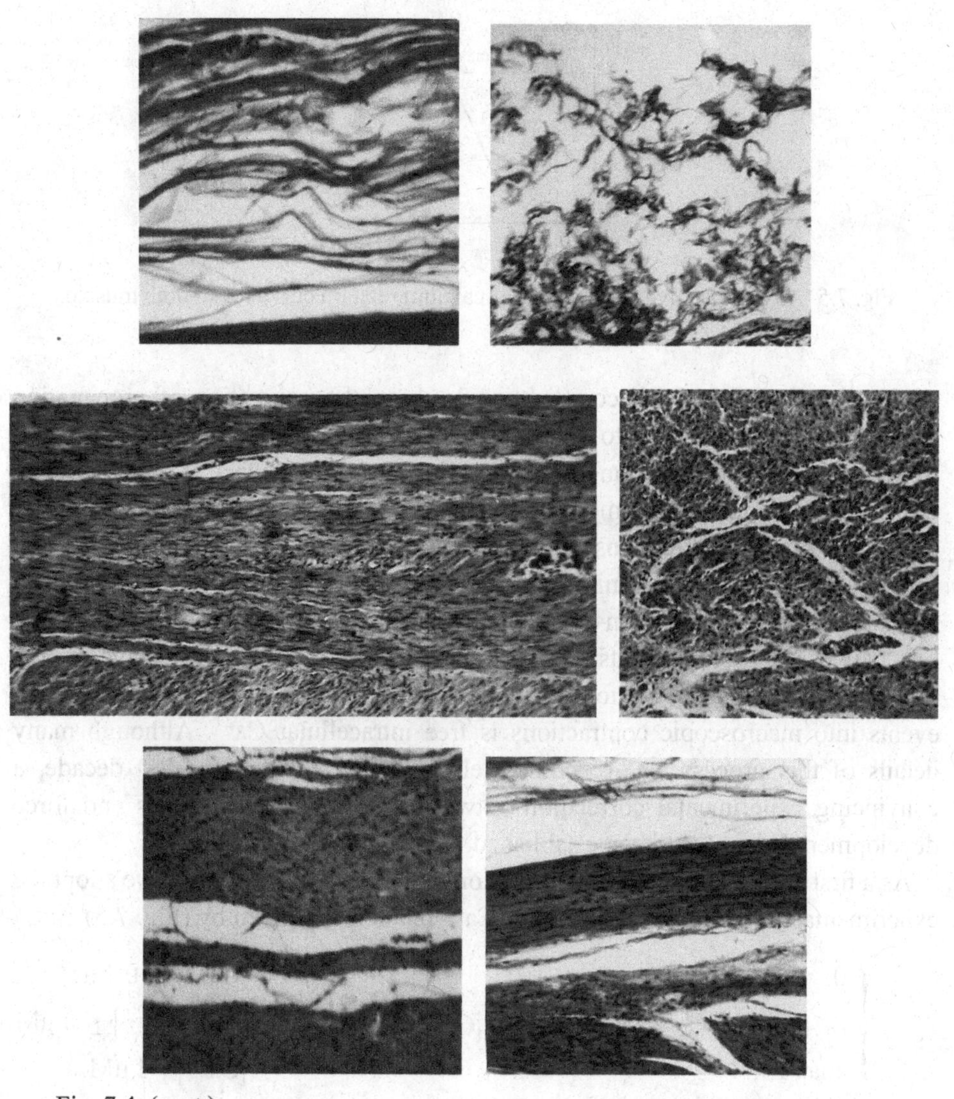

Fig. 7.4 (cont.)

which this book was being written there were no experimental data available on the in-plane active behaviour of the wall of the stomach. The main problem is the need to keep specimens physiologically viable and stable, in order for *in vitro* samples to reproduce myoelectrical patterns that are consistent with those observed *in vivo*. Thus, it is practically impossible to sustain and to control simultaneously the slow-wave, spiking and contractile activity of smooth muscle syncytia.

Current theories of the motility of the gastrointestinal tract suggest that there are reciprocal mechanical relationships between the longitudinal and circular muscle syncytia. Contractions of the longitudinal muscle layer always precede contractions

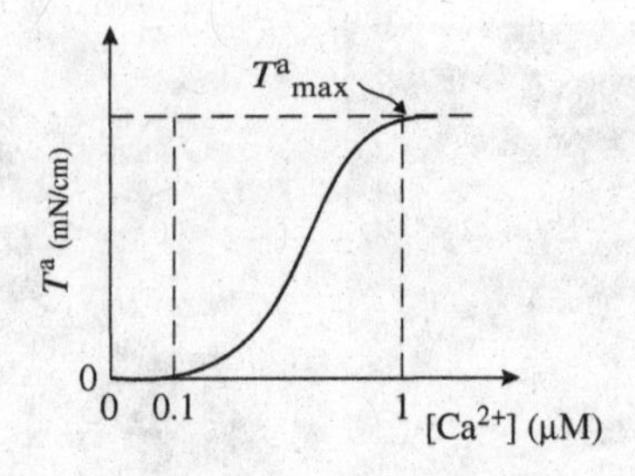

Fig. 7.5 A typical active-force–internal-calcium relation curve for smooth muscle.

of the circular layer. Such coordination leads to the generation of propagating peristaltic waves, in contrast to a non-propagating spastic type of activity that results from simultaneous contraction of both muscle syncytia. This fact, together with the fine fibrillar structure of the smooth muscle, suggests that active forces are produced only in the preferred directions, either longitudinal or circumferential, and thus can be characterized in full by uniaxial tests. Further, constructive modelling requires formulation of the excitation–contraction-coupling phenomenon that provides a link between electrical events and the muscle mechanics. The key player in the dynamics of the transformation of microscopic properties of electrical excitatory events into macroscopic contractions is free intracellular Ca^{2+}. Although many details of this process have become well established within the last decade, a convincing experimental correlation between ultra-structural changes and force development has not yet been established.

As a first approximation of the excitation–contraction phenomenon we adopt the experimental active-force–intracellular-Ca^{2+} relationship given by (Fig. 7.5)

$$T^{\mathrm{a}}_{\mathrm{c,l}} = \begin{cases} 0, & [\mathrm{Ca}^{2+}] \leq 0.1\ \mu\mathrm{M}, \\ c_7 + c_8[\mathrm{Ca}^{2+}]^4 + c_9[\mathrm{Ca}^{2+}]^3 + c_{10}[\mathrm{Ca}^{2+}]^2 + c_{11}[\mathrm{Ca}^{2+}], & 0.1 < [\mathrm{Ca}^{2+}] \leq 1\ \mu\mathrm{M}, \\ \max T^{\mathrm{a}}, & [\mathrm{Ca}^{2+}] > 1\ \mu\mathrm{M}, \end{cases} \tag{7.5}$$

where $[Ca^{2+}]$ is the intracellular concentration of calcium ions.

The myoelectrical processes in the tissue of the stomach are governed by voltage-dependent Ca^{2+} channels (L- and T-type), mixed Ca^{2+}–K^+ and K^+ channels and chloride channels. Following the general principles of the Hodgkin–Huxley formalism, the system of equations of the dynamics of the membrane potential $V_{c,l}$ is described as

$$\lambda C_{\mathrm{m}} \frac{\mathrm{d}V_{\mathrm{c,l}}}{\mathrm{d}t} = -\sum_j \tilde{I}_j, \tag{7.6}$$

where $\tilde{\lambda}$ is the numerical parameter, C_m is the muscle-cell membrane capacitance, and $\tilde{I}_j$ are the fast and slow inward Ca^{2+}, Ca^{2+}-activated K^+, voltage-dependent K^+ and leak Cl^- currents given by

$$\begin{aligned}
\tilde{I}^{f}_{Ca} &= \tilde{g}^{f}_{Ca}\tilde{m}^{3}_{I}\tilde{h}\left(V_{c,l} - \tilde{V}_{Ca}\right),\\
\tilde{I}^{s}_{Ca} &= \tilde{g}^{s}_{Ca}\tilde{x}_{Ca}\left(V_{c,l} - \tilde{V}_{Ca}\right),\\
\tilde{I}_{K} &= \tilde{g}_{K}\tilde{n}^{4}\left(V_{c,l} - \tilde{V}_{Ca}\right),\\
\tilde{I}_{Ca-K} &= \frac{\tilde{g}^{f}_{Ca-K}\left[Ca^{2+}\right]\left(V_{c,l} - \tilde{V}_{Ca}\right)}{0.5 + \left[Ca^{2+}\right]},\\
\tilde{I}_{Cl} &= \tilde{g}_{Cl}\left(V_{c,l} - \tilde{V}_{Ca}\right).
\end{aligned} \tag{7.7}$$

Here $\widetilde{V}_{Ca}$, $\widetilde{V}_{K}$ and $\widetilde{V}_{Cl}$ are the reversal potentials, $\tilde{g}^{f}_{Ca}, \tilde{g}^{s}_{Ca}, \tilde{g}_{K}, \tilde{g}_{Ca-K}$ and $\tilde{g}_{Cl}$ are the maximal conductances for the respective ion currents and $\tilde{m}_I, \tilde{h}, \tilde{n}$ and $\tilde{x}_{Ca}$ are dynamic variables described by

$$\begin{aligned}
\widetilde{m}_I &= \frac{\widetilde{\alpha}_m}{\widetilde{\alpha}_m + \widetilde{\beta}_m},\\
\tilde{\lambda}\tilde{h}\frac{d\tilde{h}}{dt} &= \tilde{\alpha}_h(1 - \tilde{h}) - \tilde{\beta}_h\tilde{h},\\
\tilde{\lambda}\tilde{h}\frac{d\tilde{n}}{dt} &= \tilde{\alpha}_n(1 - \tilde{n}) - \tilde{\beta}_n\tilde{n},\\
\tilde{\lambda}\tau_{xCa}\frac{d\tilde{x}_{Ca}}{dt} &= \frac{1}{\exp(-0.15(V_{c,l} + 50))} - \tilde{x}_{Ca},\\
\tilde{\lambda}\frac{d\left[Ca^{2+}\right]}{dt} &= \wp_{Ca}\tilde{x}_{Ca}\left(\tilde{V}_{Ca} - V_{c,l}\right) - \left[Ca^{2+}\right],
\end{aligned} \tag{7.8}$$

where the activation $\widetilde{\alpha}_y$ and deactivation $\widetilde{\beta}_y$ ($y = m, n, h$) parameters of ion channels satisfy the following empirical relations:

$$\begin{aligned}
\widetilde{\alpha}_m &= \frac{0.1\left(50 - \widetilde{V}_{c,l}\right)}{\exp\left(5 - 0.1\widetilde{V}_{c,l}\right) - 1}, & \widetilde{\beta}_m &= 4\exp\left(\frac{25 - \widetilde{V}_{c,l}}{18}\right),\\
\widetilde{\alpha}_h &= 0.07\exp\left(\frac{25 - 0.1\widetilde{V}_{c,l}}{20}\right), & \widetilde{\beta}_h &= \left[1 + \exp\left(5.5 - 0.1\widetilde{V}_{c,l}\right)\right]^{-1},\\
\widetilde{\alpha}_n &= \frac{0.01\left(55 - \widetilde{V}_{c,l}\right)}{\exp\left(5.5 - 0.1\widetilde{V}_{c,l}\right) - 1}, & \widetilde{\beta}_n &= 0.125\exp\left(\frac{45 - \widetilde{V}_{c,l}}{80}\right).
\end{aligned} \tag{7.9}$$

Here $\widetilde{V}_{c,l} = (127V_{c,l} + 8265)/105$, τ_{xCa} is the time constant and $\wp_{Ca}$ is the parameter referring to the dynamics of calcium channels; $\hbar$ is a numerical constant.

The evolution of L-type voltage-dependent Ca^{2+} channels is defined by

$$\widetilde{g}^{s}_{Ca} = \delta(V_{c,l})\widetilde{g}^{s}_{Ca}, \tag{7.10}$$

where

$$\delta(V_{c,l}) = \begin{cases} 1, & \text{for } \delta(V_{c,l}) \geq p(c,l), \\ 0, & \text{otherwise.} \end{cases}$$

The pacemaker–ICC dynamics V_i is given by

$$\alpha C_m \frac{dV_i}{dt} = -\sum_j I_j + I_{ext(i)}, \tag{7.11}$$

where C_m is the membrane capacitance, I_j ($j = Ca^{2+}, Ca^{2+}\text{–}K^+, Na^+, K^+, Cl^-$) are ion currents carried through different ion channels, $I_{ext(i)} = V_i/R_{ICC}$ is the external membrane current and R_{ICC} is the input cellular resistance. The ion currents are defined by

$$\begin{aligned} I_{Ca} &= \frac{g_{Ca(i)} z_{Ca} (V_i - V_{Ca})}{1 + \vartheta_{Ca}[Ca^{2+}]_i}, \\ I_{Ca-K} &= \frac{g_{Ca-K(i)} \rho_\infty (V_i - V_{Ca-K})}{0.5 + [Ca^{2+}]_i}, \\ I_{Na} &= g_{Na(i)} m^3_{Na} h_{Na} (V_i - V_{Na}), \\ I_K &= g_{K(i)} n^4_K (V_i - V_K), \\ I_{Cl} &= g_{Cl(i)} (V_i - V_{Cl}), \end{aligned} \tag{7.12}$$

where $V_{Ca}, V_{Ca-K}, V_{Na}, V_K$ and V_{Cl} are the reversal potentials and $g_{Ca(i)}, g_{Ca-K(i)}, g_{Na(i)}, g_{K(i)}$ and $g_{Cl(i)}$ are the maximal conductances of voltage-dependent Ca^{2+} channels (N-type), Ca^{2+}-activated K^+ channels, Na^+ channels, K^+ channels and leak Cl^- channels, ϑ_{Ca} is the parameter of calcium inhibition of the Ca^{2+} channels, $[Ca^{2+}]_i$ is the intracellular (ICC) concentration of free calcium that yields

$$\frac{d[Ca^{2+}]_i}{dt} = \frac{0.2 z_{Ca} (V_i - V_{Ca})}{1 + \vartheta_{Ca}[Ca^{2+}]_i} - 0.3[Ca^{2+}]_i \tag{7.13}$$

and z_{Ca}, ρ_∞, m_{Na}, h_{Na} and n_K represent dynamic variables of the ion channels given by

$$\begin{aligned}
dz_{Ca}/dt &= (z_\infty - z_{Ca})/\tau_z, \\
dh_{Na}/dt &= \lambda_h (h_\infty - h_{Na})/\tau_h, \\
dn_K/dt &= \lambda_n (n_\infty - n_K)/\tau_n, \\
m_{Na} &= m_\infty(V_i), \\
\rho_\infty &= [1 + \exp(0.15(V_i + 47))]^{-1}.
\end{aligned} \tag{7.14}$$

In the above $m_\infty, h_\infty, n_\infty$ and z_∞ are calculated as

$$\begin{aligned}
y_\infty &= \frac{\alpha_{y\infty}}{\alpha_{y\infty} + \beta_{y\infty}} \quad (y = m, h, n), \\
z_\infty &= [1 + \exp(-0.15(V_i + 42))]^{-1},
\end{aligned} \tag{7.15}$$

where

$$\begin{aligned}
\alpha_{m\infty} &= \frac{0.12(V_i + 27)}{1 - \exp(-(V_i + 27)/8)}, \qquad \beta_{m\infty} = 4\exp(-(V_i + 47)/15), \\
\alpha_{h\infty} &= 0.07\exp(-(V_i + 47)/17), \qquad \beta_{h\infty} = [1 + \exp(-(V_i + 22)/8)]^{-1}, \\
\alpha_{n\infty} &= \frac{0.012(V_i + 12)}{1 - \exp(-(V_i + 12)/8)}, \qquad \beta_{n\infty} = 0.125\exp(-(V_i + 20)/67).
\end{aligned}$$

The discharge of the pacemaker initiates the electrical wave of depolarization V_1^s in the longitudinal muscle syncytium. Electrophysiological extracellular recordings of the dynamics of the propagation of V_1^s revealed anisotropic electrical properties of the syncytium. The dynamics of V_1^s is described by Eq. (4.42),

$$C_m \frac{\partial V_1^s}{\partial t} = I_{m1}(\alpha_1, \alpha_2) + I_{m2}(\alpha_1 - \alpha_1', \alpha_2 - \alpha_2') + I_{ion}, \tag{7.16}$$

where I_{m1} and I_{m2} are the transmembrane currents described by Eqs. (4.38) and (4.41). The intracellular and extracellular conductivities $\hat{g}_{i_{(o)}}$ of the syncytium is defined by

$$\hat{g}_{i_{(o)}} := 1/R^m_{i(o)}, \tag{7.17}$$

where $R^m_{i_{(o)}}$ is the intracellular (subscript i) and extracellular (o) smooth-muscle-membrane resistance. According to Ohm's law

$$R^m_{i(o)} = \frac{R^{ms}_{i(o)}\lambda_{c,l}}{\tilde{S}_{s,l}}, \tag{7.18}$$

where $\lambda_{c,l}$ are the stretch ratios and $\tilde{S}_{c,l}$ are the cross-sectional areas of smooth muscle layers and $R^{ms}_{i(o)}$ is the specific smooth-muscle-membrane resistance. By substituting Eq. (7.16) into (7.15) and assuming that $\tilde{S}_{\mathrm{c,l}}$ is constant throughout deformation, we obtain

$$\hat{g}_{\mathrm{i(o)}} = \frac{\tilde{S}_{\mathrm{c,l}}}{R^{\mathrm{m}}_{\mathrm{i(o)}}\lambda_{\mathrm{c,l}}} := \frac{\hat{g}^*_{\mathrm{i(o)}}}{\lambda_{\mathrm{c,l}}}, \tag{7.19}$$

where $\hat{g}^*_{\mathrm{i(o)}}$ have the meanings of maximal intracellular and interstitial-space conductivities. By substituting (7.17) into (4.38) and (4.41), we find

$$\begin{aligned}
I_{\mathrm{m}1}(\alpha_1,\alpha_2) &= M_{\mathrm{vs}}\left\{\frac{2(\mu_{\alpha_2}-\mu_{\alpha_1})}{(1+\mu_{\alpha_1})(1+\mu_{\alpha_1})}\tan^{-1}\left(\frac{\mathrm{d}\alpha_1}{\mathrm{d}\alpha_2}\sqrt{\frac{G_{\alpha_2}}{G_{\alpha_1}}}\right)+\frac{\hat{g}^*_{\mathrm{o},\alpha_2}}{G_{\alpha_1}}\right\}\\
&\quad\times\left(\frac{\partial}{\partial\alpha_1}\left(\frac{\hat{g}^*_{\mathrm{o},\alpha_1}}{\lambda_{\mathrm{c}}}\frac{\partial V_{\mathrm{l}}^{\mathrm{s}}}{\partial\alpha_1}\right)+\frac{\partial}{\partial\alpha_2}\left(\frac{\hat{g}^*_{\mathrm{o},\alpha_2}}{\lambda_{\mathrm{l}}}\frac{\partial V_{\mathrm{l}}^{\mathrm{s}}}{\partial\alpha_2}\right)\right),\\
I_{m2}(\alpha_1,\alpha_2) &= M_{\mathrm{vs}}\iint\limits_S \frac{\mu_{\alpha_1}-\mu_{\alpha_2}}{2\pi(1+\mu_{\alpha_1})(1+\mu_{\alpha_1})}\frac{(\alpha_2-\alpha_2')/G_{\bar{s}_2}-(\alpha_1-\alpha_1')/G_{\alpha_1}}{\left[(\alpha_1-\alpha_1')/G_{\bar{s}_1}-(\alpha_2-\alpha_2')/G_{\alpha_2}\right]^2}\\
&\quad\times\left(\frac{\partial}{\partial\alpha_1}\left(\frac{\hat{g}^*_{\mathrm{o},\alpha_1}}{\lambda_{\mathrm{c}}}\frac{\partial V_{\mathrm{l}}^{\mathrm{s}}}{\partial\alpha_1}\right)+\frac{\partial}{\partial\alpha_2}\left(\frac{\hat{g}^*_{\mathrm{o},\alpha s_2}}{\lambda_{\mathrm{l}}}\frac{\partial V_{\mathrm{l}}^{\mathrm{s}}}{\partial\alpha_2}\right)\right)\mathrm{d}\alpha_1'\,\mathrm{d}\alpha_2',\\
\mu_{\alpha_1} &= \hat{g}^*_{\mathrm{o},\alpha_1}/\hat{g}^*_{\mathrm{i},\alpha_1}, \qquad \mu_{\alpha_2} = \hat{g}^*_{\mathrm{o},\alpha_2}/\hat{g}^*_{\mathrm{i},\alpha_2},\\
G_{\alpha_1} &= \frac{\hat{g}^*_{\mathrm{o},\alpha_1}+\hat{g}^*_{\mathrm{i},\alpha_1}}{\lambda_{\mathrm{c}}}, \qquad G_{\alpha_2} = \frac{\hat{g}^*_{\mathrm{o},\alpha_2}+\hat{g}^*_{\mathrm{i},\alpha_2}}{\lambda_{\mathrm{l}}}, \qquad G=\sqrt{G_{\alpha_1}G_{\alpha_2}}.
\end{aligned} \tag{7.20}$$

where α_1 and α_2 are the Lagrange coordinates of the longitudinal and circular smooth muscle syncytia, respectively, and the meanings of other parameters are as described above.

The total ion current I_{ion} is given by

$$I_{\mathrm{ion}} = \tilde{g}_{\mathrm{Na}}\hat{m}^3\hat{h}(V_{\mathrm{l}}^{\mathrm{s}}-\tilde{V}_{\mathrm{Na}})+\tilde{g}_{\mathrm{K}}\hat{n}^4(V_{\mathrm{l}}^{\mathrm{s}}-\tilde{V}_{\mathrm{K}})+\tilde{g}_{\mathrm{Cl}}(V_{\mathrm{l}}^{\mathrm{s}}-\tilde{V}_{\mathrm{Cl}}), \tag{7.21}$$

where $\tilde{g}_{\mathrm{Na}}, \tilde{g}_{\mathrm{K}}$ and $\tilde{g}_{\mathrm{Cl}}$ represent maximal conductances, whereas $\tilde{V}_{\mathrm{Na}}, \tilde{V}_{\mathrm{K}}$ and $\tilde{V}_{\mathrm{Cl}}$ denote reversal potentials of Na^+, K^+ and Cl^- currents. The dynamics of change in the probability variables $\hat{m}, \hat{h}$ and $\hat{n}$ of opening of the ion gates are obtained from

$$\begin{aligned}
\frac{\mathrm{d}\hat{m}}{dt} &= \hat{\alpha}_m(1-\hat{m})-\hat{\beta}_m\hat{m},\\
\frac{\mathrm{d}\hat{h}}{\mathrm{d}t} &= \hat{\alpha}_h(1-\hat{h})-\hat{\beta}_h\hat{h},\\
\frac{\mathrm{d}\hat{n}}{\mathrm{d}t} &= \hat{\alpha}_n(1-\hat{n})-\hat{\beta}_n\hat{n}.
\end{aligned} \tag{7.22}$$

The activation $\hat{\alpha}_y$ and deactivation $\hat{\beta}_y$ $(y = \hat{m}, \hat{h}, \hat{n})$ parameters are given by

$$\hat{\alpha}_m = \frac{0.005(V_1^s - \tilde{V}_m)}{\exp(0.1(V_1^s - \tilde{V}_m)) - 1}, \qquad \hat{\beta}_m = 0.2\exp\left(\frac{V_1^s + \tilde{V}_m}{38}\right),$$

$$\hat{\alpha}_h = 0.014\exp - \left(\frac{\tilde{V}_h + V_1^s}{20}\right), \qquad \hat{\beta}_h = \frac{0.2}{1 + \exp(0.2(\tilde{V}_h - V_1^s))}, \tag{7.23}$$

$$\hat{\alpha}_n = \frac{0.006(V_1^s - \tilde{V}_n)}{\exp(0.1(V_1^s - \tilde{V}_n)) - 1}, \qquad \hat{\beta}_n = 0.75\exp(\tilde{V}_n - V_1^s).$$

In contrast to the longitudinal smooth muscle syncytium, the circular layer possesses properties of electrical isotropy. The dynamics of the propagation of the electrical wave V_c^s along it satisfies Eq. (4.43),

$$C_m \frac{\partial V_c^s}{\partial t} = \frac{M_{vs}}{1 + \mu_{\alpha_1}} \left\{ \frac{\partial}{\partial \alpha_1}\left(\frac{\hat{g}^*_{0,\alpha_1}}{\lambda_c} \frac{\partial V_c^s}{\partial \alpha_1}\right) + \frac{\partial}{\partial \alpha_2}\left(\frac{\hat{g}^*_{0,\alpha_1}}{\lambda_1} \frac{\partial V_c^s}{\partial \alpha_2}\right) \right\} - I_{ion}, \tag{7.24}$$

where Eqs. (7.21)–(7.23) are used to calculate I_{ion}. In the above formulas V_1^s should be substituted by V_c^s.

Equations (7.1)–(7.24), complemented by initial and boundary conditions, constitute the mathematical model of the biomechanical properties of the wall of the stomach and describe

(i) the 'passive' nonlinear mechanical properties of the tissue under uniaxial and in-plane biaxial stress–strain states;
(ii) slow-wave myoelectrical processes;
(iii) the activity of pacemaker cells represented by the ICC;
(iv) the generation and propagation of the wave of depolarization in smooth muscle syncytia;
(v) the production of spikes on the crests of slow waves;
(vi) the dynamics of the turnover of free intracellular calcium; and
(vii) the development of 'active' forces of contraction.

The commonly used initial conditions assume that the pacemaker cell discharges electrical signals of given amplitude $\overset{0}{V}_i$ and duration t_i^d; the myoelectrical potentials of the circular and longitudinal smooth muscle syncytia attain equilibrium values $V_{c,l}^r$; both syncytia are in an unexcitable state; and the intracellular concentrations of calcium ions in the ICC, $[\overset{0}{Ca}{}^{2+}]$, and muscle cells, $[\overset{0}{Ca}{}^{2+}]$, are known. Thus, at $t=0$,

$$V_i = \begin{cases} 0, & 0 < t < t_i^d, \\ \overset{0}{V}_i & t \geq t_i^d, \end{cases} \qquad V_{c,l} = V_{c,l}^r, \qquad V_{c,l}^s = 0,$$

$$[Ca^{2+}]_i = [\overset{0}{Ca}{}^{2+}]_i, \qquad [Ca^{2+}] = [\overset{0}{Ca}{}^{2+}] \tag{7.25}$$

and the dynamic variables of the various ion channels are defined by

$$\hat{m} = \hat{m}_\infty, \quad \hat{h} = \hat{h}_\infty, \quad \hat{n} = \hat{n}_\infty, \quad z = z_\infty, \quad h_{\text{Na}} = h_{\text{Na}\infty},$$
$$n_{\text{Na}} = n_{\text{Na}\infty}, \quad \tilde{h} = \tilde{h}_\infty, \quad \tilde{n} = \tilde{n}_\infty, \quad \tilde{x}_{\text{Ca}} = \tilde{x}_{\text{Ca}\infty}.$$

Boundary conditions should be specified depending on a particular problem.

7.3 A one-dimensional model of gastric muscle

The rest of the book will deal with various biomechanical applications of the theory of soft shells to study the motility of the digestive tract. To understand biological phenomena that underlie complex processes such as peristalsis, bolus propulsion and intestinal dysrythmia, we begin with an analysis of the basic myoelectrical phenomena. We proceed from a one-dimensional biomechanical model of gastric smooth muscle. We shall not distinguish between the longitudinal and circular smooth muscle; therefore, subscripts l and c can be omitted.

Consider a bundle of isolated gastric smooth muscle cells embedded in a connective tissue stroma. The cells are connected end-to-end by tight junctions to form a homogeneous electromechanically active biological continuum – a soft fibre of length L (Fig. 7.6). The governing equation of motion of the fibre is given by (see Eq. (6.85))

$$\rho \frac{\partial v}{\partial t} = \frac{\partial}{\partial \alpha} T \quad (0 \leq \alpha \leq L), \tag{7.26}$$

where ρ represents density, v velocity, T force, α the Lagrange coordinate of the smooth muscle fibre and t is time. Assuming that the fibre possesses viscoelastic mechanical properties, for the total force T we have

$$T = k_{\text{v}} \frac{\partial(\lambda - 1)}{\partial t} + T^{\text{a}}([\text{Ca}^{2+}]) + T^{\text{p}}(\lambda), \tag{7.27}$$

where T^{a} and T^{p} are the active and passive components that satisfy Eqs. (7.2) and (7.5), λ is the stretch ratio, $[\text{Ca}^{2+}]$ is the concentration of intracellular calcium ions

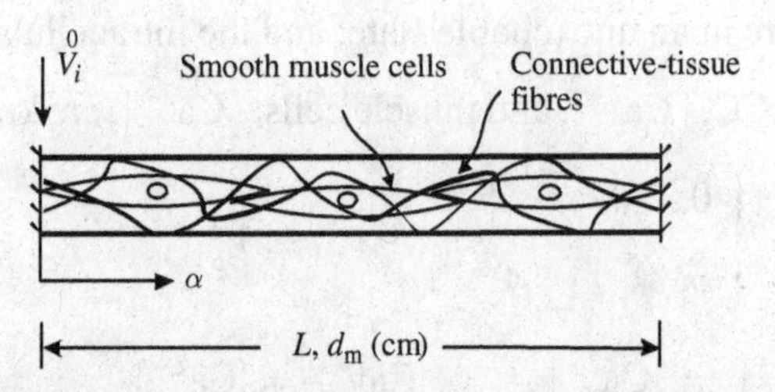

Fig. 7.6 A generalized one-dimensional 'fibre' model of the wall of the abdominal organs.

given by Eq. (7.8) and k_v is the viscosity. By substituting Eq. (7.27) into (7.26) we obtain

$$\rho \frac{\partial v}{\partial t} = \frac{\partial}{\partial \alpha}\left(k_v \frac{\partial(\lambda - 1)}{\partial t} + T^a([\mathrm{Ca}^{2+}]) + T^p(\lambda) \right). \tag{7.28}$$

The electrical cable properties of the fibre are described by the modified Hodgkin–Huxley equations

$$C_m \frac{\partial V}{\partial t} = \frac{d_m}{R_s} \frac{\partial}{\partial \alpha}\left(\lambda(\alpha) \frac{\partial V}{\partial \alpha} \right) - \sum I_{ion}, \tag{7.29}$$

where the dynamics of ion currents, I_{ion}, satisfy Eqs. (7.21)–(7.23), d_m is the diameter and R_s is the specific resistance of the smooth muscle fibre. The meanings of other parameters are as described above.

Myogenic electrical events, either slow waves or spikes, are a result of the activity of spatially distributed autonomous oscillators. They are divided into pools according to their natural frequencies. Two oscillators communicate weakly, such that the phase of one of them is sensitive to the phase of the other, if they have nearly equal frequencies; two oscillators are strongly connected if they have equal frequencies, and two oscillators are disconnected if they have essentially different frequencies. Let each oscillator be silent. It transforms to a firing state as a result of the propagating electrical wave of depolarization, V^s. Its dynamics is described by Eqs. (7.6)–(7.10). Discharges of pacemaker cells – the cells of Cajal, V_i, are described by Eqs. (7.11)–(7.15).

Assume that the fibre is initially in the resting state,

$$t = 0: \quad V_i = V(\alpha, 0) = 0, \quad v(\alpha, 0) = 0, \quad \left[Ca^{2+}\right] = \left[\overset{0}{Ca}{}^{2+}\right]. \tag{7.30}$$

The excitation is provided by action potentials generated by the ICC:

$$V_i = \begin{cases} 0, & 0 < t < t_i^d, \\ \overset{0}{V}_i, & t \geq t_i^d \end{cases} \tag{7.31}$$

The ends of the fibre are clamped and remain unexcitable throughout,

$$t > 0: \quad V(0, t) = V(L, t) = 0, \quad v(0, t) = v(L, t) = 0. \tag{7.32}$$

All mathematical problems presented in the book are solved using the ABS Technologies© computational platform. It employs a hybrid finite-difference scheme and a finite-element method of second-order accuracy, with respect to spatial and time variables. The platform is available upon request at enquiries@abstechnologies.co.uk

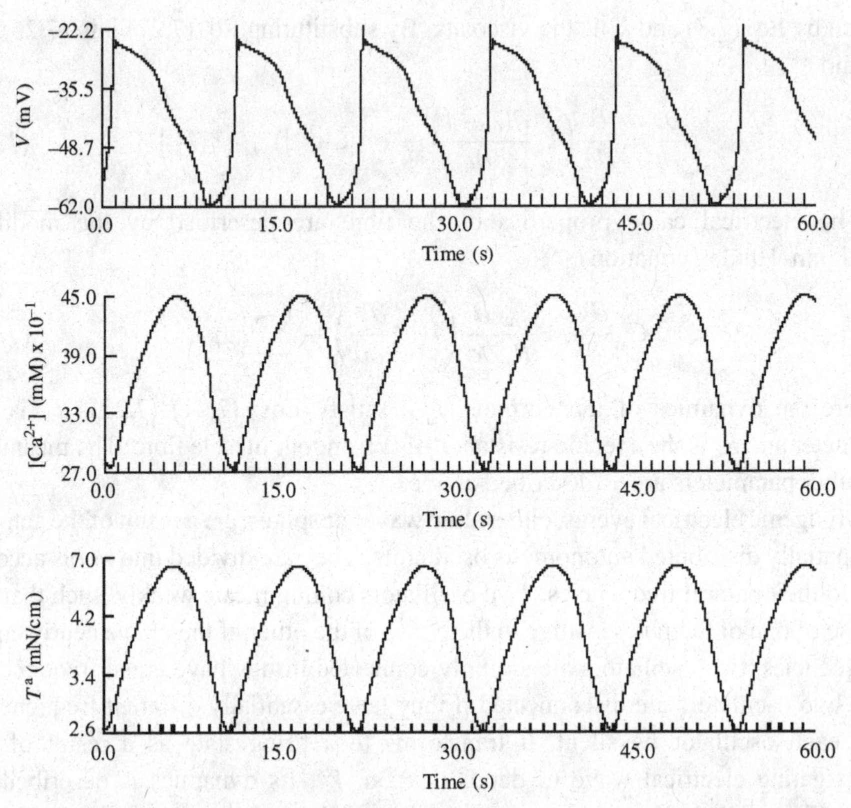

Fig. 7.7 Normal electromechanical activity of the 'fibre'. Traces from top to bottom indicate slow-wave dynamics, changes in intracellular calcium concentration and active force.

7.3.1 Myoelectrical activity

Although gastric smooth muscle exhibits a wide variety of mechanical activity that ranges from phasic to slow sustained contractions, the electrical activity underlying these events shows remarkable similarities. It consists of slow waves and action potentials (spikes). A typical pattern for slow potential changes, oscillations in internal calcium concentration and phasic contractions is shown in Fig. 7.7. The resting membrane potential of gastric muscle is −62 mV. Slow waves of constant amplitude, $V^s = 40$ mV, have a frequency 0.1–0.12 Hz. The maximum rate of depolarization is 20 mV/s and that of repolarization is −6 mV/s. The initial spike of amplitude 6 mV is omnipresent at the beginning on the top of slow waves. The conduction velocity of the wave of excitation along the muscle fibre is calculated to be 7 mm/s.

The most ubiquitous signalling molecule that regulates excitability and provides electromechanical conjugation in smooth muscle is the calcium ion. The increase in

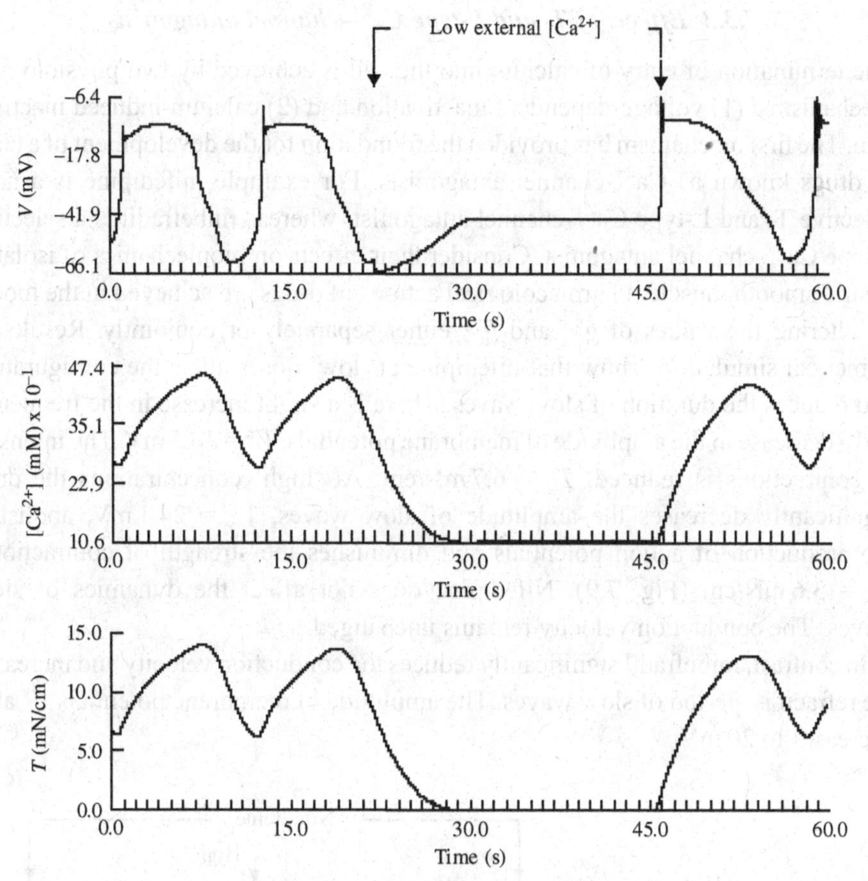

Fig. 7.8 The response of the 'fibre' to changes in external Ca^{2+} concentration.

intracellular concentration of Ca^{2+} is tightly modulated by voltage-dependent and -independent ion channels and release mechanisms of intracellular stores. The rise in $[Ca^{2+}]$ to 4.5 μM leads to phasic contractions of intensity $T^a = 6.9$ mN/cm. The contractions are concomitant with oscillations in intracellular Ca^{2+} concentration and are normally preceded by slow waves.

7.3.2 Decrease in external Ca^{2+} concentration

Slow-wave oscillations cease in a calcium-free environment, $[Ca^{2+}]_0 = 0$. The smooth muscle membrane becomes hyperpolarized at the constant level $V = -49$ mV. The concentration of free intracellular calcium of $[Ca^{2+}] = 1$ μM is insufficient to induce and sustain mechanical contractions. The muscle fibre remains relaxed. Only with an increase in external calcium concentration to its original value $[Ca^{2+}] = 4.5$ μM does the fibre regain its myoelectrical and mechanical activity (Fig. 7.8).

7.3.3 Effects of T- and L-type Ca^{2+}-channel antagonists

The termination of entry of calcium into the cell is achieved by two physiological mechanisms: (1) voltage-dependent inactivation and (2) calcium-induced inactivation. The first mechanism has provided the foundation for the development of a class of drugs known as Ca^{2+}-channel antagonists. For example, nifedipine is a non-selective T- and L-type Ca^{2+}-channel antagonist, whereas mibefradil is a selective T-type Ca^{2+}-channel antagonist. Consider their effects on biomechanics of isolated gastric smooth muscle. Pharmacological actions of drugs are achieved in the model by altering the values of $\tilde{g}^{s}_{Ca}$ and $\tilde{g}^{f}_{Ca}$ either separately or conjointly. Results of numerical simulations show that nifedipine at 'low' doses alters the configuration and reduces the duration of slow waves. There is a slight increase in the frequency and a decrease in the amplitude of membrane potentials, $V^{s} = 36.5$ mV. The intensity of contractions is reduced, $T^{a} = 6.7$ mN/cm. At 'high' concentrations the drug significantly decreases the amplitude of slow waves, $V^{s} = 24.1$ mV, abolishes the production of action potentials and diminishes the strength of contractions, $T^{a} = 3.6$ mN/cm (Fig. 7.9). Nifedipine does not affect the dynamics of slow waves. The conduction velocity remains unchanged.

In contrast, mibefradil significantly reduces the conduction velocity and increases the refractory period of slow waves. The amplitude of membrane potentials V^{s} also decreases to 20 mV.

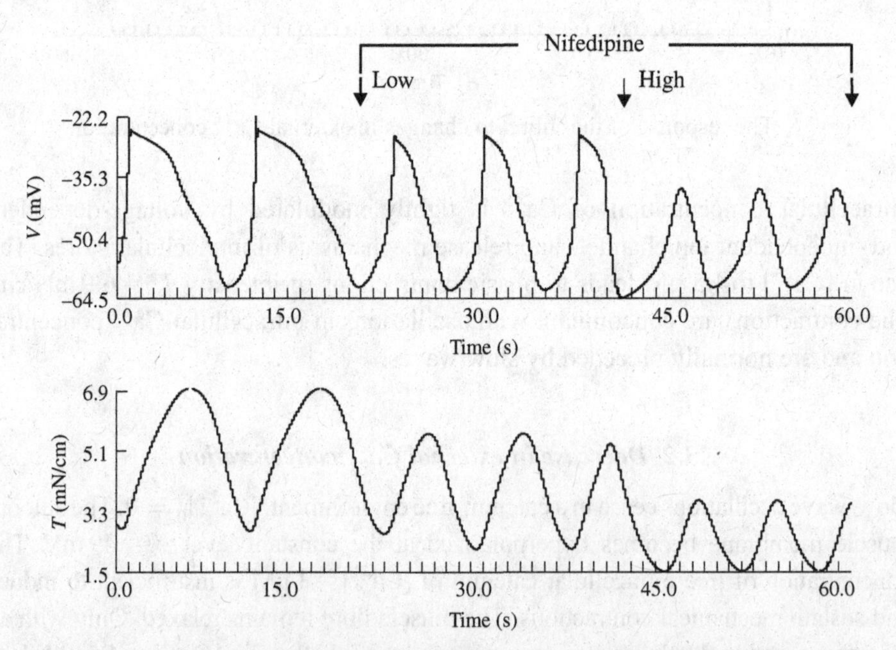

Fig. 7.9 Dose-dependent responses of the 'fibre' to nifedipine and mibefradil, non-selective and selective Ca^{2+}-channel antagonists.

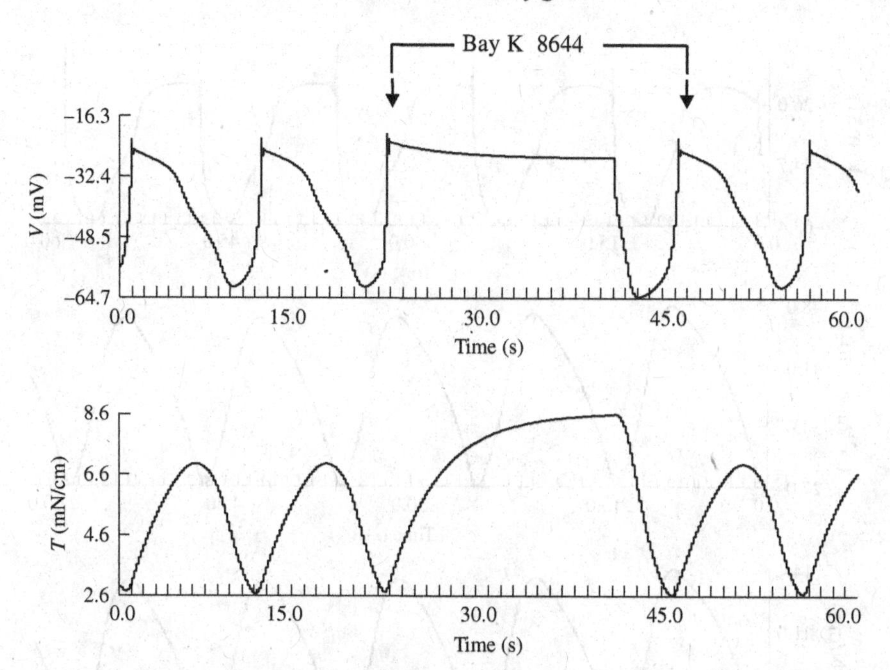

Fig. 7.10 Effect of Bay K 8644, a Ca^{2+}-channel agonist, on the electromechancial activity of the 'fibre'.

Entry of calcium into the cell can also be increased by the process of facilitation. For example, Bay K 8644 – a weakly selective L-type Ca^{2+}-channel agonist, increases the tone and causes spastic, tonic-type contraction of the fibre, max $T^a = 8.46$ mN/cm (Fig. 7.10). As a result, the slow-wave activity is completely abolished and the muscle undergoes hyperpolarization, $V^s = -27.1$ mV. Only after washout of the substance does the fibre regain its normal physiological activity.

7.3.4 Acetylcholine-induced myoelectrical responses

Acetylcholine (ACh) is a major excitatory neurotransmitter in the gastrointestinal tract. The result of externally applied or internally released ACh is a transient increase in the permeability of T-type Ca^{2+} channels located on the smooth muscle membrane. This effect in the model is simulated by varying the parameter $\tilde{g}^{f}_{Ca}$. Simulations show that the effects of ACh are dose-dependent. At 'low' concentrations, ACh increases the amplitude of slow waves and initial spikes by 9 mV (Fig. 7.11). No significant changes are seen in the frequency of oscillations and the resting membrane potential. 'High' concentrations of ACh cause depolarization of the membrane and an upwards shift of the resting membrane potential by 7.2 mV. These changes are concomitant with a decrease in the amplitude of slow waves, $V^s = 26$ mV.

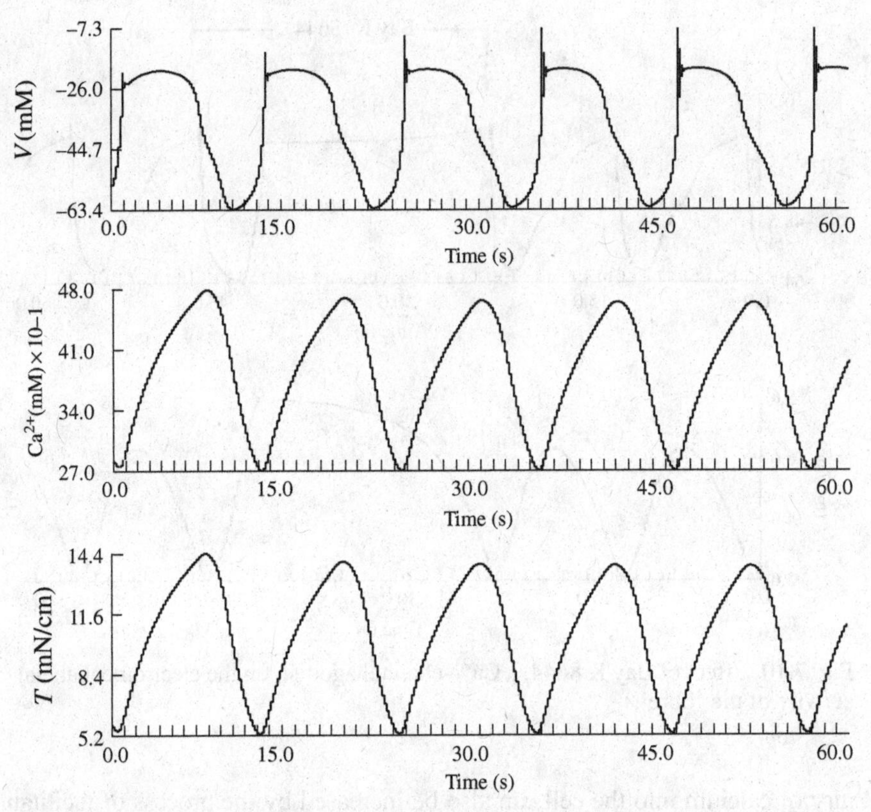

Fig. 7.11 Changes in slow-wave and mechanical activity of the gastric wall after application of acetylcholine.

7.3.5 Effect of chloride-channel antagonist

The effect of 4,4′-diisothiocyano-2,2′-stilbene disulphonic acid – a selective Cl^-- channel blocker, on the electromechanical activity of smooth muscle fibre is achieved by decreasing the value of $\tilde{g}_{Cl}$. The drug abolishes the production of slow waves, action potentials and phasic contractions. The fibre becomes silent.

7.3.6 Effect of selective K^+-channel antagonist

Tetraethylammonium chloride is a selective K^+-channel antagonist. Its pharmacological effect is achieved in the model by varying $\tilde{g}_K$. At 'low' concentrations the compound increased the amplitude and the plateau level of slow waves. These changes correlate with a rise in the strength of phasic contractions, $T^a = 7.9$ mN/cm (Fig. 7.12). Tetraethylammonium chloride causes the generation of high-frequency action potentials on the crests of slow waves. The increase in their amplitude and frequency is dose-dependent. Spikes of maximum amplitude 42 mV are produced at

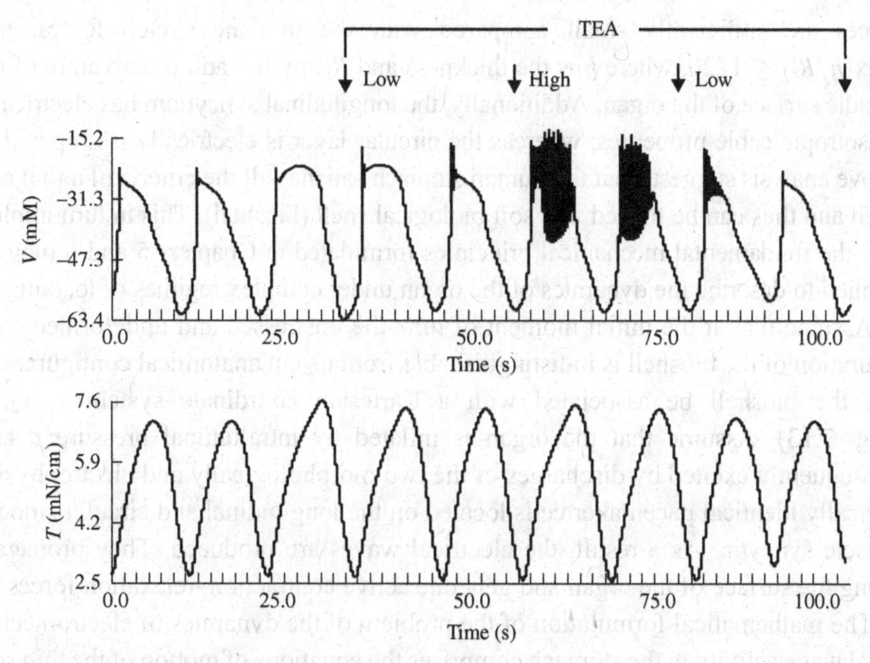

Fig. 7.12 Biomechanics of the 'fibre' in the presence of tetraethylammonium chloride, a selective K^+-channel agonist.

'high' doses of tetraethylammonium chloride. There is a concomitant increase in strength of contractions, max $T^a = 12.2$ mN/cm.

7.4 The stomach as a soft biological shell

A natural way to address the question of selection and development of an adequate mathematical model of the stomach is to analyse the anatomical, physiological and mechanical data from the biomechanical perspective. Starting with the microscopic structure of the tissue, it can be treated as a three-phase biocomposite: the 'passive' phase is presented by inactive smooth muscle syncytia; the 'active' phase is given by mechanochemically active smooth muscle fibres that generate forces of contraction; and the 'supporting' phase is comprised of collagen and elastin fibres arranged in a two-dimensional network with minimal interfibre connectivity.

At the macroscopic level the wall of the organ is formed of two distinct smooth muscle layers (syncytia) embedded into a network of connective tissue. The syncytia and the network display an orthogonal type of weaving. The wall of the stomach is a nonlinear, viscoelastic, transversely anisotropic medium with axes of anisotropy oriented along the direction of the longitudinal and circular smooth muscle fibres. It does not resist compression forces and possesses zero flexural rigidity. The shear forces are sufficiently small compared with the in-plane stretch forces, and $\max(h/R_i) \leq 1/20$, where h is the thickness and R_i are the radii of curvature of the middle surface of the organ. Additionally, the longitudinal syncytium has electrically anisotropic cable properties, whereas the circular layer is electrically isotropic. The above analysis suggests that the human stomach satisfies all the criteria of a thin soft shell and thus can be treated as a soft biological shell (bioshell). This in turn implies that the fundamental mechanical principles formulated in Chapters 5 and 6 may be applied to describe the dynamics of the organ under complex regimes of loading.

Assume that at the initial moment of time the unstressed and undeformed configuration of the bioshell is indistinguishable from its cut anatomical configuration. Let the bioshell be associated with a Cartesian coordinate system x_1, x_2, x_3 (Fig. 7.13). Assume that the organ is inflated by intraluminal pressure p and subsequently excited by discharges of the two morphologically and electrophysiologically identical pacemaker cells located on the longitudinal and circular smooth muscle syncytia. As a result, the electrical waves are produced. They propagate along the surface of the organ and generate active contraction–relaxation forces.

The mathematical formulation of the problem of the dynamics of electromechanical wave activity in the stomach comprises the equations of motion of the thin soft shell, constitutive relations for the biocomposite and initial and boundary conditions (Miftakhov, 1988). The equations of motion of the soft biological shell are given by Eqs. (6.97),

$$
\begin{aligned}
\rho\frac{\partial v_{x_1}}{\partial t} &= \frac{\partial}{\partial\alpha_1}\left[\left(k_{\mathrm{v}}\frac{\partial(\lambda_{\mathrm{c}}-1)}{\partial t}+T^{\mathrm{a}}([\mathrm{Ca}^{2+}])+T^{\mathrm{p}}(\lambda_{\mathrm{c}},\lambda_{\mathrm{l}})\right)e_{1x_1}\sqrt{a_{22}}\right] \\
&\quad+\frac{\partial}{\partial\alpha_2}\left[\left(k_{\mathrm{v}}\frac{\partial(\lambda_{\mathrm{l}}-1)}{\partial t}+T^{\mathrm{a}}([\mathrm{Ca}^{2+}])+T^{\mathrm{p}}(\lambda_{\mathrm{c}},\lambda_{\mathrm{l}})\right)e_{2x_1}\sqrt{a_{11}}\right]+p_1\sqrt{a}\bar{m}_{x_1}, \\
\rho\frac{\partial v_{x_2}}{\partial t} &= \frac{\partial}{\partial\alpha_1}\left[\left(k_{\mathrm{v}}\frac{\partial(\lambda_{\mathrm{c}}-1)}{\partial t}+T^{\mathrm{a}}([\mathrm{Ca}^{2+}])+T^{\mathrm{p}}(\lambda_{\mathrm{c}},\lambda_{\mathrm{l}})\right)e_{1x_2}\sqrt{a_{22}}\right] \\
&\quad+\frac{\partial}{\partial\alpha_2}\left[\left(k_{\mathrm{v}}\frac{\partial(\lambda_{\mathrm{l}}-1)}{\partial t}+T^{\mathrm{a}}([\mathrm{Ca}^{2+}])+T^{\mathrm{p}}(\lambda_{\mathrm{c}},\lambda_{\mathrm{l}})\right)e_{2x_2}\sqrt{a_{11}}\right]+p_2\sqrt{a}\bar{m}_{x_2}, \\
\rho\frac{\partial v_{x_3}}{\partial t} &= \frac{\partial}{\partial\alpha_1}\left[\left(k_{\mathrm{v}}\frac{\partial(\lambda_{\mathrm{c}}-1)}{\partial t}+T^{\mathrm{a}}([\mathrm{Ca}^{2+}])+T^{\mathrm{p}}(\lambda_{\mathrm{c}},\lambda_{\mathrm{l}})\right)e_{1x_3}\sqrt{a_{22}}\right] \\
&\quad+\frac{\partial}{\partial\alpha_2}\left[\left(k_{\mathrm{v}}\frac{\partial(\lambda_{\mathrm{l}}-1)}{\partial t}+T^{\mathrm{a}}([\mathrm{Ca}^{2+}])+T^{\mathrm{p}}(\lambda_{\mathrm{c}},\lambda_{\mathrm{l}})\right)e_{2x_3}\sqrt{a_{11}}\right]+p_3\sqrt{a}\bar{m}_{x_3},
\end{aligned}
\tag{7.33}
$$

where v_{x_1}, v_{x_2} and v_{x_3} are components of the velocity vector ($v_{x_i} = \mathrm{d}x_i/dt$), λ_c and λ_l are the stretch ratios along the circular and longitudinal smooth muscle fibres, k_v is the viscosity parameter, and the meanings of other parameters are as described above.

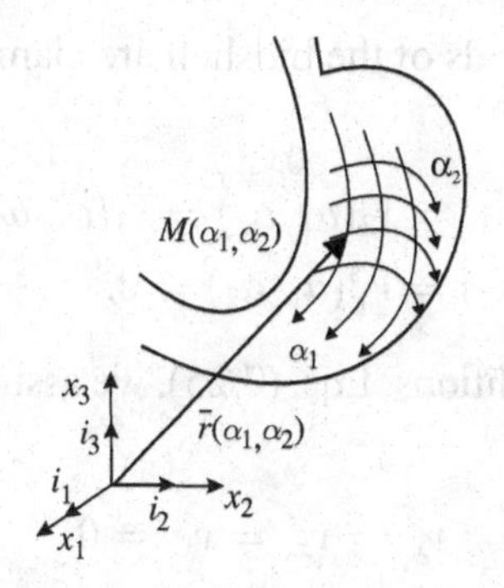

Fig. 7.13 The human stomach as a soft biological shell of complex geometry.

The components and the determinant of the metric tensor **A** are

$$a_{ij} = \frac{\partial x_1}{\partial \alpha_i}\frac{\partial x_1}{\partial \alpha_j} + \frac{\partial x_2}{\partial \alpha_i}\frac{\partial x_2}{\partial \alpha_j} + \frac{\partial x_3}{\partial \alpha_i}\frac{\partial x_3}{\partial \alpha_j}, \qquad (7.34)$$
$$a = a_{11}a_{22} - a_{12}^2,$$

and the cosines of the outward normal $\overline{m}$ to S are given by

$$e_{ix_1} = \frac{1}{\sqrt{a_{ii}}}\frac{\partial x_1}{\partial \alpha_i}, \qquad e_{ix_2} = \frac{1}{\sqrt{a_{ii}}}\frac{\partial x_2}{\partial \alpha_i}, \qquad e_{ix_3} = \frac{1}{\sqrt{a_{ii}}}\frac{\partial x_3}{\partial \alpha_i},$$
$$\bar{m}_{x_1} = (e_{1x_2}e_{2x_2} - e_{1x_3}e_{2x_3})\sqrt{a_{11}a_{22}}/\sqrt{a}, \qquad (7.35)$$
$$\bar{m}_{x_2} = (e_{1x_3}e_{2x_1} - e_{1x_1}e_{2x_3})\sqrt{a_{11}a_{22}}/\sqrt{a},$$
$$\bar{m}_{x_3} = (e_{1x_1}e_{2x_2} - e_{1x_2}e_{2x_1})\sqrt{a_{11}a_{22}}/\sqrt{a}.$$

Assume that the intragastric pressure changes according to the adiabatic law

$$p = \overset{0}{p}\,\Delta \breve{V}^{1.41}, \qquad (7.36)$$

here $\overset{0}{p}$ is the initial pressure and $\Delta\breve{V} = \breve{V}/\overset{0}{\breve{V}}$ is the ratio of a current intraluminal volume $\breve{V}$ to its initial value $\overset{0}{\breve{V}}$.

The characteristic feature of the bioshell is the possibility of the simultaneous coexistence of smooth (biaxially stressed), wrinkled (uniaxially stressed) and unstressed zones. Thus, from Corollaries 2 and 4–6 in Chapter 5 it follows that if both λ_c and $\lambda_l > 1.0$ ($T_{l,c} > 0$) then Eqs. (7.3)–(7.5) are used to calculate forces in the wall. If wrinkles develop, then either $\lambda_l \leq 1.0$ and $\lambda_c > 1.0$ ($T_l = 0$ $T_c > 0$), or $\lambda_c \leq 1.0$ and $\lambda_l > 1.0$ ($T_l > 0$, $T_c = 0$). The wrinkled zone is substituted by an ironed-out zone made of a set of unbound muscle and connective-tissue fibres aligned in the direction of the applied positive tension. Forces are calculated using Eqs. (7.2) and (7.5). This is necessary for conservation of the smoothness of the surface and preservation of the continuity of in-plane membrane forces.

The cardiac and pyloric ends of the bioshell are clamped and remain unexcitable throughout,

$$\begin{aligned} x_1(\alpha_1,\alpha_2) = x_2(\alpha_1,\alpha_2) = x_3(\alpha_1,\alpha_2) = 0, \\ V_c^s(\alpha_1,\alpha_2) = V_l^s(\alpha_1,\alpha_2) = 0. \end{aligned} \tag{7.37}$$

In addition to the initial conditions, Eqs. (7.25), we assume that the stomach is in the state of equilibrium

$$v_{x_1} = v_{x_2} = v_{x_3} = 0. \tag{7.38}$$

Excitation is provided by discharges of pacemaker cells

$$\begin{aligned} V_{i(l)}(\alpha_1,\alpha_2) &= \begin{cases} 0, & 0<t<t_i^d, \\ \overset{0}{V}_i, & t \geq t_i^d, \end{cases} \\ V_{i(c)}(\alpha_1,\alpha_2) &= \begin{cases} 0, & \Delta t<t<t_i^d, \\ \overset{0}{V}_i, & t \geq t_i^d + \Delta t, \end{cases} \end{aligned} \tag{7.39}$$

where Δt is the time interval between discharges on the longitudinal and circular smooth muscle layers.

The combined nonlinear system of partial and ordinary differential equations (7.6), (7.11), (7.16), (7.20), (7.24) and (7.33), the initial and boundary conditions Eqs. (7.25) and (7.37)–(7.39), and supplementary relations given by Eqs. (7.2)–(7.5), (7.7)–(7.9), (7.12)–(7.15), (7.21)–(7.23) and (7.34)–(7.36) presents the closed mathematical problem of complex loading of the human stomach.

7.4.1 Inflation of the stomach

The strain distribution in the wall of the bioshell inflated by internal pressure $p=20$ kPa in the state of dynamic equilibrium is shown in Fig. 7.14. The maximal distension is observed in the cardio-fundal region along the greater curvature of the organ, where $\lambda_l=1.28$ and $\lambda_c=1.34$ are recorded. A part of the fundus bordering the body of the stomach experiences maximal biaxial extension with $\lambda_l=1.3$ and $\lambda_c=1.38$. Uniaxial deformation and wrinkling along the longitudinal axis of the shell, $\lambda_l=1.2$, $\lambda_c=0$, is observed along the lesser curvature in the cardia and the body. In the antro-pyloric region at the greater curvature, $\lambda_l=0$ and $\lambda_c=1.21$ are registered. The fundus and the body of the stomach are biaxially distended throughout, while a change from biaxial to uniaxial stretching is seen in the cardia and antro-pyloric areas. The intragastric volume $\breve{V}$ varies in the range 0.11–0.12 l.

Analysis of the total tension distribution in the bioshell shows that $\max T_l = \max T_c = 14.5$ mN/cm is produced in the fundus and the body. A maximum

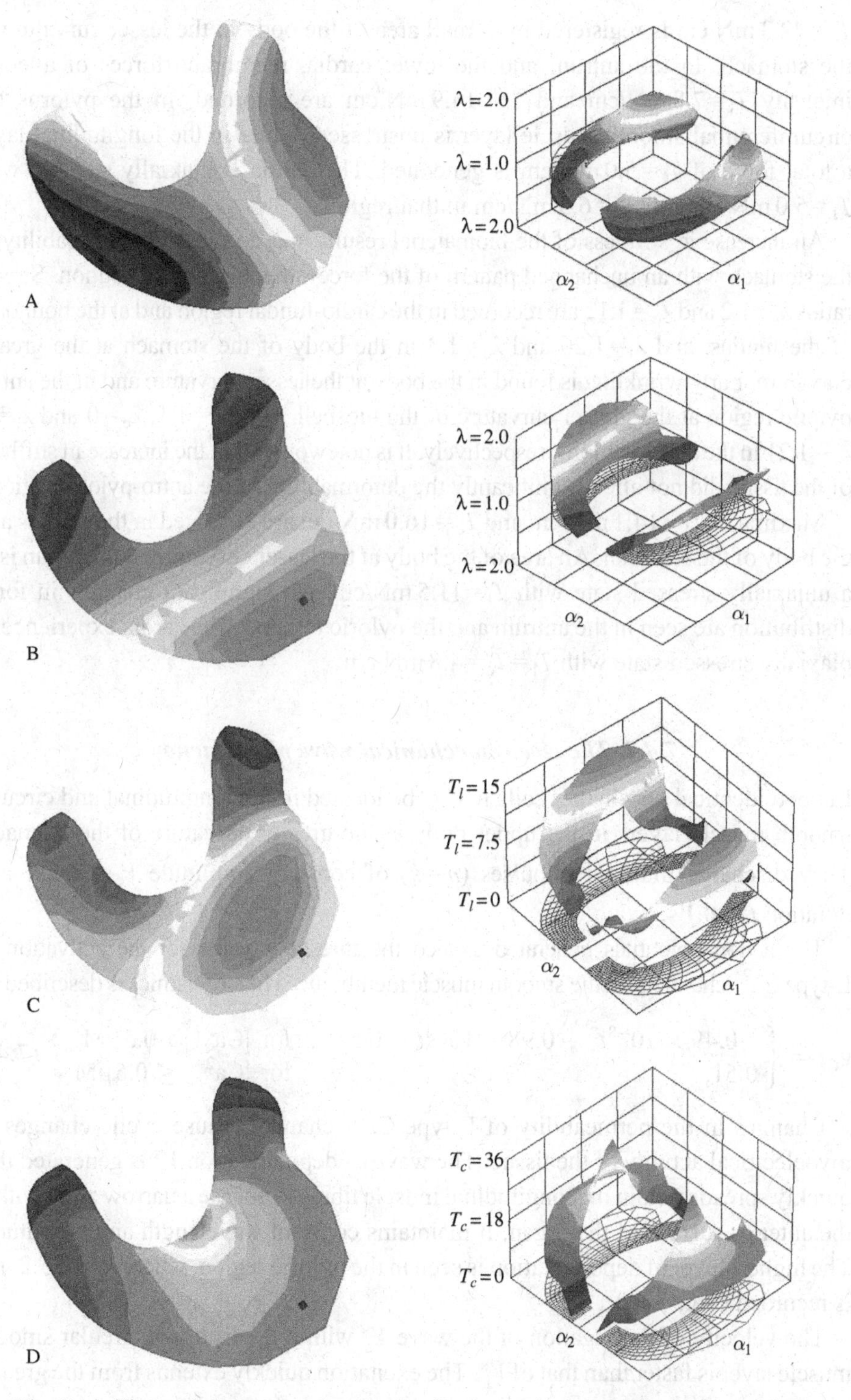

Fig. 7.14 The total-force–stretch-ratio distribution in the longitudinal, (A) and (C), and circular, (B) and (D), smooth muscle syncytia of the bioshell (stomach). Results are presented for the intact stomach and the open envelope, respectively.

$T_c = 42.3$ mN/cm is registered in a small area of the body at the lesser curvature of the stomach. In the antrum and the lower cardia, membrane forces of average intensity $T_l = 7.8$ mN/cm and $T_c = 14.9$ mN/cm are recorded. In the pylorus the circumferential smooth muscle layer is unstressed, while in the longitudinal layer a total force of $T_l = 5.0$ mN/cm is generated. The cardia is biaxially stressed with $T_l = 5.0$ mN/cm and $T_c = 6.3$ mN/cm in that region.

An increase in stiffness of the biomaterial results in a decrease in deformability of the stomach with an unchanged pattern of the force–stretch-ratio distribution. Stretch ratios $\lambda_l = 1.2$ and $\lambda_c = 1.12$ are recorded in the cardio-fundal region and at the boundary of the fundus, and $\lambda_l = 1.26$ and $\lambda_c = 1.3$ in the body of the stomach at the greater curvature. Early wrinkling is found in the body at the lesser curvature and in the antro-pyloric region at the greater curvature of the bioshell, with $\lambda_l = 1.1$, $\lambda_c = 0$ and $\lambda_l = 0$, $\lambda_c = 1.21$ in the above regions, respectively. It is noteworthy that the increase in stiffness of the tissue did not affect significantly the deformability of the antro-pyloric region.

Maxima of $T_l = 14.1$ mN/cm and $T_c = 16.0$ mN/cm are observed in the fundus and the body of the stomach. An area of the body at the lesser curvature of the organ is in a uniaxially stressed state with $T_l = 11.5$ mN/cm. No significant changes in force distribution are seen in the antrum and the pyloric regions. The cardia experiences a biaxially stressed state with $T_l = T_c = 4.8$ mN/cm.

7.4.2 The electromechanical wave phenomenon

Let two identical pacemaker cells ICC_{IM} be located in the longitudinal and circular smooth muscle layers in the upper body at the greater curvature of the stomach. They discharge multiple impulses ($n = 5$) of constant amplitude $\overset{0}{V}_i = 96$ mV and duration $t_d = 0.1$ s.

The action potentials generated exceed the threshold value for the activation of L-type Ca^{2+} channels on the smooth muscle membrane. Their dynamics is described by

$$\tilde{g}^{f}_{Ca} = \begin{cases} -0.49 \times 10^{-8} t^2 + 0.98 \times 10^{-4} t + 0.51, & \text{for } [Ca^{2+}] > 0.5\,\mu M, \\ 0.51, & \text{for } [Ca^{2+}] \leq 0.5\,\mu M. \end{cases} \tag{7.40}$$

Changes in the permeability of L-type Ca^{2+} channels cause cyclic changes in myoelectrical activity of the tissue. The wave of depolarization V_l^s is generated that quickly spreads within the longitudinal muscle fibres to encase a narrow zone within the anterior surface of the organ. It maintains constant wavelength and amplitude. The highest level of depolarization is seen in the pyloric region, where $V_l^s = 25.2$ mV is recorded (Fig. 7.15).

The velocity of propagation of the wave V_c^s within the isotropic circular smooth muscle layer is faster than that of V_l^s. The excitation quickly extends from the greater

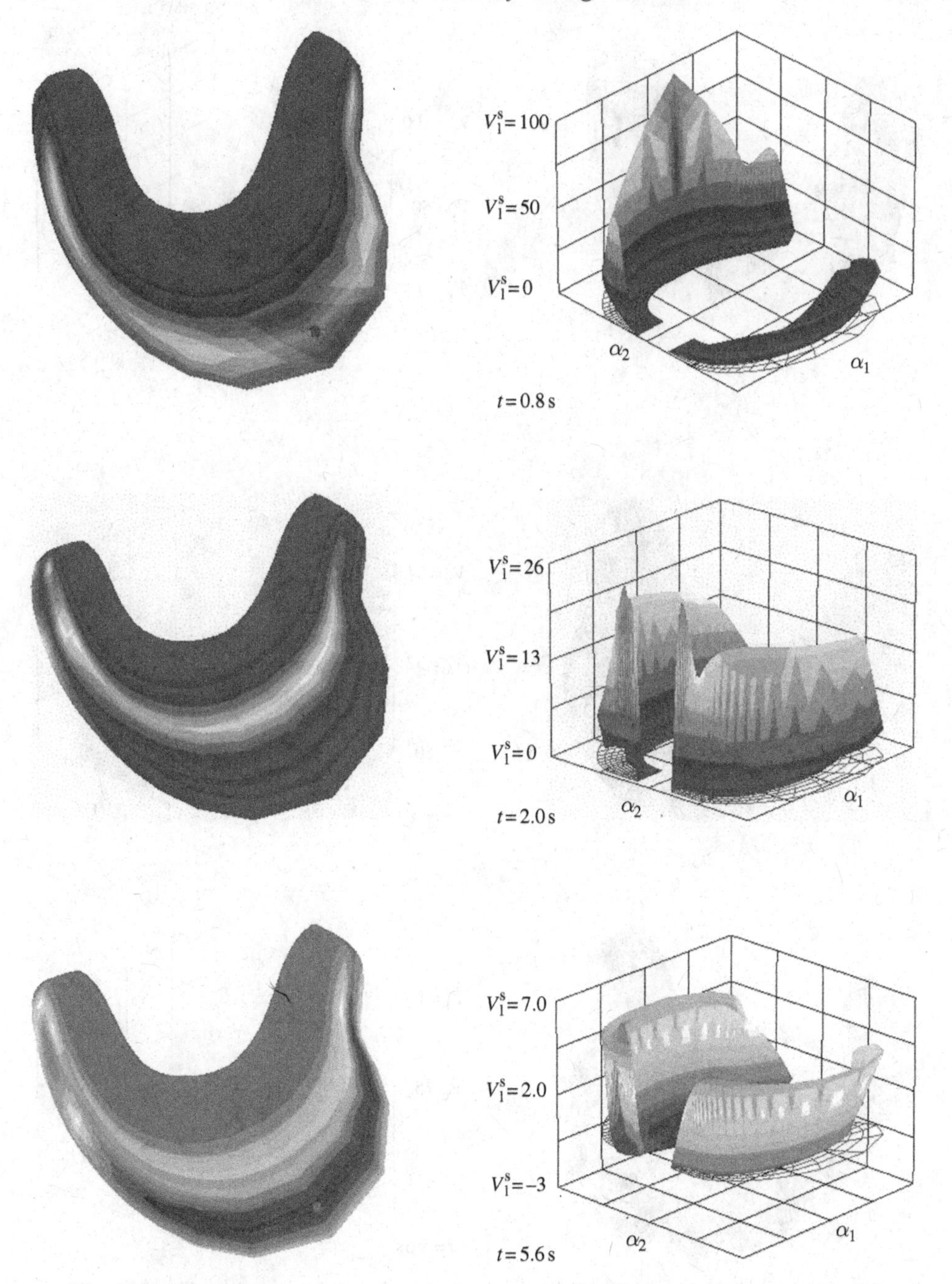

Fig. 7.15 Propagation of the electrical waves of depolarization $V^s_{(l,c)}$ within the longitudinal (l) and circular (c) muscle layers of the stomach.

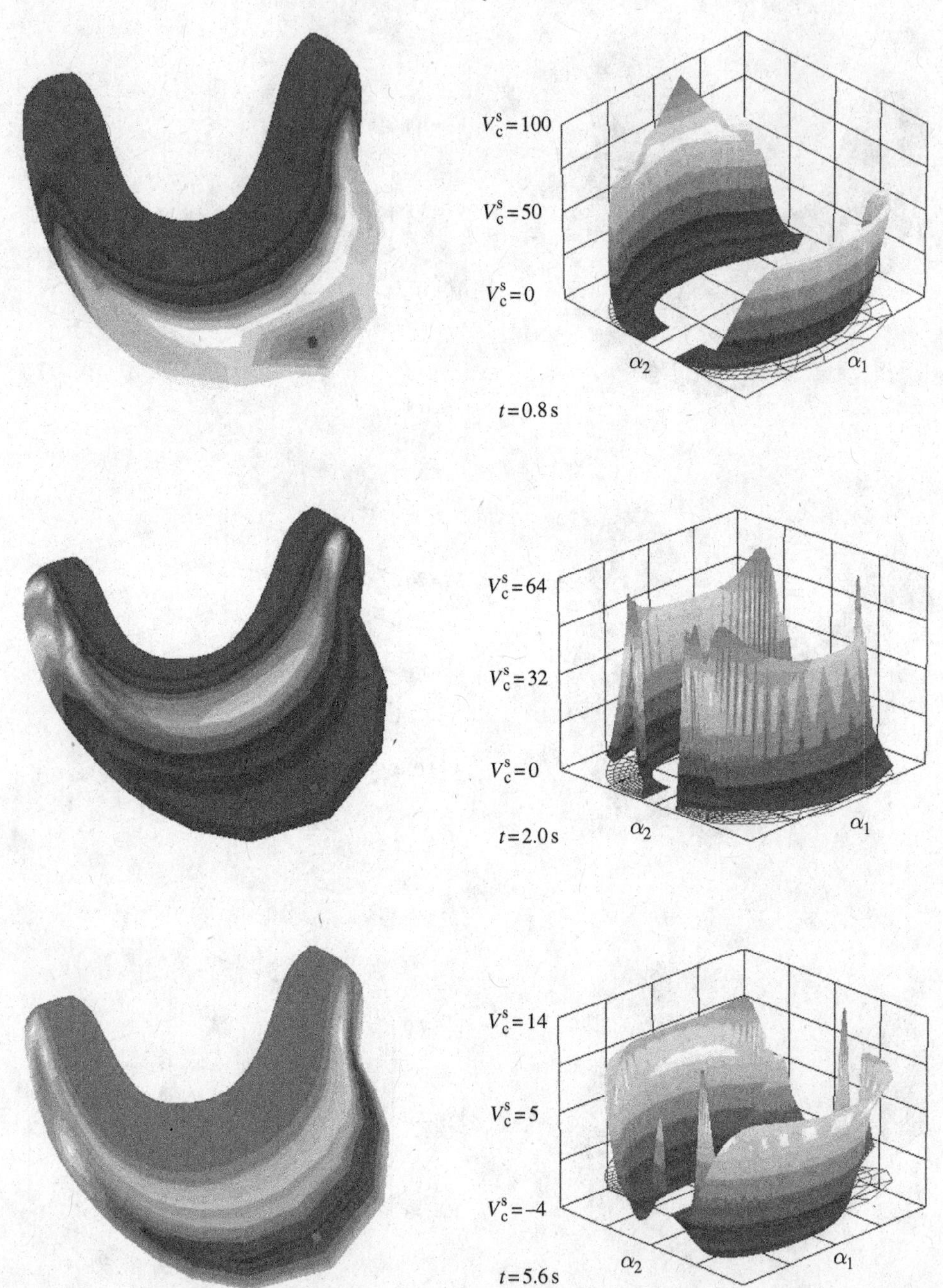
$V_c^s = 100$
$V_c^s = 50$
$V_c^s = 0$
α_2
α_1
$t = 0.8$ s
$V_c^s = 64$
$V_c^s = 32$
$V_c^s = 0$
α_2
α_1
$t = 2.0$ s
$V_c^s = 14$
$V_c^s = 5$
$V_c^s = -4$
α_2
α_1
$t = 5.6$ s

Fig. 7.15 (cont.)

curvature of the body of the organ towards its lesser curvature. The cardiac and pyloric regions experience extensive depolarization, max $V_{c}^{s} = 43.3$ mV.

The wave $V_{(c,l)}^{s}$ activates voltage-dependent calcium channels on the smooth muscle membrane, resulting in a rapid influx of extracellular Ca^{2+} into cells. A rise in the free cytosolic calcium-ion concentration leads to activation of the cascade of mechanochemical reactions with production of the active forces of contraction. The contractions are concomitant in phase and time with the dynamics of calcium-concentration oscillations. The active forces in the fundus and the body of the stomach are $T_{l}^{a} = 6.1\,\mathrm{mN/cm}$ and $T_{c}^{a} = 7.4\,\mathrm{mN/cm}$ (Fig. 7.16). The cardia and the antro-pyloric regions experience less intense tensions, $T_{l}^{a} = 3.0\text{–}3.6\,\mathrm{mN/cm}$ and $T_{c}^{a} = 3.7\text{–}4.4\,\mathrm{mN/cm}$.

The most intense contractions are produced by the longitudinal syncytium in the region of the body of the stomach, max $T_{l}^{a} = 7.9\,\mathrm{mN/cm}$. There is a small zone at the lesser curvature in the body where max $T_{c}^{a} = 10.5\,\mathrm{mN/cm}$. The fundus undergoes uniform contractions in both smooth muscle layers, $T_{l}^{a} = 6.9\,\mathrm{mN/cm}$ and $T_{c}^{a} = 8.4\,\mathrm{mN/cm}$.

Analysis of deformations in the wall of the stomach shows that the existing wrinkles in the antro-pyloric regions expand at the greater curvature to the distal part of the body of the organ. Uniaxial stretching in the longitudinal direction persists in the cardia. Additionally, new wrinkles oriented along the longitudinal axis at the lesser curvature of the bioshell are formed.

The pattern of total force distribution in the organ is similar to that observed in the state of dynamic equilibrium. An increase in the intensity of tension is consistent with the process of generation of the active forces of contraction by the muscle syncytia (Fig. 7.16). Maxima of $T_{l} = 15.1$ mN/cm and $T_{c} = 22.4$ mN/cm are recorded in the body, and max $T_{l} = 2.7$ mN/cm and max $T_{c} = 5.3$ mN/cm in the cardia and the pylorus of the stomach.

7.4.3 The chronaxiae of pacemaker discharges

Consider the effect of a time delay, Δt, between discharges of pacemaker cells located in the longitudinal and circular smooth muscle syncytia on the dynamics of the force–stretch-ratio distribution in the bioshell. Let the number of excitatory impulses, their amplitude and their duration be as described above. Assume that the pacemaker located in the longitudinal muscle syncytium fires first, followed by a discharge of the cell of the circular muscle syncytium.

The delay, $\Delta t = 0.75$ s, in activation results in early wrinkling of the bioshell that begins along the lesser curvature. The delay in excitation of muscle layers results in a reciprocal contraction–relaxation relation between the two muscle syncytia. The observed electromechanical pattern resembles peristaltic movements (Fig. 7.17).

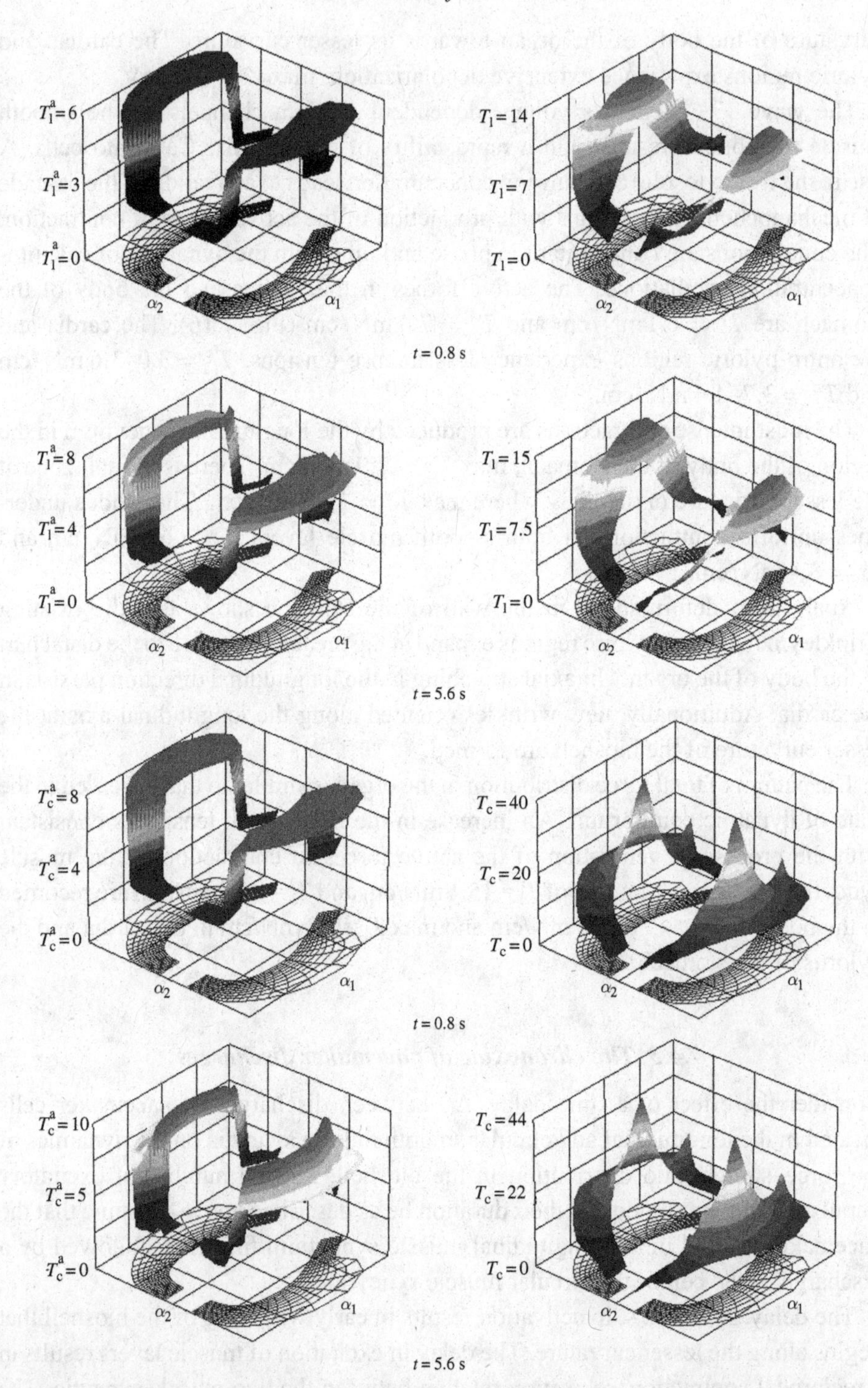

Fig. 7.16 Active, $T^a_{l,c}$, and total, $T_{l,c}$, force distributions in longitudinal (l) and circular (c) muscle layers.

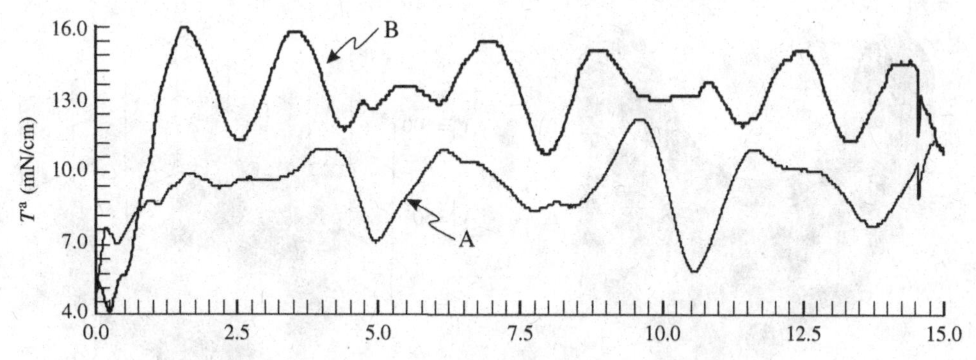

Fig. 7.17 Dynamics of active force development in the longitudinal (A) and circular (B) muscle layers of the bioshell. Recordings were made at a point on the anterior wall in projection of the body of the stomach.

There are no significant quantitative changes in the force–stretch states in the bioshell from those described above.

7.4.4 Multiple pacemakers

Consider three identical pacemaker cells spatially distributed along the anterior wall of the stomach. Assume that the leading pacemaker is located in the upper part of the body, and the other two are located in the antrum and the pyloric regions of the stomach, respectively. The cells discharge simultaneously five impulses of constant amplitude $\overset{0}{V}_{\mathrm{i}} = 96$ mV and duration $t_{\mathrm{d}} = 0.1$ s. The time lag between discharges is constant, $t_i = 0.75$ s.

As a result of excitation the organ undergoes higher levels of depolarization, $V_{\mathrm{l}}^{\mathrm{s}} \approx V_{\mathrm{c}}^{\mathrm{s}} \approx 55$ mV (Fig. 7.18). In time the longitudinal smooth muscle layer of the anterior wall of the stomach in the projection of the body, antrum and pylorus experiences excitation, $V_{\mathrm{l}}^{\mathrm{s}} = 6.8$ mV, while the circumferential layer undergoes hyperpolarization, $V_{\mathrm{c}}^{\mathrm{s}} = -1.2$ mV. There is a focal area of high depolarization in the pyloric region, where $V_{\mathrm{c}}^{\mathrm{s}} = 17.1$ mV is recorded.

Intense contractions, $T_{\mathrm{l}}^{\mathrm{a}} = 7.4\ \mathrm{mN/cm}$ and $T_{\mathrm{c}}^{\mathrm{a}} = 9.2\ \mathrm{mN/cm}$, are produced in the fundus, body and antrum. First longitudinal wrinkles occur in the cardia and pyloric regions along the lesser curvature. They disappear with the redistribution of forces in the circular syncytium. The strongest contractions are generated in the body at the lesser curvature of the bioshell, max $T_{\mathrm{l}}^{\mathrm{a}} = 7\ \mathrm{mN/cm}$ and max $T_{\mathrm{c}}^{\mathrm{a}} = 8.7\ \mathrm{mN/cm}$.

The network of connective tissue allows equal force distribution in all regions of the stomach. There are no zones of force gradients in any region of the organ. Only a focal area in the body at the lesser curvature is overstressed, max $T_{\mathrm{c}} = 44$ mN/cm. Other anatomical regions undergo even biaxial loading with average intensities: in the cardia $T_{\mathrm{l}} = T_{\mathrm{c}} = 2$–3 mN/cm, in the fundus $T_{\mathrm{l}} = 6.5$ mN/cm and $T_{\mathrm{c}} = 9.4$ mN/cm,

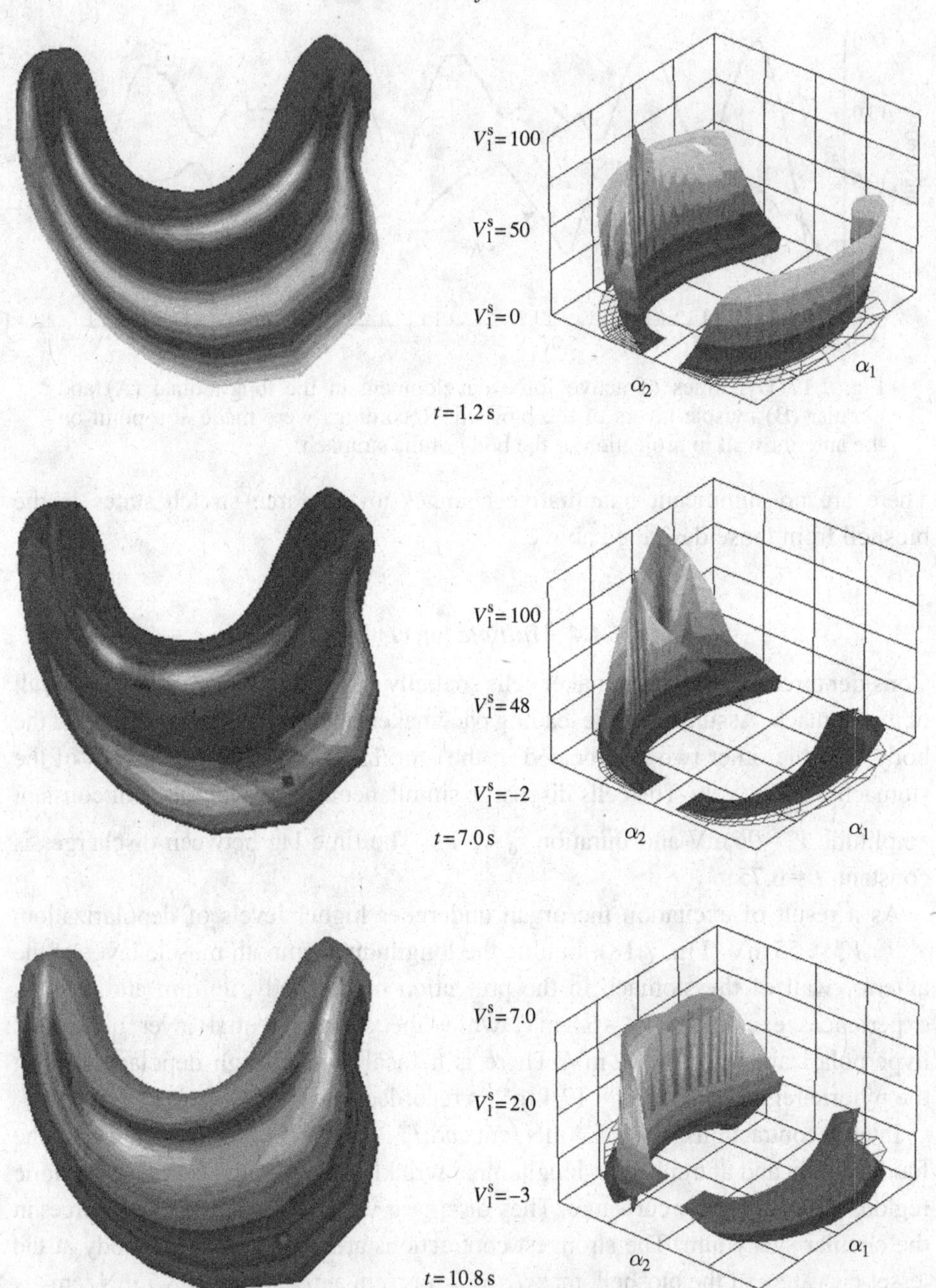

Fig. 7.18 Electromechanical responses of the human stomach to high-frequency stimulation by multiple pacemakers located in various regions of the organ.

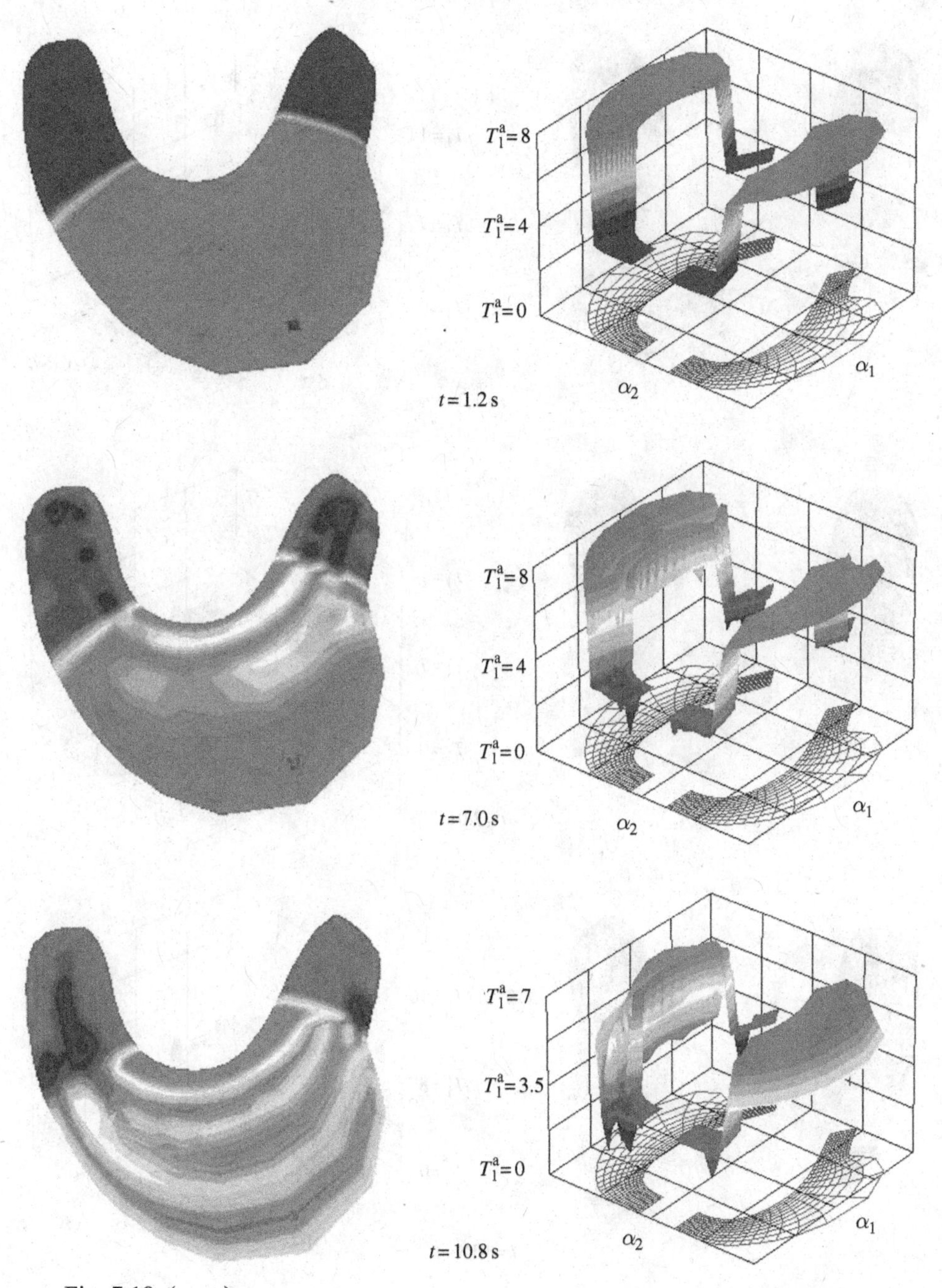

$T_1^a=8$
$T_1^a=4$
$T_1^a=0$
α_1
α_2
$t=1.2$ s
$T_1^a=8$
$T_1^a=4$
$T_1^a=0$
α_1
α_2
$t=7.0$ s
$T_1^a=7$
$T_1^a=3.5$
$T_1^a=0$
α_1
α_2
$t=10.8$ s

Fig. 7.18 (cont.)

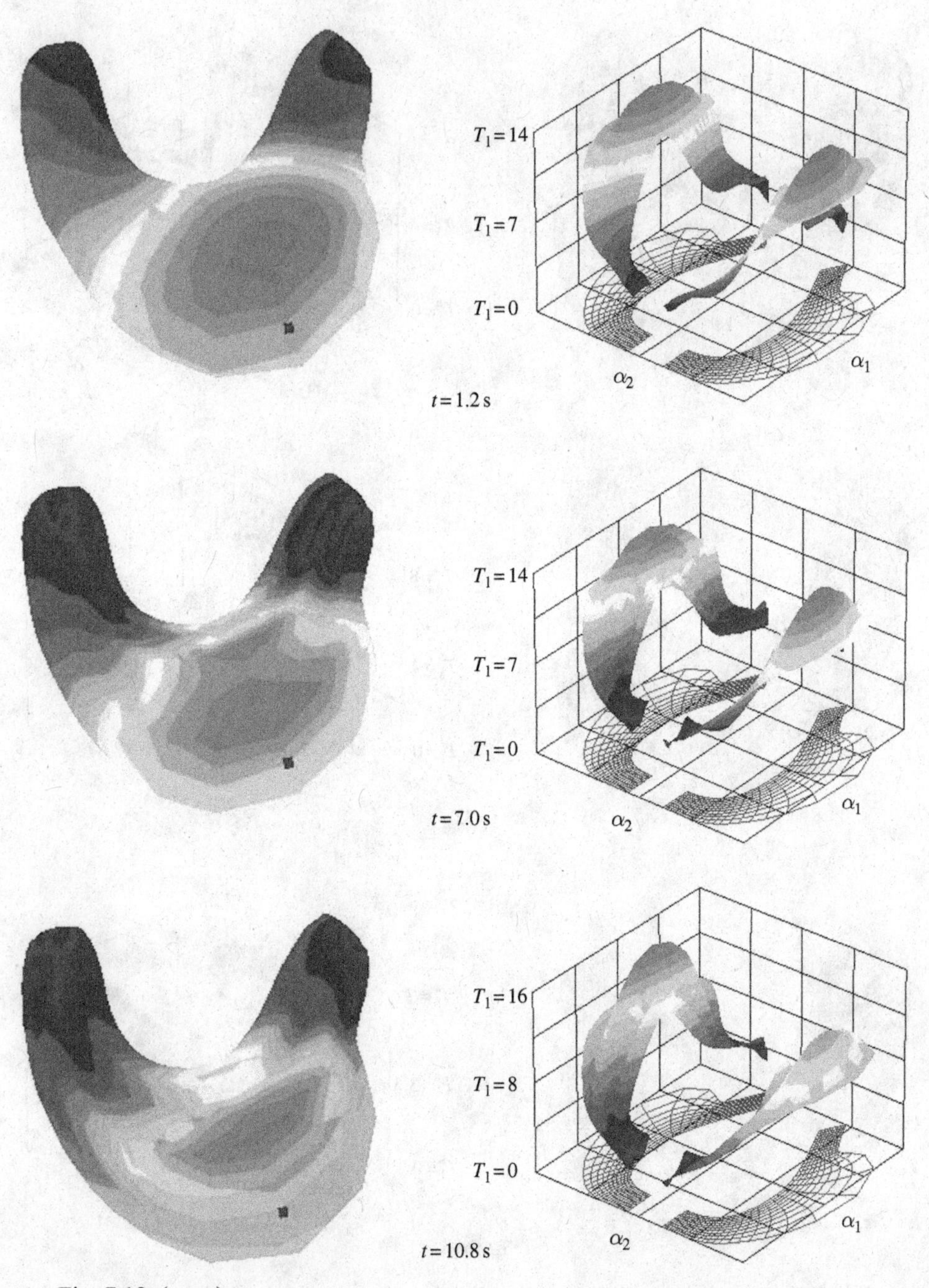
$T_1 = 14$
$T_1 = 7$
$T_1 = 0$
α_2
α_1
$t = 1.2$ s
$T_1 = 14$
$T_1 = 7$
$T_1 = 0$
α_2
α_1
$t = 7.0$ s
$T_1 = 16$
$T_1 = 8$
$T_1 = 0$
α_2
α_1
$t = 10.8$ s

Fig. 7.18 (cont.)

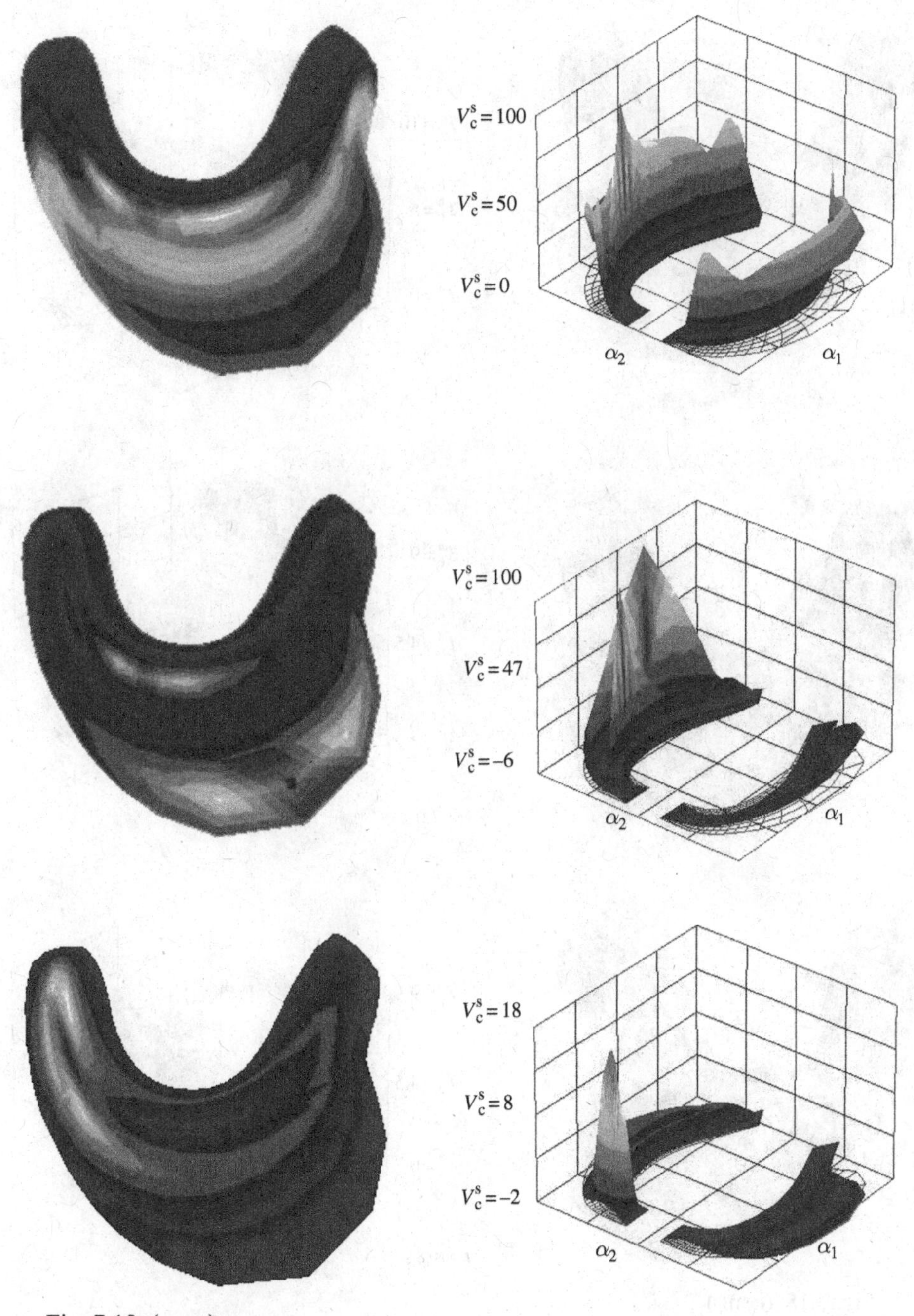
$V_c^s = 100$
$V_c^s = 50$
$V_c^s = 0$
α_2
α_1
$V_c^s = 100$
$V_c^s = 47$
$V_c^s = -6$
α_2
α_1
$V_c^s = 18$
$V_c^s = 8$
$V_c^s = -2$
α_2
α_1

Fig. 7.18 (cont.)

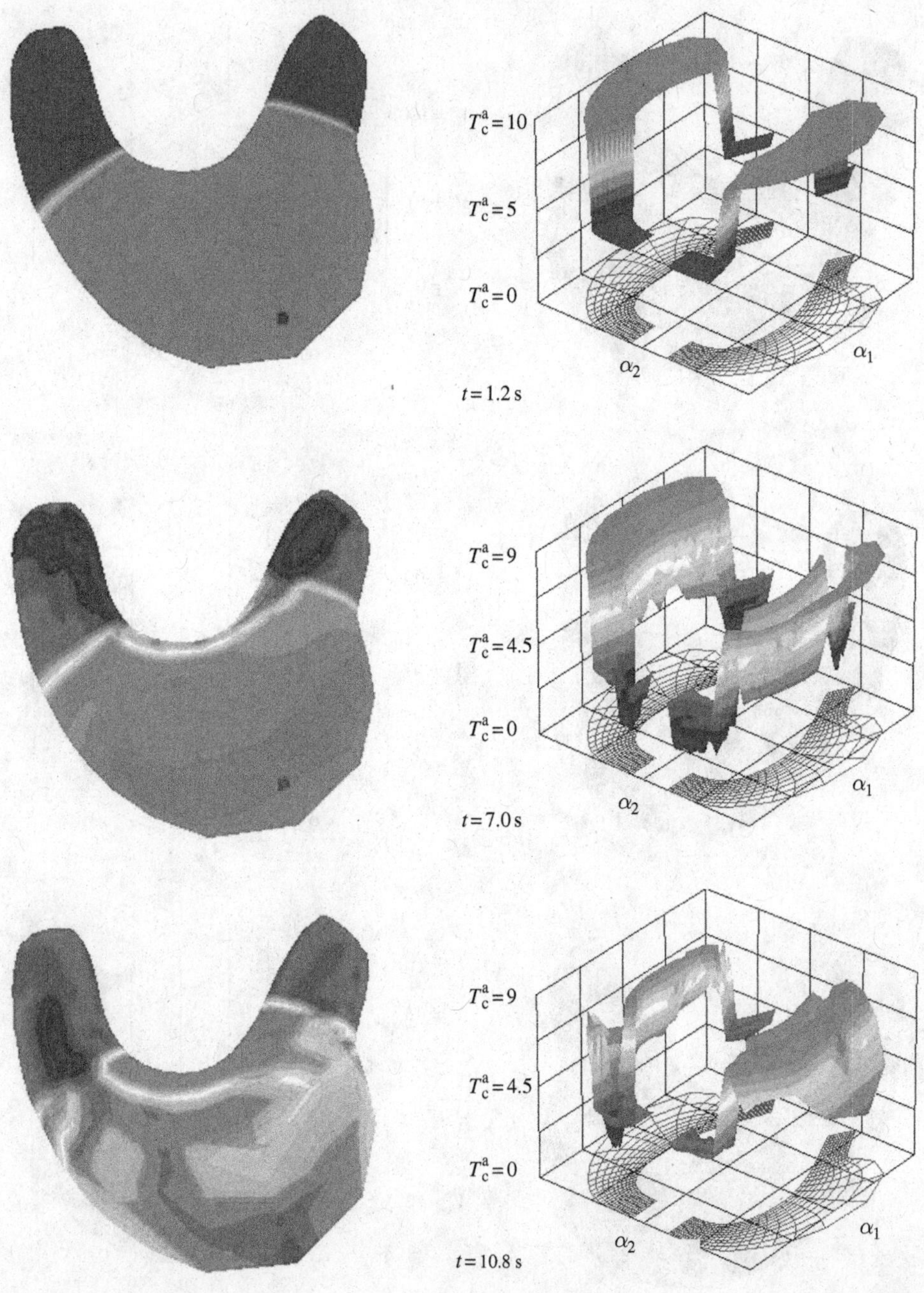
$T_c^a = 10$
$T_c^a = 5$
$T_c^a = 0$
α_2
α_1
$t = 1.2$ s
$T_c^a = 9$
$T_c^a = 4.5$
$T_c^a = 0$
α_2
α_1
$t = 7.0$ s
$T_c^a = 9$
$T_c^a = 4.5$
$T_c^a = 0$
α_2
α_1
$t = 10.8$ s

Fig. 7.18 (cont.)

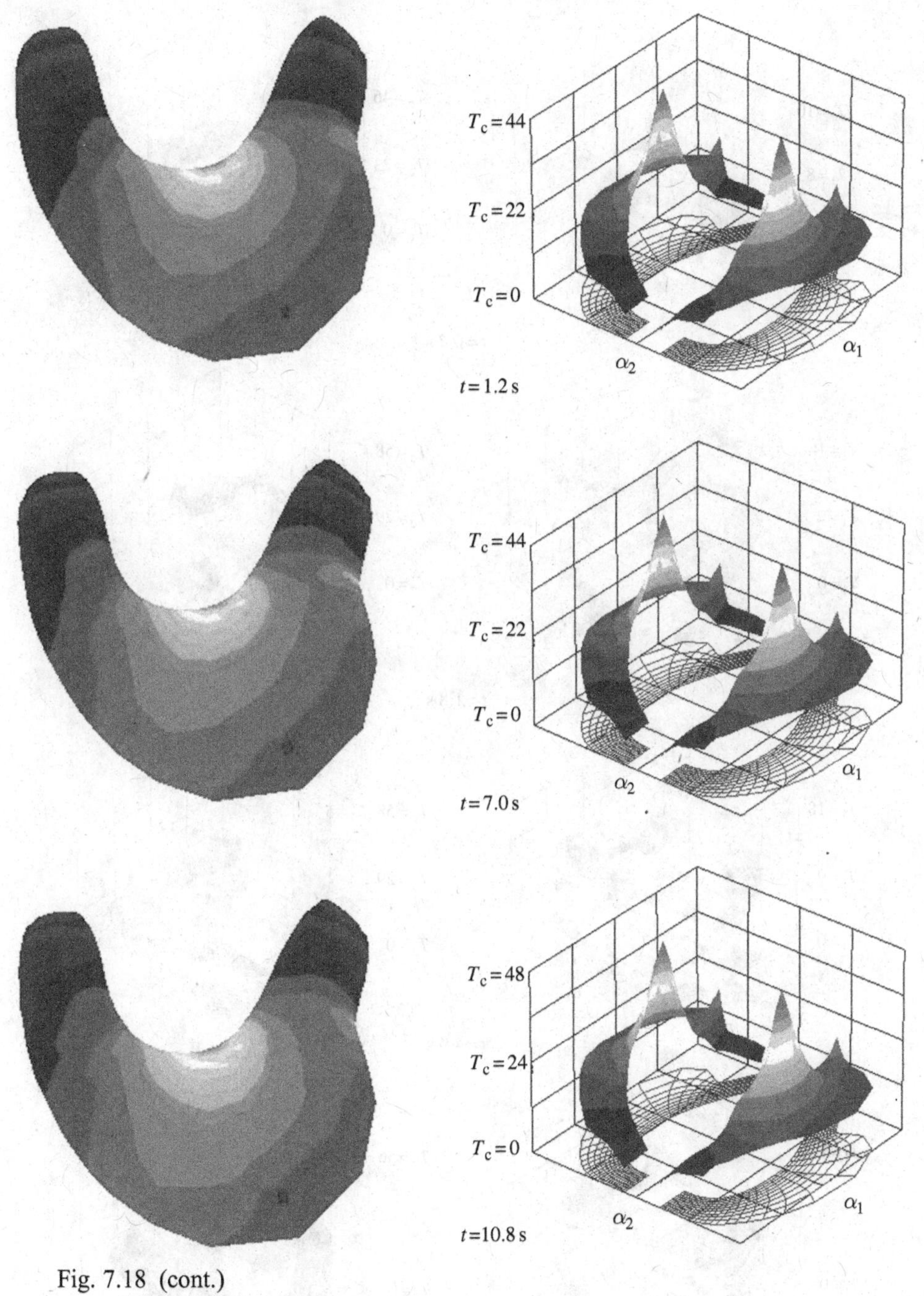

$T_c=44$
$T_c=22$
$T_c=0$
α_1
α_2
$t=1.2$ s
$T_c=44$
$T_c=22$
$T_c=0$
α_1
α_2
$t=7.0$ s
$T_c=48$
$T_c=24$
$T_c=0$
α_1
α_2
$t=10.8$ s

Fig. 7.18 (cont.)

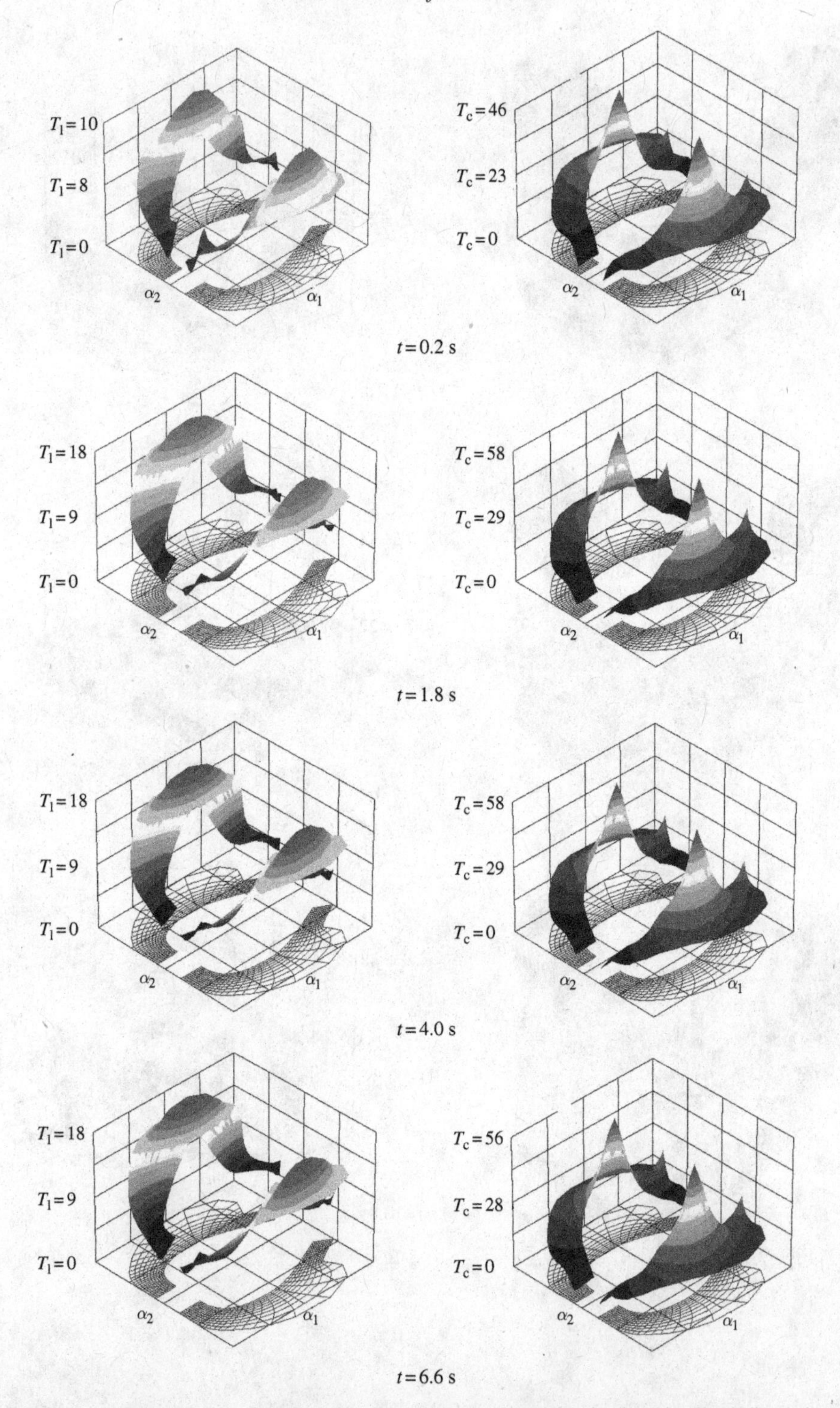

Fig. 7.19 Total $T_{l,c}$ and active $T^{a}_{l,c}$ force development in the longitudinal (l) and circular (c) smooth muscle syncytia of the stomach after application of metoclopramide.

in the body $T_1 = 10$ mN/cm and $T_c = 15$ mN/cm and $T_1 = 5$ mN/cm and $T_c = 9$ mN/cm in the antrum–pylorus.

7.4.5 *Pharmacology of myoelectrical activity*

Consider the effects of prokinetic drugs – metoclopramide and azithromycin – on the biomechanics of the stomach. Metoclopramide enhances the action of ACh in the gastrointestinal tract by blocking dopamine receptors. Azithromycin is a macrolide antibiotic that acts on motilin receptors.

The effect of metoclopramide is simulated by varying the maximal conductance of T-type Ca^{2+} channels, $\tilde{g}^{f}_{Ca}$ (see Section 7.3.4). Application of metoclopramide increases the muscle tone of the stomach. A rise in the level of total forces is mainly due to an increase in the intensity of active forces of contractions. Thus, the maximal value of the total force in the fundus is $T_1 = 15$ mN/cm, $T_c = 26.2$ mN/cm, that in the body is $T_1 = 17.8$ mN/cm, $T_c = 57.5$ mN/cm and that in the antrum is $T_1 = 12.8$ mN/cm, $T_c = 22.1$ mN/cm (Fig. 7.19), with the pattern of force distribution resembling those described above (Section 7.4.2).

The rise in intragastric pressure results in quick adaptive relaxation of the bioshell followed by a period of shortening of the organ along the lesser curvature. Periodic waves of contraction–relaxation cause strong peristaltic movements of the stomach. The general increase in smooth-muscle tone results in a decrease in intragastric volume as earlier: $\breve{V} = 0.75$ l.

Metoclopramide does not affect the propagation of electrical waves of depolarization in the syncytia. The amplitude of slow waves is slightly increased, with no significant change in their frequency.

The effect of azithromycin is modelled by a conjoint increase in the maximal conductances of T- and L-type Ca^{2+} channels, $\tilde{g}^{f}_{Ca}$ and $\tilde{g}^{s}_{Ca}$. Treatment of the stomach with the drug has a profound effect on the contractility of the longitudinal smooth muscle layer. It experiences tonic-type contractions. The circumferential layer, though, shows strong phasic activity. No significant differences are recorded in force–stretch-ratio distributions compared with those elicited by the action of metoclopramide.

Exercises

1. Consider the mathematical model of self-oscillatory myoelectrical activity of smooth muscle syncytium given by Eqs. (7.6)–(7.10). Discuss the nature of the dynamics of the system. Show that stable equilibria and periodic solutions exist.
2. The interstitial cells of Cajal (ICC) were first described by Ramon Cajal (1852–1934) more than a hundred years ago. However, only recently have researchers discovered their significance in gastrointestinal motility. What are the roles of the ICC in regulating gastrointestinal motility?

3. The morphology and electrophysiological characteristics of neuronal cells are crucial in mathematical modelling of the neural circuitry of the enteric nervous plexus. What are the histomorphological and synaptic relationships between the ICC and the neurons of the enteric nervous plexus in the human stomach? State the neurotransmitter types involved.
4. What is the functional significance of the distribution and variation in histological structures of the ICC in different regions of the gastrointestinal tract?
5. The ICC discharge high-amplitude action potentials at a constant frequency. Using the Hodgkin–Huxley formalism, a mathematical description of their pacemaker activity was proposed. Construct the phase portrait and the cusp surface of the dynamic system given by Eqs. (7.11)–(7.15).
6. Show that the system of equations (7.6)–(7.15) is Hamiltonian.
7. Fitting a mathematical model with data on geometry, electrophysiology, biochemistry and mechanical properties remains the main problem in the biomechanics of living tissues. The large variability in stomach location and size among individuals is one of the hurdles a modeller faces in defining its actual geometry. Design an experimental protocol for the acquisition of geometrical data of the human stomach.
8. What is the morphological evidence that strongly supports the model of the longitudinal and circular smooth muscle layers as electrically active bisyncytia?
9. Gastric dysrhythmia is a pathological condition that is observed in patients with dyspepsia, anorexia nervosa, gastro-oesophageal-reflux disease, motion sickness, pregnancy, diabetes, etc. and is linked to symptoms such as nausea and vomiting. The pathological basis of the condition is disturbances in myoelectrical activity. Dysrhythmias are classified into tachygastria (frequency higher than normal), bradygastria (frequency lower than normal) and arrhythmia (no rhythmic activity). Use ABS Technologies© software to simulate the effect of vasopressin on gastric electromechanical activity. (*Hint*: vasopressin increases the free intracellular Ca^{2+} concentration in dispersed smooth muscle cells and induces their hypomotility.) Discuss a possible association of the reduced gastric motility and bradygastria.
10. One of the possible causes of gastric arrhythmia is discoordinated firing activity of multiple ectopic pacemakers. Use ABS Technologies© software to study the effect of spatially distributed and disconcordant pacemaker cells (ICC) on myoelectrical and mechanical activity in the stomach.
11. Gastroparesis is a debilitating motility disorder that affects diabetic patients. Use ABS Technologies© software to simulate the effect of long-standing hyperglycaemia on gastric motility. (*Hint*: learn about the effects of hyperglycaemia on the electrical and biomechanical properties of the smooth muscle and neuronal cells.)
12. Myopathies, e.g. amyloidosis, muscular dystrophies and familial visceral myopathies, can affect muscle layers of the upper gastrointestinal tract and lead to significantly delayed gastroduodenal emptying. Study the effect of muscle stiffening on the stress–strain distribution in the stomach.
13. The electromechanical activity of the gastric muscle is very sensitive to changes in external calcium concentration. Identify the intracellular mechanisms (pathways) that are involved.

8

Biomechanics of the small intestine

8.1 Anatomical and physiological background

The small intestine is a long cylindrical tube that extends from the stomach to the caecum of the colon. The absolute length of the small bowel generally makes up to 80% to 90% of the entire gut length. In the abdomen most of the intestine is loosely suspended by the mesentery and it is looped upon itself. The diameter of the intestine is not constant but gradually decreases from the proximal to the distal part. For example, the diameter of the duodenum is 25–35 mm, that of the jejunum is ~30 mm and that of the ileum is 20–25 mm.

The intestinal wall is a biological composite formed of four layers: mucosa, submucosa, muscular and serosa. The mucosa is the innermost layer and its primary function is to digest and absorb nutrients.

The submucosa consists mainly of connective tissue and serves a purely mechanical function. Septa of connective-tissue fibres carrying nerves, blood and lymphatic vessels penetrate into the muscle layer and form a fibrillary three-dimensional network. It maintains a stable organization of the wall and allows the intestine to undergo reversible changes in length and diameter, offering remarkable properties of stiffness and elasticity.

The muscle coat is made of two smooth muscle layers – a thick (inner) layer of circumferentially oriented smooth muscle cells and a thin (outer) layer of longitudinally oriented muscle elements. The two layers are distinct and separate, although there are intermediate bundles that pass from one layer to the other. The smooth muscle cells form planes and run orthogonal to one another. They form cell–stroma junctions that are of mechanical significance, i.e. equivocal stress–strain distribution during the reaction of contraction–relaxation. The thickness of the muscle layers, h, varies greatly between individuals and according to the anatomical part of the organ ($h \approx 0.5$–0.7 mm).

The serosa is composed of a thin sheet of epithelial cells and connective tissue.

Electromechanical processes in the small intestine have similar basic physiological principles. They require the neuroanatomical integrity of constructive elements – gap junction continuity, an intact enteric nervous plexus and uninterrupted neurotransmission. However, the smooth muscle layers of the stomach, small intestine and colon are electrically isolated. Therefore, one should expect that these organs have a few settled differences, e.g. the frequency of slow waves is higher in the small intestine, $\nu = 0.15$–0.2 Hz than it is in the stomach, they last ~2 s and their amplitude varies in the range 15–30 mV). The slow wave recorded from the small intestine has a sinusoidal configuration with rapid depolarization and a slow repolarization phase. It is a result of coordinated function mainly of L- and T-type Ca^{2+} channels, Ca^{2+}-activated K^+ channels, potential-sensitive K^+ channels and Cl^- channels. Spike bursts occur on the crests of slow waves only in response to neurohumoural stimulation. Their occurrence is essential for the development of contractions.

The variety of mechanical activity of the small intestine – pendular movements, segmentations, peristalsis – is regulated by intrinsic reflexes originating in the enteric nervous system. A basic neuroanatomical circuit is morphologically and functionally uniform and thus can be viewed as a functional unit. The minimum length of the functional unit where a local contraction can be visualized and recorded is 1–2 cm. Being structurally combined together and arranged by gating mechanisms, they respond as an entity. Gating mechanisms are provided by the enteric nervous system that determines the distance, velocity and intensity of propagation of electromechanical waves.

Contractile events in the fasting intestine reveal a three-phase stereotypical motor pattern that is repeated every 90 minutes. Phase 1 is of spiking and motor quiescence, followed by phase 2 of irregular spiking activity and intermittent contractions, whereas the final phase, 3, of the cycle consists of high-frequency action-potential production and powerful regular contractions, occurring at the frequency of the slow waves. The cardinal element of the fasting motor complex is its migratory nature. The velocity of its propagation depends on the anatomical region of the intestine and the phase. Thus, in the duodenum the velocity is maximal during phase 3 (8.5 ± 2.4 cm/s), while in the ileum – the terminal part of the small intestine – it is only 4.0 ± 1.0 cm/s.

The role of pacemaker cells and intermediaries in the signal transduction between the enteric nervous plexus and smooth muscle layers is played by the ICC. The leading role in pacemaker activity belongs to the cells located in the enteric nervous plexus (ICC_{MY}). They discharge action potentials $V_i = 90$–100 mV having durations of 2–3 s. Cells of Cajal that are distributed throughout the circular and longitudinal smooth muscle layers also produce high-amplitude, $V_i = 70$–85 mV, and short-duration, 2–4 s, action potentials, albeit at a lower frequency than do the ICC_{MY}.

8.2 A one-dimensional model of intestinal muscle

We shall start our analysis of biomechanical phenomena in the small intestine from a one-dimensional model of intestinal smooth muscle fibre. All our considerations will be based on the approach developed in Chapter 6. Mathematically, the problem leads to the governing system of equations that consists of Eqs. (7.28) and (7.29), which describe the electromechanical wave processes in the smooth muscle fibre; supplementary Eqs. (7.21)–(7.23) for the dynamics of ion currents; Eqs. (7.6)–(7.10) for the propagation of the wave of depolarization within the fibre; and Eqs. (7.11)–(7.15), which define the dynamics of pacemaker activity, ICC_{MY}. Let the intestinal muscle fibre be initially in the resting state (Eq. (7.30)). It is excited by the discharge ICC_{MY} given by Eq. (7.31). The ends of the fibre are clamped and remain unexcitable throughout (Eq. (7.32)).

8.2.1 Myoelectrical activity

Changes in the permeability of the T-type Ca^{2+} channels (Eq. 7.40) induce alterations in the dynamics of membrane potential oscillations (Fig. 8.1). Initially, slow waves have a constant amplitude $V^s = 25$ mV and a frequency of 0.18 Hz. The depolarization phase lasts 1.6 s. It is followed by a short plateau of duration 0.4 s,

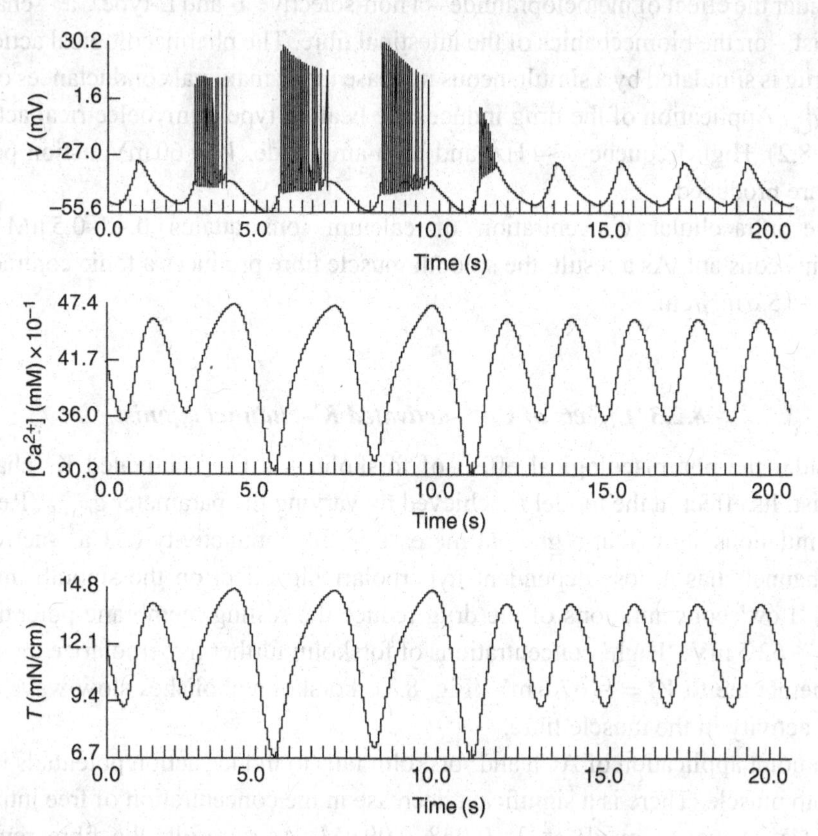

Fig. 8.1 Myoelectrical activity in a normal intestinal smooth muscle 'fibre'. Used with permission from World Scientific Publishing Company.

and finally decreases slowly to the resting value $V_r^s = -51$ mV. The flux of Ca^{2+} ions ($\max[Ca^{2+}] = 0.49$ μM) triggers regular rhythmic contractions of intensity $T^a = 4.8$ mN/cm. The maximum total force generated by the muscle fibre is $T = 14$ mN/cm.

Over time, the smooth muscle begins to fire action potentials $V^s = 56$–72 mV at a frequency $\simeq 17$ Hz. The dynamics of Ca^{2+}-ion influx coincides in phase with the depolarization process, i.e. the concentration of intracellular calcium rises concomitantly with the production of spikes. It achieves a maximum of $[Ca^{2+}] = 0.47$ μM immediately after the firing has ended. As a result, phasic contractions with the maximum total force $T = 15.1$ mN/cm are generated by the intestinal muscle fibre.

With a decrease in $\tilde{g}_{Ca}^f \leq 0.65\ \mathrm{mSm/cm^2}$, the system transforms to an irregular bursting mode and later it reverts to the slow-wave regime.

8.2.2 Effects of non-selective Ca^{2+}-channel agonists

Consider the effect of metoclopramide – a non-selective T- and L-type Ca^{2+}-channel agonist – on the biomechanics of the intestinal fibre. The pharmacological action of the drug is simulated by a simultaneous increase in the maximal conductances of $\tilde{g}_{Ca}^s$ and $\tilde{g}_{Ca}^f$. Application of the drug induces the beating type of myoelectrical activity (Fig. 8.2). High-frequency, ~4 Hz, and high-amplitude, $V^s = 60$ mV, action potentials are produced.

The intracellular concentration of calcium ions attains 0.48–0.5 μM and remains constant. As a result, the smooth muscle fibre produces a tonic contraction of $T = 15.6$ mN/cm.

8.2.3 Effects of Ca^{2+}-activated K^+-channel agonist

Consider the pharmacological effect of forskolin – a Ca^{2+}-activated K^+-channel agonist. Its effect in the model is achieved by varying the parameter $\tilde{g}_{Ca-K}$. Results of simulations show that a gradual increase in the conductivity of Ca^{2+}-activated K^+ channels has a dose-dependent hyperpolarizing effect on the smooth muscle fibre. 'Low' concentrations of the drug reduce the resting membrane potential to $V_r^s = -62.5$ mV. 'High' concentrations of forskolin further hyperpolarize the muscle membrane to $V_r^s = -67.6$ mV (Fig. 8.3). Forskolin abolishes slow-wave electrical activity in the muscle fibre.

Conjoint application of ACh and forskolin fails to induce action potentials in the smooth muscle. There is a significant decrease in the concentration of free intracellular Ca^{2+} ions, to $\max[Ca^{2+}] = 0.048$–0.09 μM. As a result, the fibre remains hyperpolarized and mechanically inactive throughout.

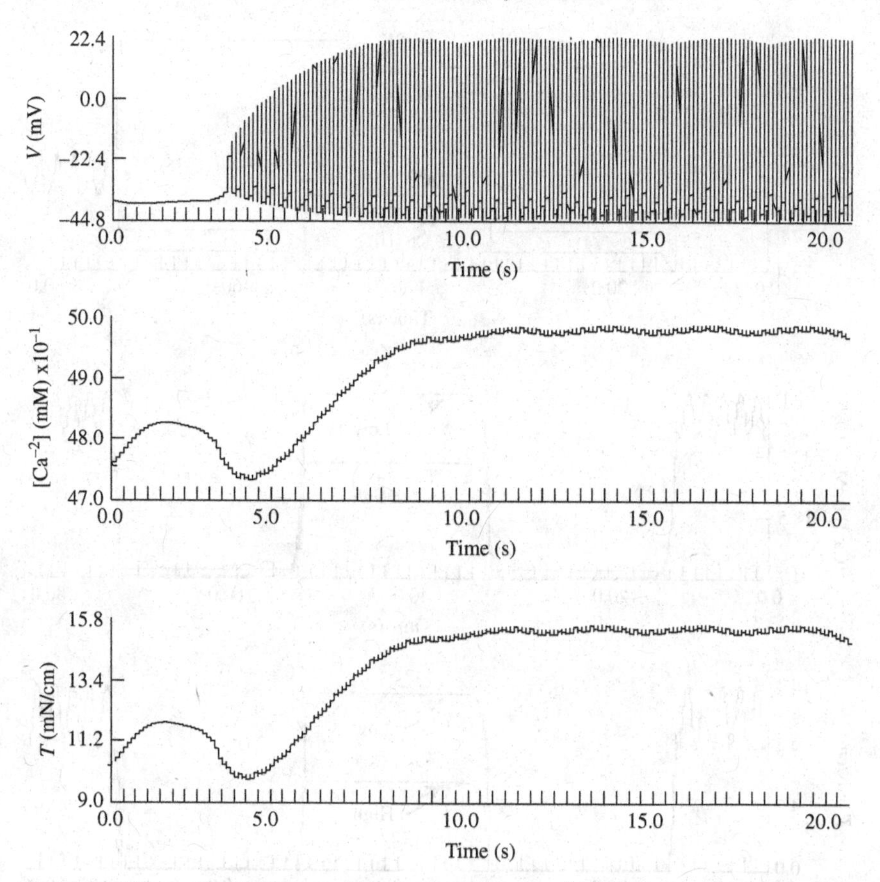

Fig. 8.2 Effect of metoclopramide on electromechanical activity of the intestinal 'fibre'. Used with permission from World Scientific Publishing Company.

Simultaneous treatment of the muscle fibre with forskolin and a concurrent increase in the concentration of extracellular potassium ions produces a strong depolarizing effect – the membrane potential V^s rises to −21.5 mV. The depolarization leads to an influx of extracellular calcium and activation of the contractile proteins, with the generation of a force of intensity $T = 16.4$ mN/cm. After the excess extracellular K^+ has been removed the fibre returns to the hyperpolarized state.

8.2.4 Response to a selective K^+-channel agonist

Lemakalim is a selective K^+-channel agonist. The action of the drug in the model is achieved by varying the parameter $\tilde{g}_K$. Results show that an increase in conductivity of K^+-channels depolarizes the membrane, $V^s = -19.2$ mV, that remains at this level throughout (Fig. 8.4).

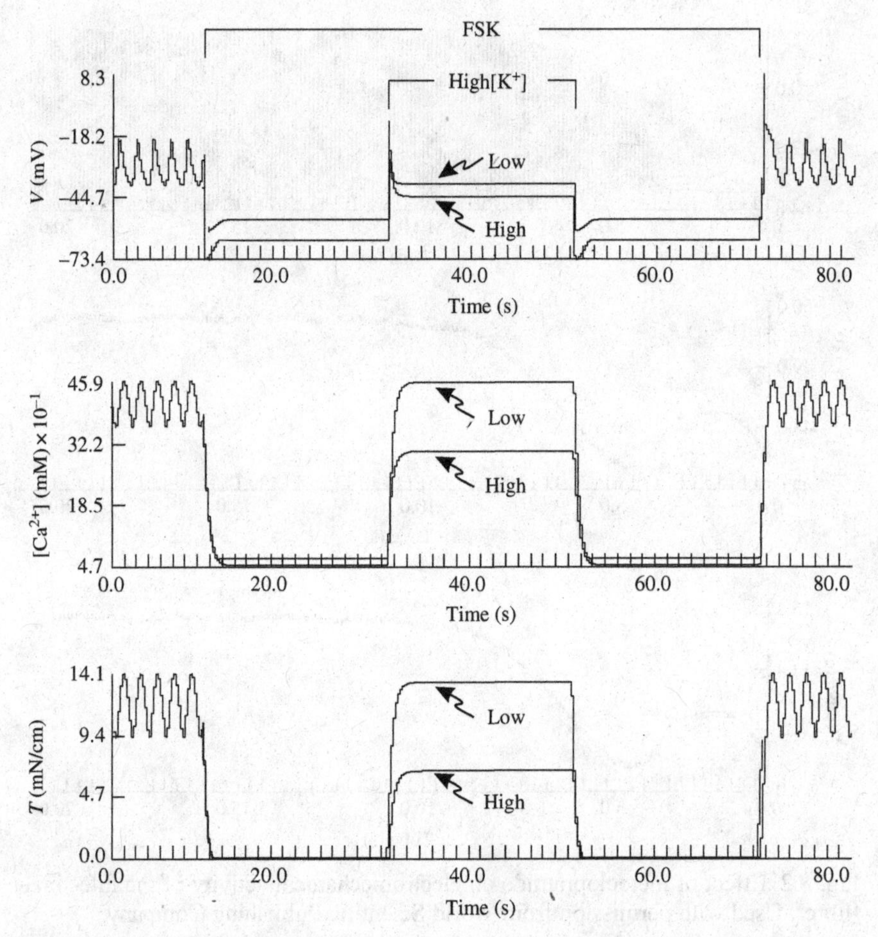

Fig. 8.3 Effect of forskolin and changes in extracellular K^+ concentration on electromechanical responses of the intestinal 'fibre'.

The dynamics of cytosolic Ca^{2+} changes and active force production corresponds to the dynamics of $V^s(t)$. Lemakalim abolishes phasic contractile activity in the fibre. Instead, tonic-type contractions with $T = 17.2$ mN/cm are generated.

Addition of lemakalim to the muscle fibre pre-exposed to a high extracellular potassium-ion concentration causes its slight hyperpolarization. The concentration of free intracellular calcium and the total force rise to 0.52 μM and 17 mN/cm, respectively.

Conjoint application of lemakalim and external ACh induces bursting in the intestinal muscle fibre. It generates high-amplitude action potentials $V^s = 68–72$ mV of high frequency $\nu = 6–8$ Hz. Subsequent introduction of forskolin hyperpolarizes the syncytium and completely abolishes its electromechanical activity.

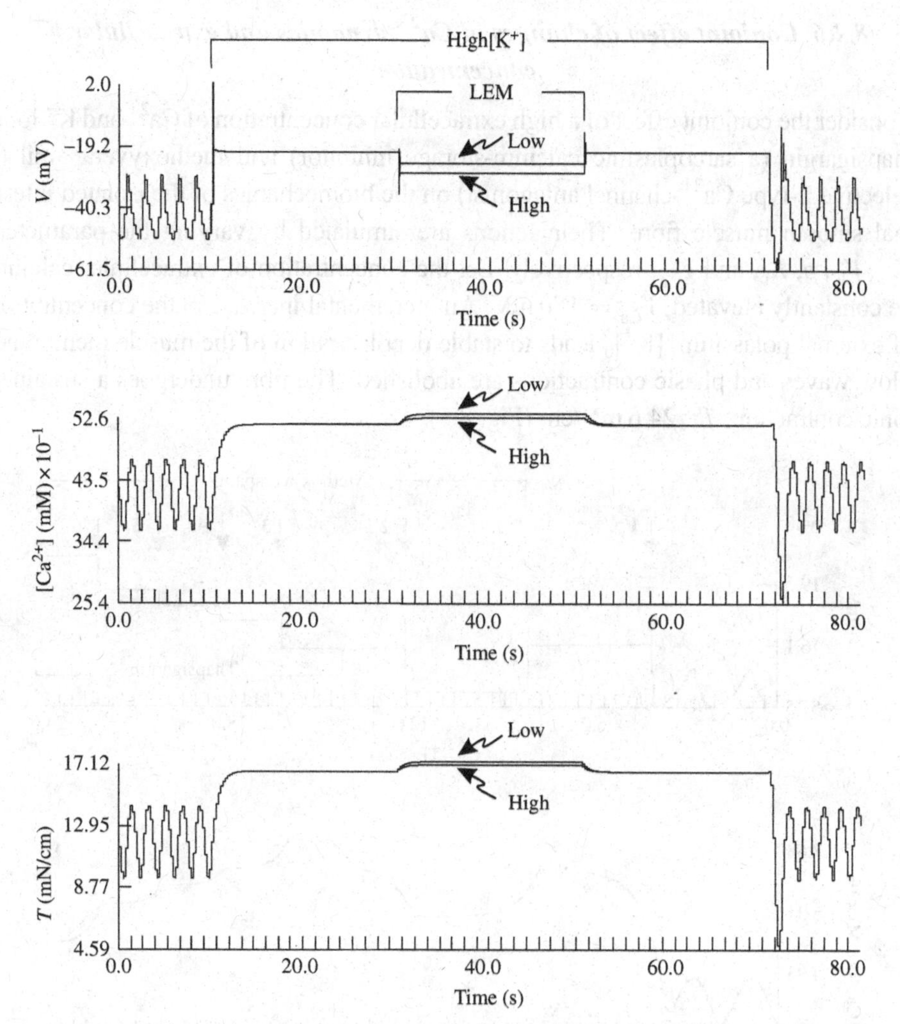

Fig. 8.4 Dose-dependent effects of lemakalim and extracellular K^+ on the biomechanics of the intestinal smooth muscle 'fibre'. Used with permission from World Scientific Publishing Company.

8.2.5 *Effect of selective K^+-channel antagonist*

Consider the pharmacological effects of phencyclidine – a selective K^+-channel antagonist – on myoelectrical activity of the muscle fibre. The pharmacological effect of the compound in the model is simulated by setting $\tilde{g}_K = 0$. Phencyclidine entirely abolishes slow waves and depolarizes the smooth muscle membrane, $V^s = -16.9$ mV. The depolarization process is accompanied by an increase in concentration of free intracellular Ca^{2+} ($\max[Ca^{2+}] = 0.48$ μM) and the development of a tonic contraction of intensity $T = 15.4$ mN/cm.

8.2.6 Conjoint effect of changes in Ca^{2+} dynamics and extracellular K^{+} concentrations

Consider the conjoint effect of a high extracellular concentration of Ca^{2+} and K^{+} ions, thapsigargin (a sarcoplasmic calcium-storage inhibitor) and methoxyverapamil (a selective L-type Ca^{2+}-channel antagonist) on the biomechanics of the isolated intestinal smooth muscle fibre. Their actions are simulated by varying the parameters $\tilde{V}_{Ca}$, $\tilde{V}_{K}$, ρ, K_c, and $\tilde{g}^s_{Ca}$, respectively. Let the concentration of extracellular calcium be constantly elevated, $\tilde{V}_{Ca} = 150$ mV. An incremental increase in the concentration of external potassium, $[K^+]_0$ leads to stable depolarization of the muscle membrane. Slow waves and phasic contractions are abolished. The fibre undergoes a sustained tonic contraction, $T = 24.6$ mN/cm (Fig. 8.5).

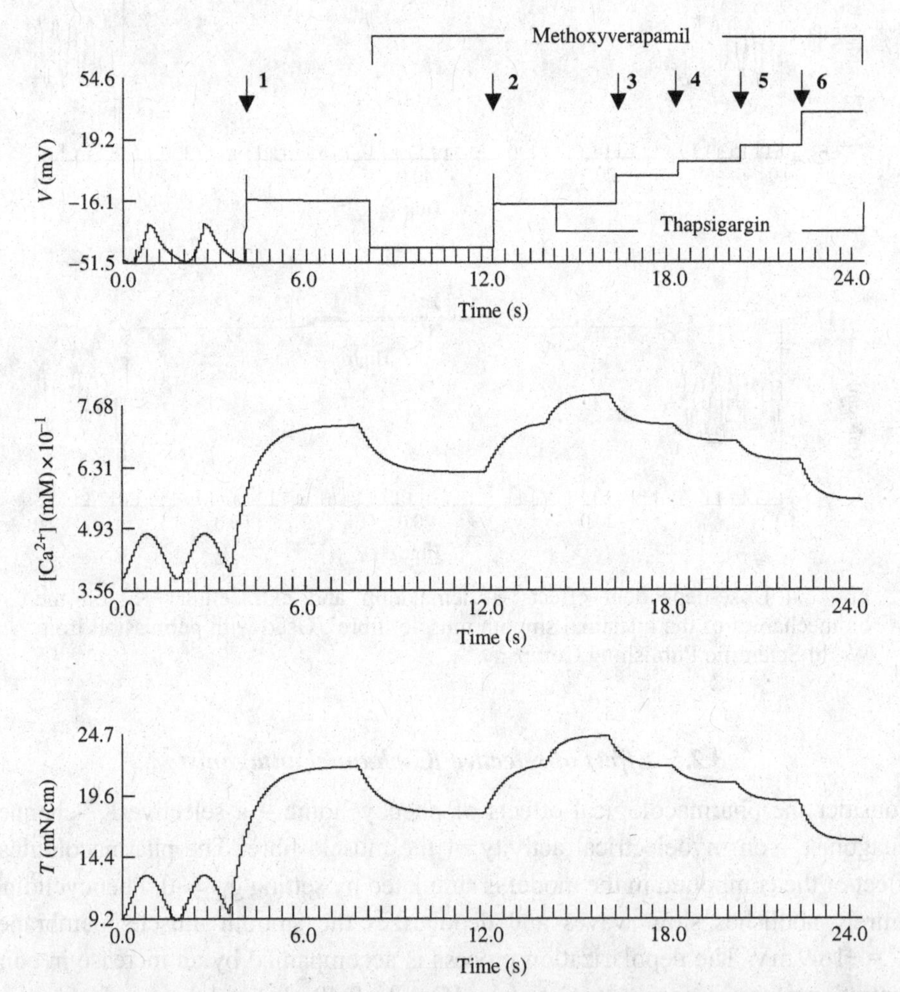

Fig. 8.5 Myoelectrical activity of the intestinal wall in the presence of high extracellular concentrations of Ca^{2+} and K^{+} ions, methoxyverapamil and thapsigargin. Used with permission from World Scientific Publishing Company.

Subsequent application of methoxyverapamil hyperpolarizes the fibre further, $V^s = -43.6$ mV. It also decreases the inward calcium current. As a result, there is a fall in concentration to $[Ca^{2+}] = 0.53$ μM and a fall in the total force to $T = 17.5$ mN/cm.

The addition of thapsigargin and the concurrent gradual increase in $[K^+]_0$ reverse the effect of methoxyverapamil. The muscle becomes depolarized. The level of depolarization depends on $[K^+]_0$. The total force dynamics is nonlinear and depends on the concentration of thapsigargin present. At 'low' concentrations of the compound there is an increase in intracellular calcium concentration, $[Ca^{2+}] = 0.76$ μM, and in the intensity of contraction, $T = 24.7$ mN/cm. As the concentration of thapsigargin continues to rise the concentration of Ca^{2+} begins to decline and the muscle fibre relaxes, min $T = 15$ mN/cm.

8.3 The small intestine as a soft cylindrical shell

The anatomical and physiological data about the small intestine discussed in Section 8.1 can be specified by the following modelling assumptions.

(i) The small intestine is a soft cylindrical shell formed of identical overlapping myogenic functional units (loci).

(ii) Each locus is of a given length L and radius r (Fig. 8.6).

(iii) The wall of the bioshell is composed of two smooth muscle layers embedded in the connective tissue network; muscle fibres in the outer layers are oriented longitudinally to the anatomical axis of the locus, and in the inner layer they run in the circumferential direction.

(iv) The tissue possesses nonlinear viscoelastic mechanical properties, which are uniform along the bioshell.

(v) Both muscle layers are electrogenic two-dimensional bisyncytia with cable electrical properties; the longitudinal layer has anisotropic and the circular layer isotropic electrical properties.

(vi) The self-oscillatory activity of syncytia, V_l and V_c, is a result of spatially distributed oscillators; the oscillators are divided into pools according to their natural frequencies.

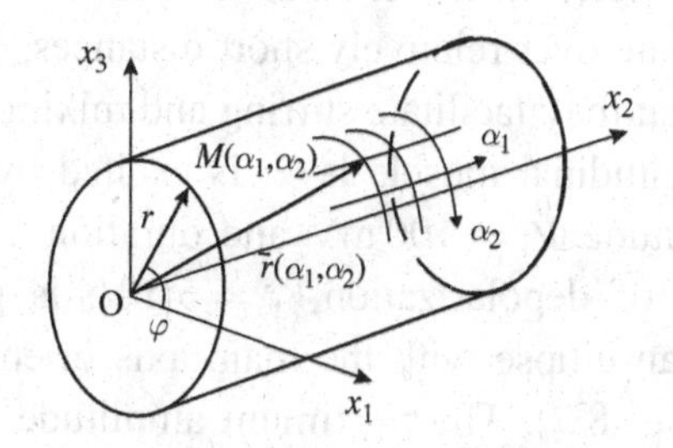

Fig. 8.6 A segment of the small intestine as a soft biological shell.

(vii) The slow-wave and spiking activity of the functional unit represents the integrated function of voltage-dependent L- and T-type Ca^{2+}, potential-sensitive K^+, Ca^{2+}-activated K^+ and leak Cl^- ion channels.

(viii) The role of pacemaker cells belongs to the ICC; their discharges, V_i, generate the propagating excitatory waves in the longitudinal and circular muscle syncytia, V_l^s and V_c^s.

(ix) V_l^s and V_c^s modulate the permeability of L-type Ca^{2+} channels; the effect is mainly chronotropic with an increase in the time of opening of the channel.

(x) Electromechanical coupling in the smooth muscle and the generation of contraction–relaxation forces are a result of the evolution of the excitatory waves; active forces of contraction, $T_{l,c}^a$, result from a multicascade process involving activation of the contractile protein system; passive forces, $T_{l,c}^p$, are explained by the mechanics of viscoelastic connective tissue stroma.

(xi) The stroma is formed from the collagen and elastin fibres arranged in a regular orthogonal net.

(xii) The bioshell is supported by intraluminal pressure p.

The system of equations that describes electromechanical processes in the bioshell – a segment of the small intestine – includes the equations of motion of the soft cylindrical bioshell, Eqs. (6.99); Eqs. (7.6)–(7.10) that describe the myoelectrical activity; Eqs. (7.11)–(7.15) to simulate the dynamics of the ICC; Eqs. (7.20)–(7.24) that model the process of propagation of the wave of excitation within electrically anisotropic longitudinal and electrically isotropic circular smooth-muscle syncytia; constitutive relations in the form Eq. (7.1); initial and boundary conditions Eqs. (7.25) and (7.37)–(7.39) (in simulations we shall always assume that the ends of the bioshell are clamped and the right boundary remains electrically unexcitable throughout); and an additional equation (7.36) that describes the dynamics of the intraluminal pressure.

8.3.1 Pendular movements

Pendular movements of the small intestine are a result of the electromechanical activity of the longitudinal muscle layer. The circular smooth muscle remains electrically and mechanically idle. Pendular movements are classified as local contractions and propagate over relatively short distances, 3–9 mm. Their physiological significance is that they facilitate stirring and mixing of intestinal content.

Assume that the longitudinal muscle layer is excited by a single discharge of the pacemaker of amplitude $\overset{0}{V}_i = 100\,\text{mV}$ and duration 1.5 s. Immediately after the excitation, a wave of depolarization, $V_l^s = 5\,\text{mV}$, is generated. Its anterior front has the shape of an ellipse with the main axis oriented in the direction of electrical anisotropy (Fig. 8.7). The maximum amplitude $V_l^s = 68.2\,\text{mV}$ is seen

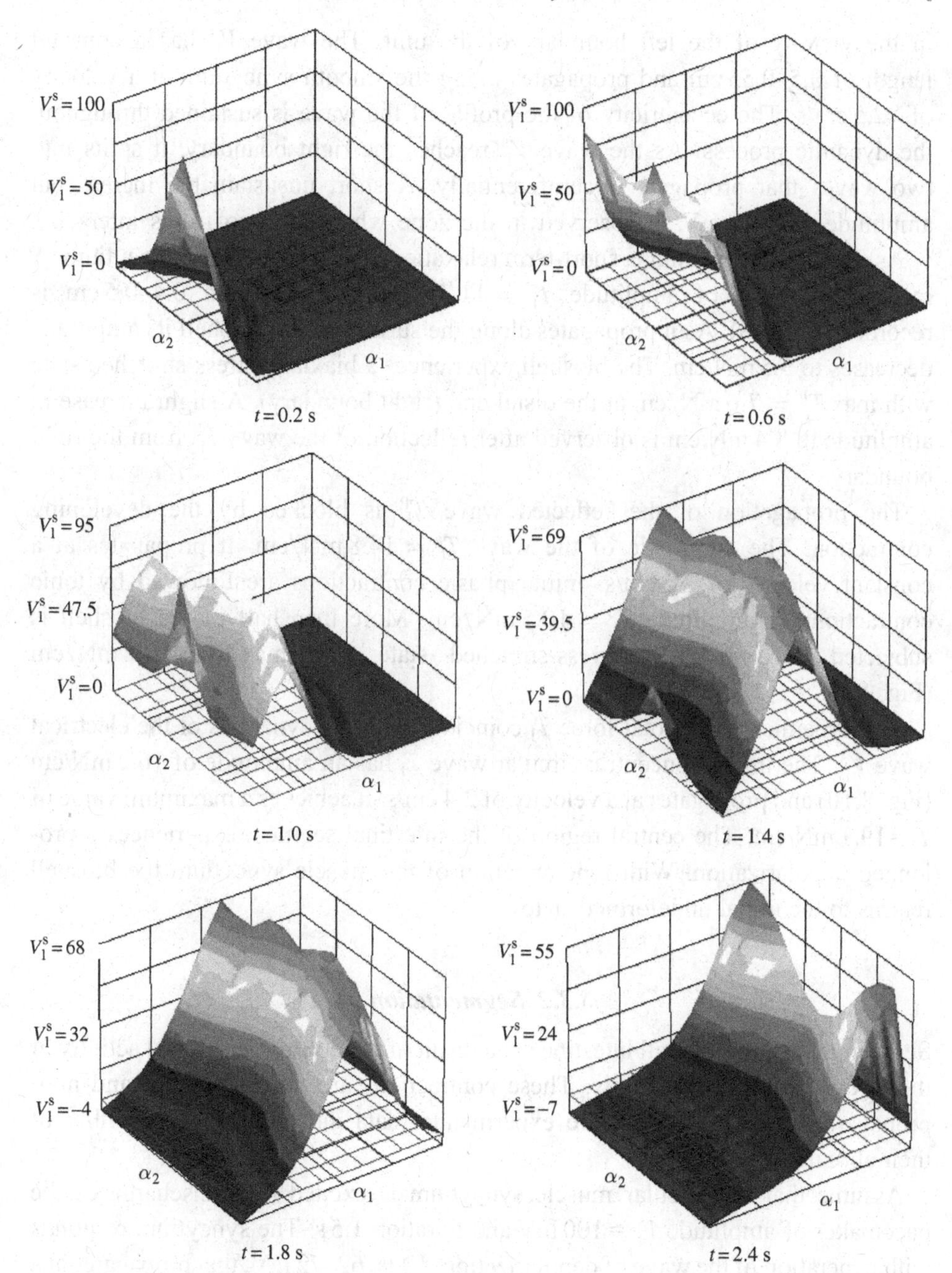

Fig. 8.7 Propagation of the wave of depolarization V_1^s within the longitudinal smooth muscle syncytium of the bioshell (intestine).

in the vicinity of the left boundary of the unit. The wave V_l^s has a constant length of 0.5–0.55 cm and propagates along the smooth syncytium at a velocity of ≈2.5 cm/s. The eccentricity of the profile of the wave is sustained throughout the dynamic process. As the wave V_l^s reaches the right boundary, it splits into two waves that propagate circumferentially. A short unsustainable increase in amplitude, to 78.8 mV, is observed in the zone where the two fronts interact.

As a result of excitation, a short-term relaxation is produced by the syncytium. A mechanical wave of amplitude $T_l^p = 11.7\,\mathrm{mN/cm}$ and length 0.4–0.5 cm is recorded (Fig. 8.8). As it propagates along the surface of the bioshell its amplitude decreases to 5.5 mN/cm. The bioshell experiences a biaxially stress-stretched state with max $T_l^p = 7.6\,\mathrm{mN/cm}$ at the distal end (right boundary). A slight increase in amplitude to 8.4 mN/cm is observed after reflection of the wave T_l^p from the right boundary.

The propagation of the reflected wave T_l^p is blocked by the developing contraction. The amplitude of the wave $T_l^a = 14.8\,\mathrm{mN/cm}$. It propagates at a constant velocity of 2.5 cm/s. Initial phasic contractions are followed by tonic contractions of amplitude $T_l^a = 10.5\,\mathrm{mN/cm}$. More than half of the bioshell is subjected to a biaxially stress-stretched state with max $T_l^a = 10.2\,\mathrm{mN/cm}$ (Fig. 8.9).

The dynamics of the total force T_l coincides with the dynamics of the electrical wave V_l^s. The initial concentric circular wave T_l has an amplitude of 16.2 mN/cm (Fig. 8.10) and propagates at a velocity of 2.4 cm/s. It achieves a maximum value of $T_l = 19.1$ mN/cm. The central region of the intestinal segment experiences a prolonged depolarization. With repolarization of the muscle syncytium the bioshell returns to the initial undeformed state.

8.3.2 Segmentation

Segmentations of the small intestine are a result of the electromechanical activity of the circular muscle layer only. These contractions are normally focal and non-propagating. However, there are experimental data indicating the possibility of their aboral propagation.

Assume that the circular muscle syncytium is excited by a discharge of the pacemaker of amplitude $V_i = 100$ mV and duration 1.5 s. The syncytium responds with generation of the wave of depolarization, $V_c^s = 62$–72 mV, that propagates at a velocity of 1.9–2.0 cm/s along the bioshell. Its anterior front has the shape of a circle (Fig. 8.11). As in the case of pendular movements, an unsustainable increase in the amplitude of V_c^s to 82 mV is observed in the vicinity of the right boundary. It is a result of the interaction of the waves V_c^s propagating towards each other in the circumferential direction. Because of the slower conduction velocity the circular

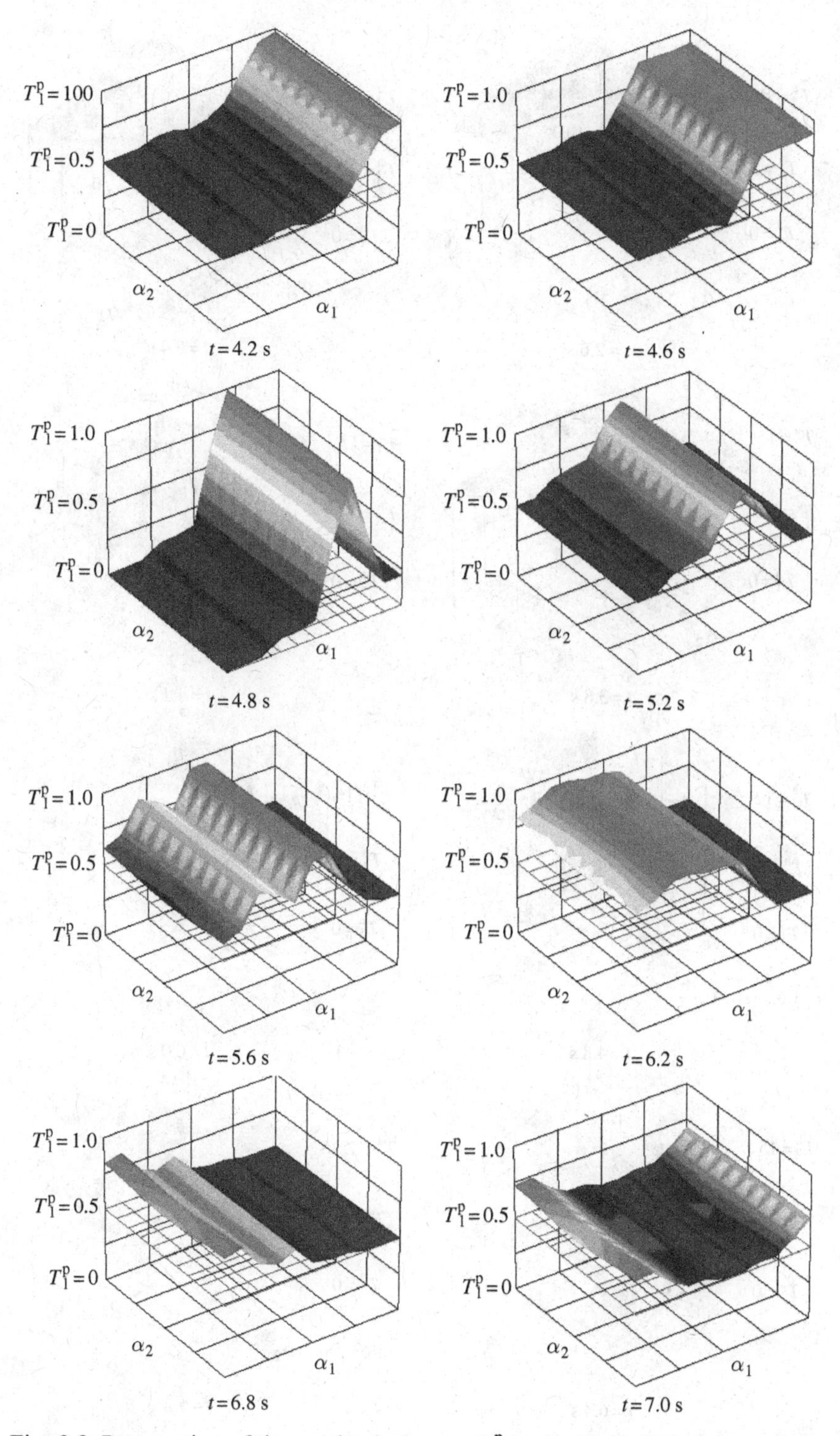

Fig. 8.8 Propagation of the mechanical wave T_1^p in the bioshell during pendular movements.

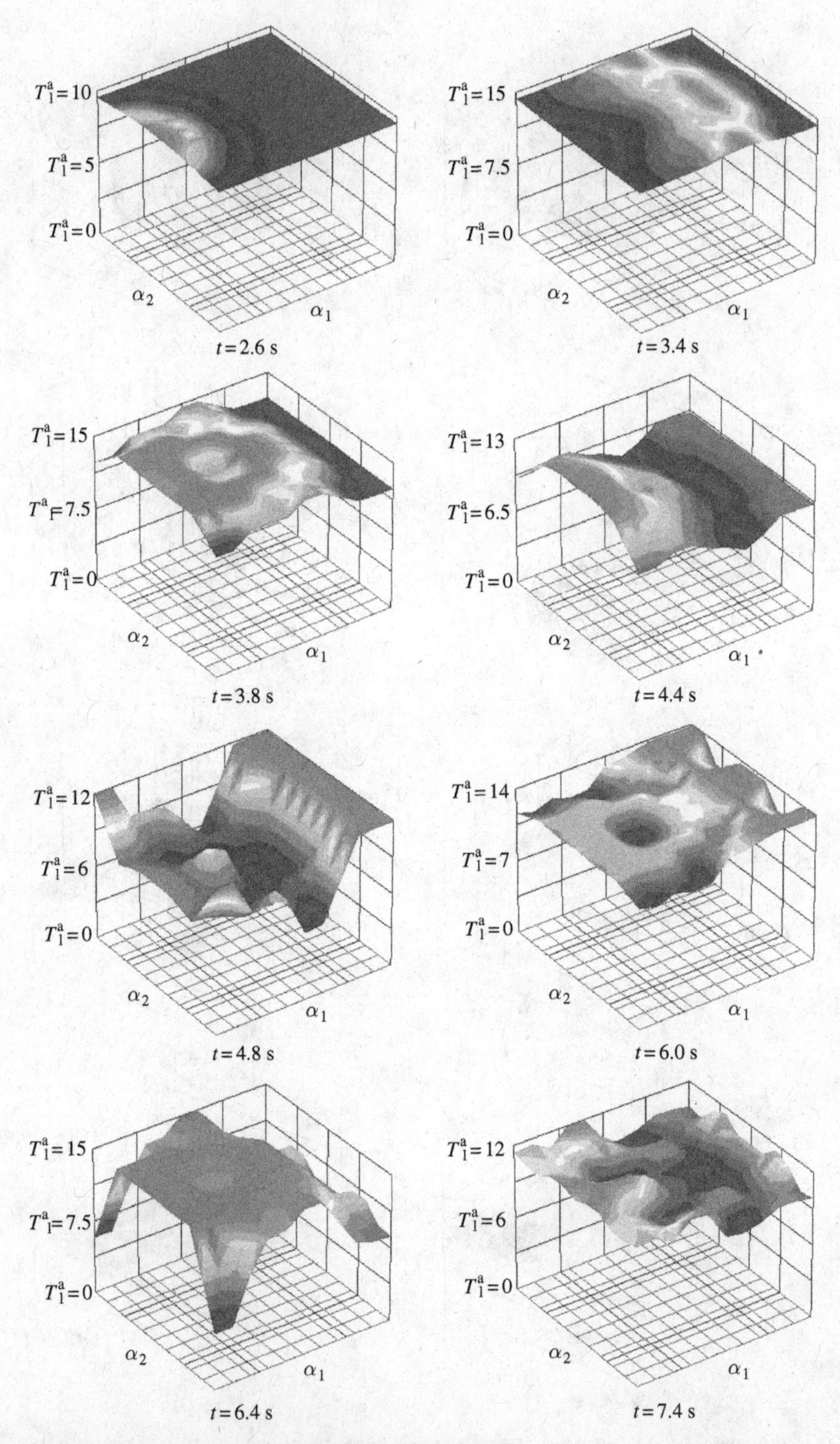

Fig. 8.9 Dynamics of the wave T_1^a in the bioshell.

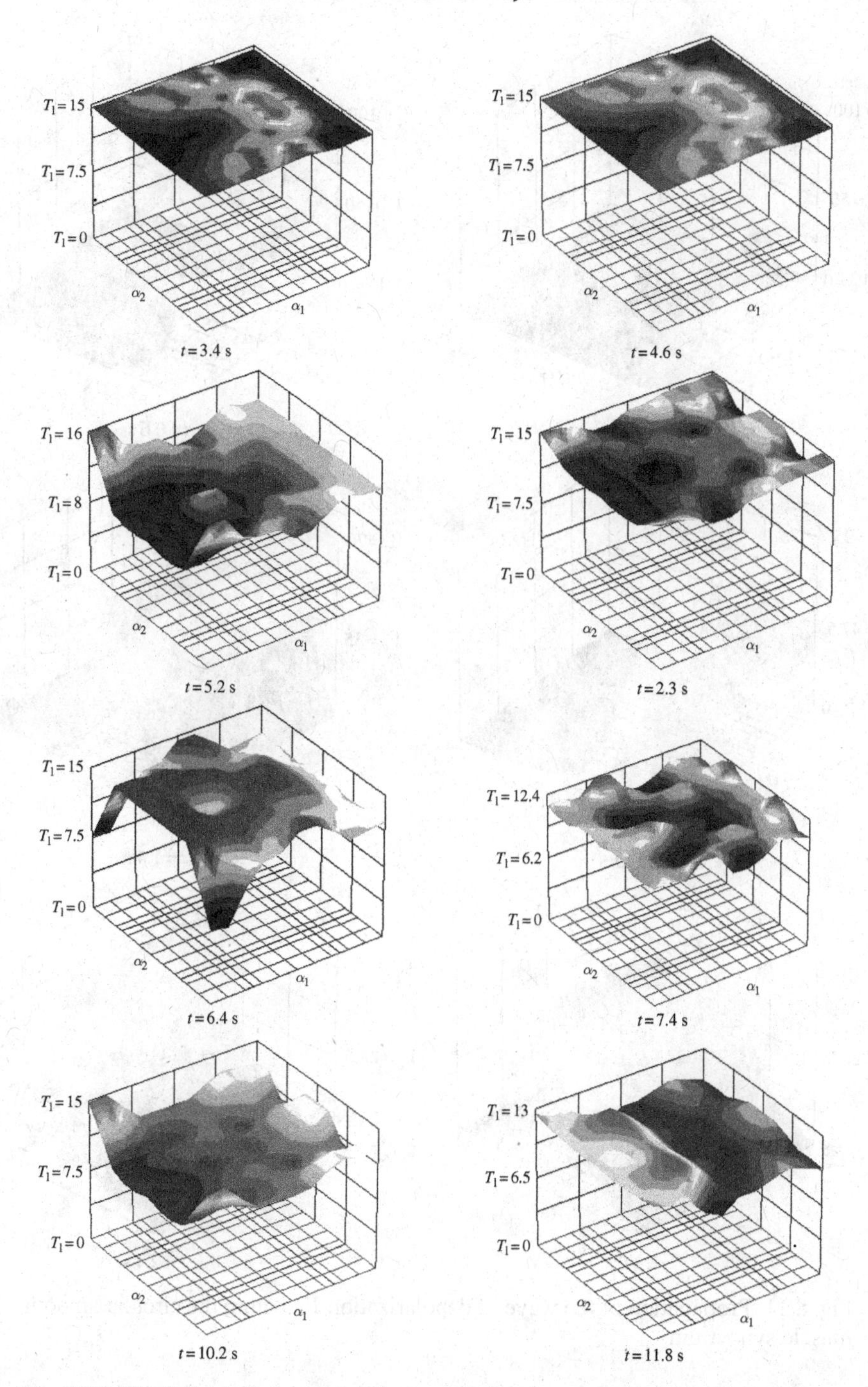

Fig. 8.10 Total force T_1 dynamics during pendular movements.

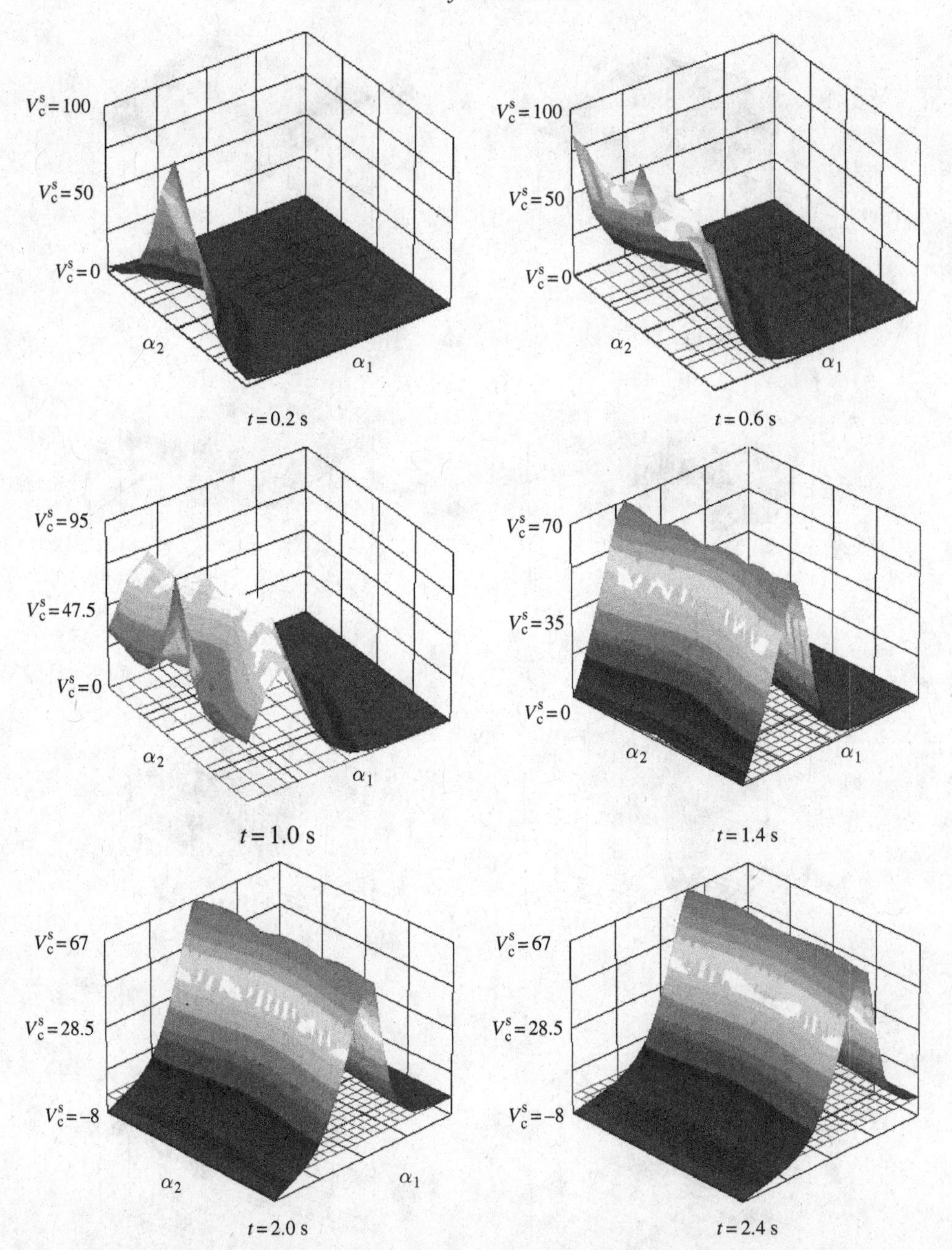

Fig. 8.11 Propagation of the wave of depolarization V_c^s within the circular smooth muscle syncytium.

muscle syncytium experiences depolarization for a longer period than does the longitudinal muscle layer.

The wave of the active forces of contraction $T_c^a = 12.2\,\text{mN/cm}$ that is produced propagates at a velocity of 1.8–1.9 cm/s (Fig. 8.12). With the development of a tonic contraction, more than half of the bioshell experiences a uniform biaxial stretching with max $T_c^a = 18.6\,\text{mN/cm}$.

The wave T_c of amplitude 28–32 mN/cm propagates in the aboral direction at a velocity of 0.5 cm/s (Fig. 8.13). The wave T_c maintains a constant length of 0.6 cm. As the wave reaches the right boundary, the wave T_c^p reflects from the boundary and begins to propagate backwards. However, it is stopped by the contraction of the muscle syncytium and the wave $T_c^a = 17.8\,\text{mN/cm}$.

The changes in configuration of the bioshell exhibit asymmetry during the first (phasic) stage of segmental contractions. Only with the development of tonic contractions is the symmetry in deformation of the proximal part of the bioshell observed.

8.3.3 Peristaltic movements

Reciprocal relations between the longitudinal and circular smooth muscle layers are essential in the development of peristalsis. The first contraction normally starts in the longitudinal syncytium. When the total force in the layer reaches a maximum, an activation of the circular muscle layer begins. This coincides with simultaneous relaxation of the longitudinal layer and vice versa.

Let two identical pacemaker cells be located in the longitudinal and circular smooth muscle syncytia. They discharge electrical impulses of amplitude $V_i = 100\,\text{mV}$ and duration 1.5 s. The time lag between discharges is specified by the dynamics of the development of the total force, T_l. The discharge of the pacemaker cell on the outer muscle layer of the bioshell initiates the excitatory wave of depolarization $V_l^s = 69\,\text{mV}$. The dynamics characteristic of the wave V_l^s and the induced mechanical wave T_l are similar to those described in Section 7.2.1. With the achievement of the maximum force of contraction the ICC_{MY} on the circular muscle syncytium discharges action potentials. They generate the electromechanical wave in the circular smooth muscle syncytium. The pattern of movements of the wall and the force dynamics resemble processes observed during pendular movements and segmentation (Fig. 8.14).

8.3.4 Self-sustained periodic activity

The frequency of discharges of ICC_{MY} and the electrical properties of the intestinal wall play a dominant role in the development of self-sustained periodic myoelectrical activity in the intestine.

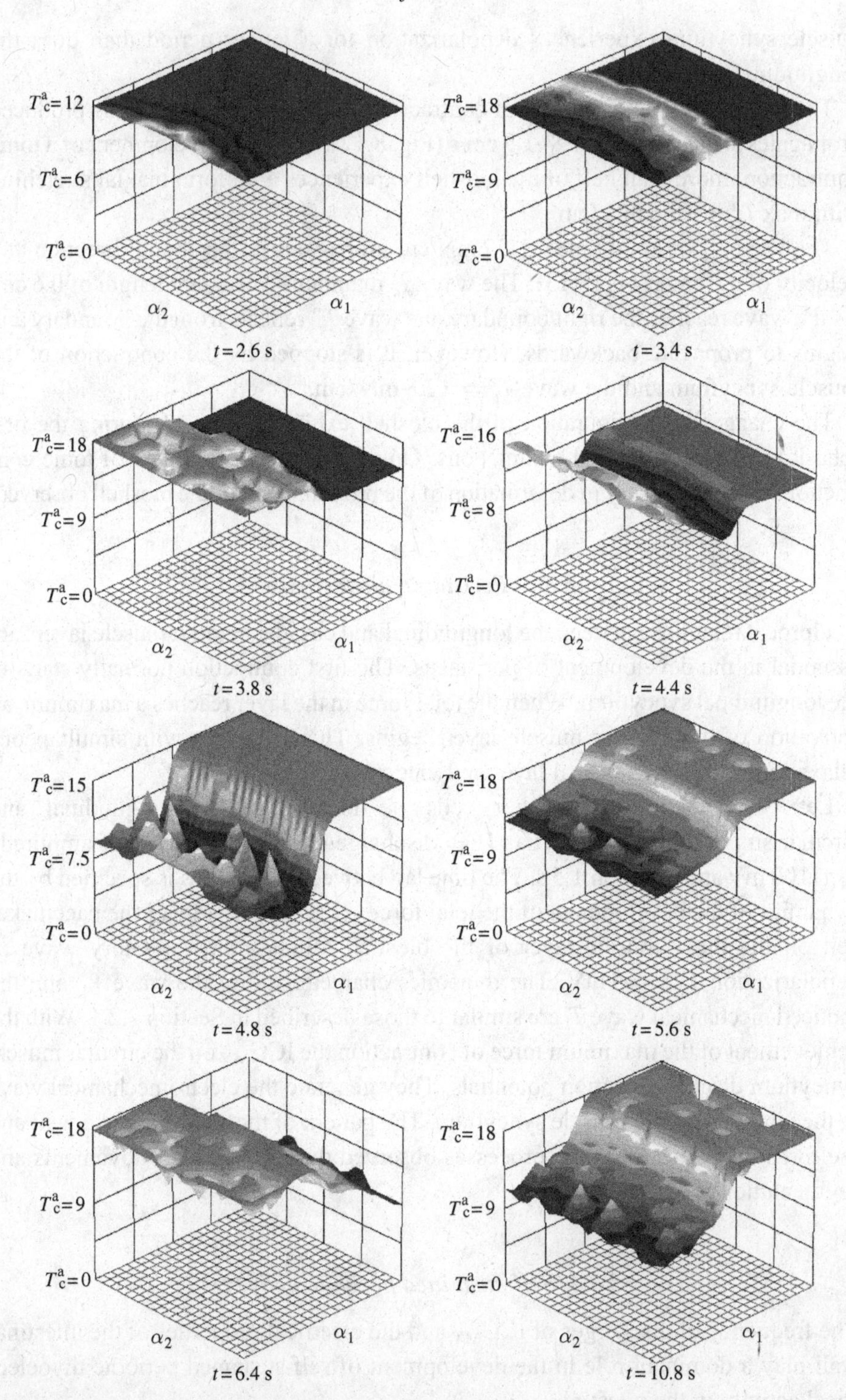

Fig. 8.12 Dynamics of the wave T^a_c in the bioshell.

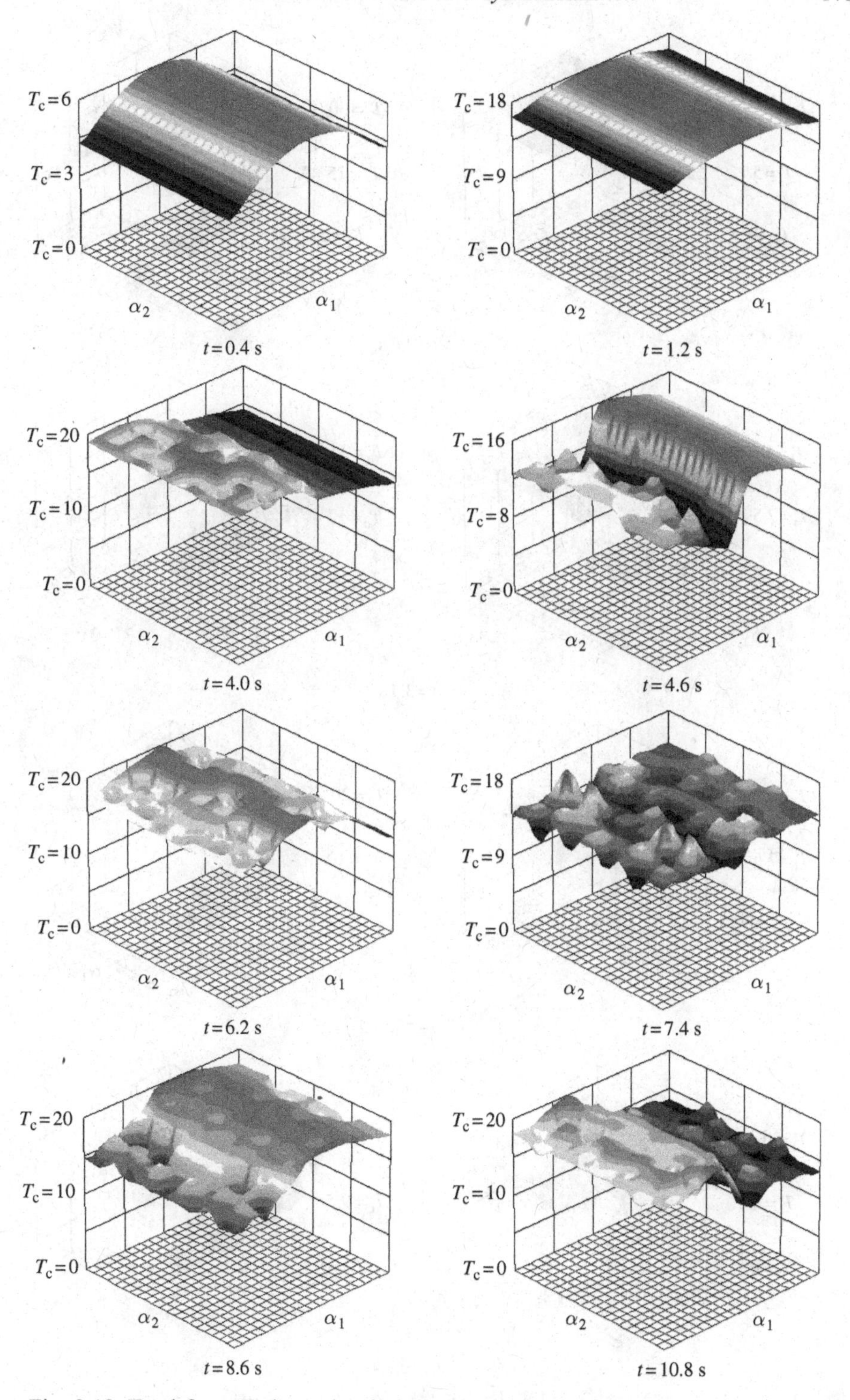

Fig. 8.13 Total-force T_c dynamics during segmental contractions.

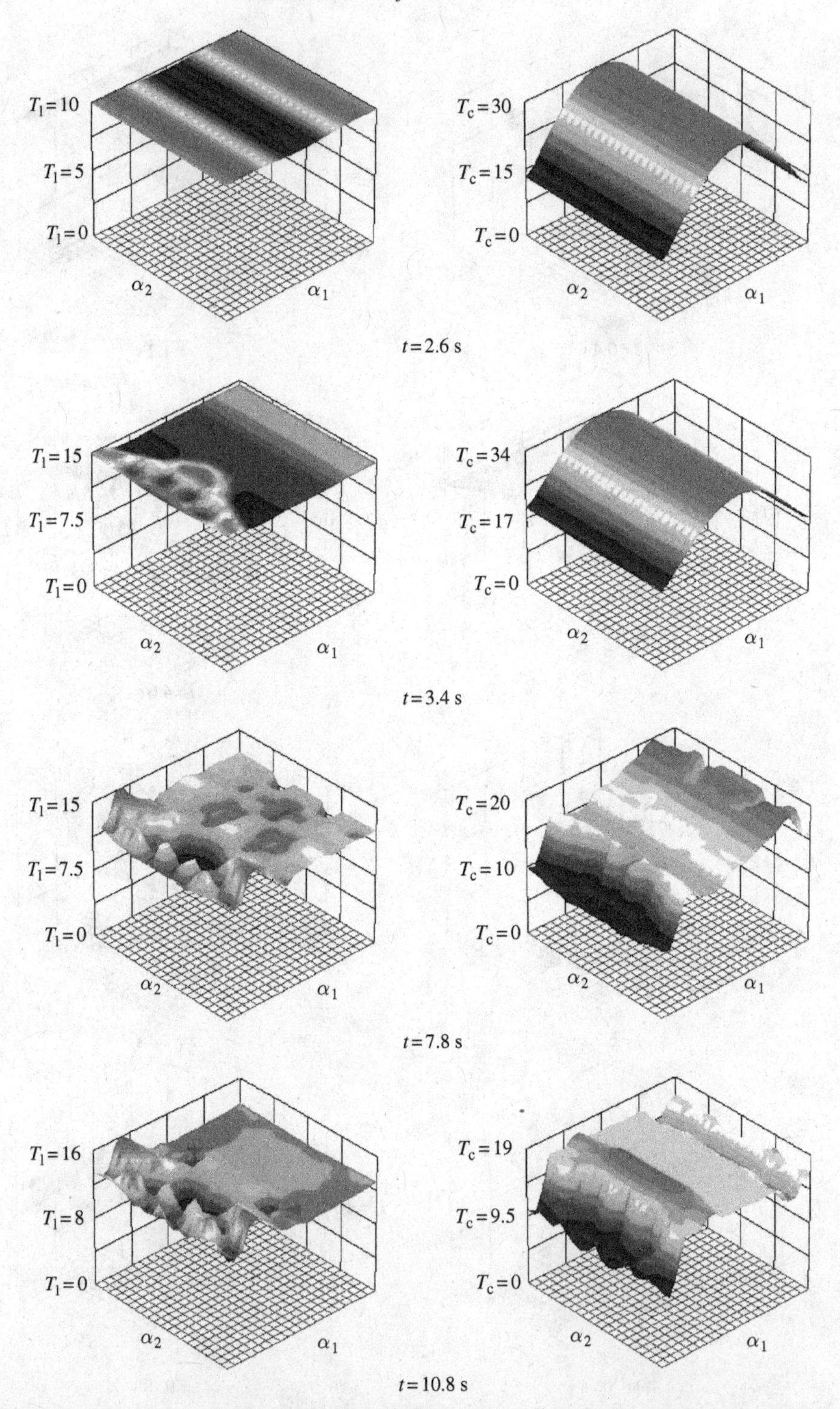

Fig. 8.14 Total-force T_l and T_c distributions in the bioshell during peristalsis.

Let the frequency of firing be $\nu = 0.16$ Hz. Following discharges of the longitudinal and circular smooth muscle syncytia the bioshell produces peristaltic movements as described above. At $t = 13.3$ s of the dynamic process the middle part of the bioshell experiences a biaxially stress-stretched state, $\lambda_l = 1.35$, $\lambda_c = 1.42$. Stretching of the syncytia causes an increase in membrane resistance (Eq. (7.18)) and as a result a decrease of the maximum amplitude of the depolarization waves, $V_l^s \approx V_c^s \approx 55.7$ mV.

This has a detrimental effect on the connectivity among myogenic oscillators in that region and on the dynamics of the propagation of succeeding waves, V_l^s. They split into two separate waves in the 'affected' part (Fig. 8.15). As the two separate waves reach the right boundary of the bioshell, their tails collide, with the generation of a new solitary wave $V_l^s = 70.8$ mV. It is strong enough to self-sustain its propagation backwards, i.e. from the electrically unexcitable right boundary towards the left boundary. As the reflected wave reaches the distended part of the bioshell it splits into two separate waves and thus produces a spiral wave of amplitude 65 mV that continues to circulate over the surface of the intestinal segment. The spiral waves provide strong connections among the spatially distributed myogenic oscillators and, therefore, support mechanical wave activity in the longitudinal syncytium.

The self-sustained spiral-wave phenomenon is produced only by the electrically anisotropic longitudinal smooth muscle syncytium. It can never be simulated by the electrically isotropic circular smooth muscle syncytium. Waves V_c^s propagate without disruption along the surface of the bioshell, after which they vanish at the right boundary.

8.3.5 Effect of lidocaine

The spiral-wave phenomenon, as described above, could be a physiological mechanism of intestinal dysrhythmia – a medical condition associated with altered motility of the small intestine. In an attempt to abort the spiral-wave formation and to convert the system back to normal, consider the effect of lidocaine. The drug blocks both fast, sodium-dependent action potentials and voltage-dependent, non-inactivating Na^+ conductance. Its pharmacological action in the model is achieved by decreasing the maximal conductance $\tilde{g}_{Na}$. At 'high' concentrations lidocaine abolishes spiral waves and mechanical activity in the bioshell. The intestinal segment returns to a resting state.

Although lidocaine 'successfully reversed' dysrhythmia in the model, its clinical application is limited owing to its narrow therapeutic index and possible high-dose-induced cardiotoxicity.

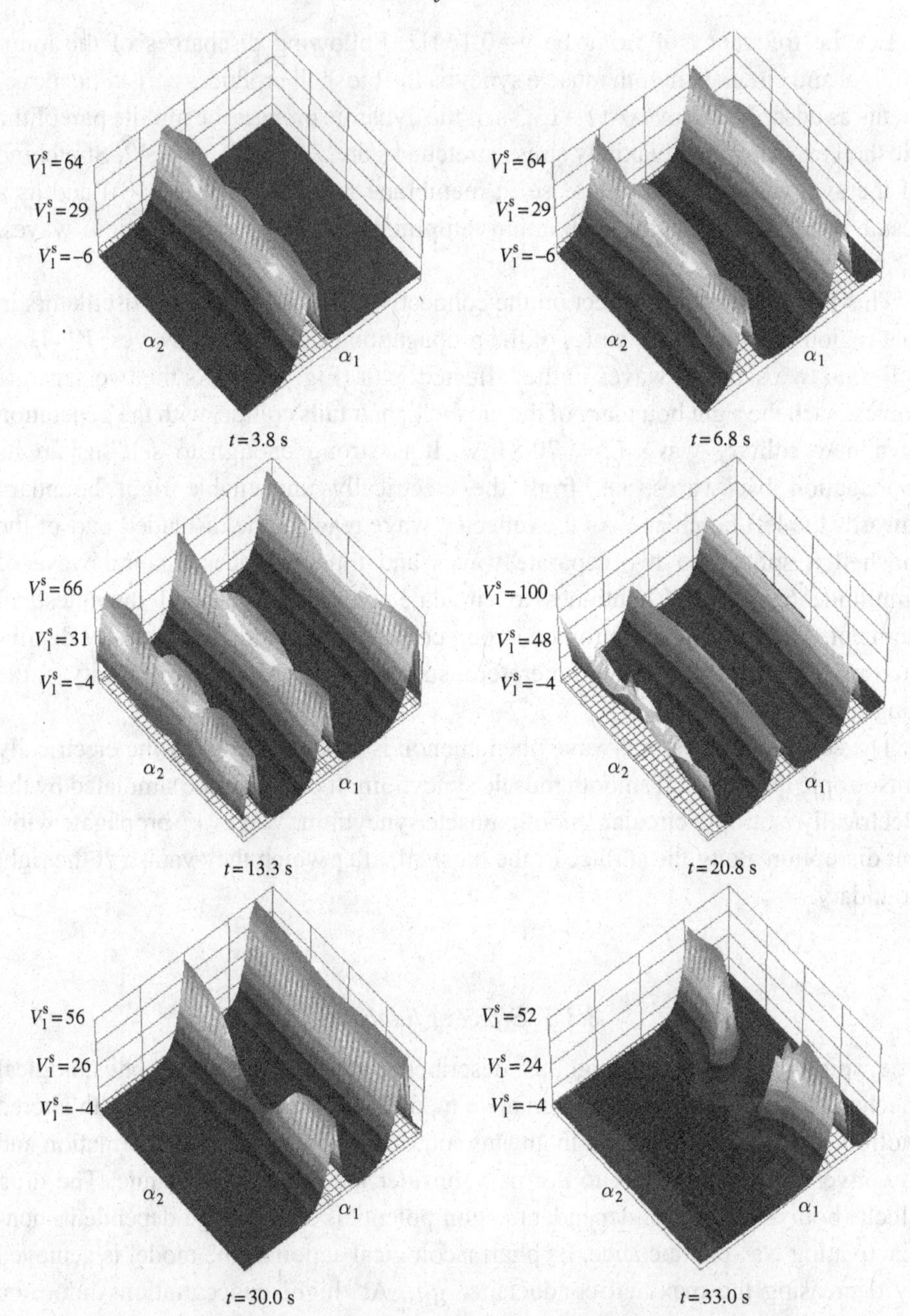

Fig. 8.15 Dynamics of self-sustained myoelectrical activity and spiral-wave formation in the longitudinal smooth muscle syncytium.

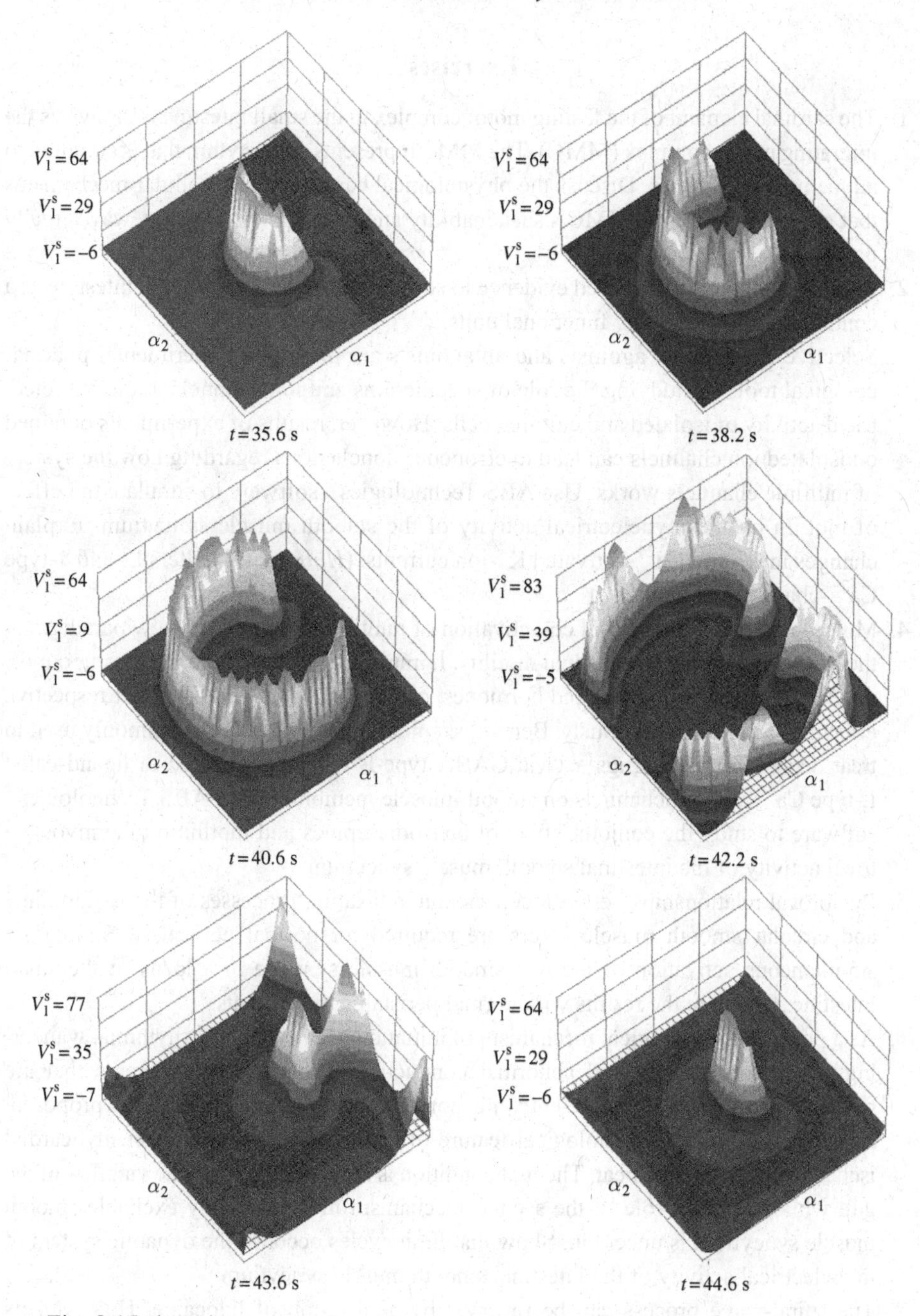

$V_1^s=64$
$V_1^s=29$
$V_1^s=-6$
α_2
α_1
$t=35.6$ s
$V_1^s=64$
$V_1^s=29$
$V_1^s=-6$
α_2
α_1
$t=38.2$ s
$V_1^s=64$
$V_1^s=9$
$V_1^s=-6$
α_2
α_1
$t=40.6$ s
$V_1^s=83$
$V_1^s=39$
$V_1^s=-5$
α_2
α_1
$t=42.2$ s
$V_1^s=77$
$V_1^s=35$
$V_1^s=-7$
α_2
α_1
$t=43.6$ s
$V_1^s=64$
$V_1^s=29$
$V_1^s=-6$
α_2
α_1
$t=44.6$ s

Fig. 8.15 (cont.)

Exercises

1. The cardinal element of the fasting motor complex in the small intestine is known as the migrating motor complex (MMC). The MMC represents a biorhythm that is common to all mammalian species. Discuss the physiological basics and intracellular mechanisms that are essential for the MMC's sustainability and propagation. Do slow waves really propagate?
2. Provide physiologically based evidence to support the model of the small intestine as a continuum of overlapping functional units.
3. Selective ion-channel agonists and antagonists are important experimental pharmacological tools in studying the role of specific ions and ion channels in the myoelectrical activity of isolated and cultured cells. However, results of experiments obtained on isolated ion channels can lead to erroneous conclusions regarding how the system of multiple channels works. Use ABS Technologies© software to simulate the effect of motilin on the myoelectrical activity of the smooth muscle syncytium. Explain changes in K^+ and Ca^{2+}-activated K^+-ion currents. (*Hint*: motilin alters L- and T-type Ca^{2+}-channel conductivity.)
4. Multiple co-transmission and co-activation of multiple receptors is a new paradigm in the physiology of gastrointestinal motility. It implies that various chemical compounds, i.e. drugs, neurotransmitters and hormones, can be present and induce their respective effects on a cell simultaneously. Benzodiazepines are drugs that are commonly used to treat depression. The drugs act via GABA type-B receptors coupled to ligand-gated L-type Ca^{2+} and Cl^- channels on smooth-muscle membrane. Use ABS Technologies© software to study the conjoint effect of benzodiazepines and motilin on the myoelectrical activity of the intestinal smooth muscle syncytium.
5. Reciprocal relationships between contraction–relaxation processes of the longitudinal and circular smooth muscle layers are required for normal peristalsis. Simulate a simultaneous activation of the two smooth muscle syncytia in a locus of the small intestine. Compare the results with normal peristaltic movements.
6. As a possible physiological mechanism of initiation of intestinal dysrhythmia, wandering pacemaker activity and abnormal conduction – the analogous events that are responsible for trigger and re-entry phenomena in the heart – have been proposed. However, the primary pathological feature of cardiac arhythmia is either myocardial ischaemia or an ectopic beat. The first condition is incompatible with the viability of the gut, while the exact role of the second mechanism in a not highly excitable smooth muscle syncytium is uncertain. Show that limit cycles occur in the dynamic system of myoelectrical activity of the intestinal smooth muscle syncytium.
7. The spiral-wave process can be reversed by application of lidocaine. However, its clinical use is limited because of possible cardiac toxicity. Use ABS Technologies© software to find other possible 'remedies'.
8. Current hypotheses regarding irritable-bowel syndrome are based on the concepts of altered visceral sensation and motor dysfunction. Visceral hypersensitivity is the

ubiquitous finding in these diseases and presents in the form of either (i) allodynia or (ii) hyperalgesia. The dysfunction of abdominal viscera is recorded primarily as altered motility patterns, which are neither constant nor specific. Propose a basic biomechanical principle underlying irritable-bowel syndrome. Use ABS Technologies© software to study the motility patterns of the small intestine in irritable-bowel syndrome.

9. Immunohistochemical and radio-ligand binding studies have demonstrated the presence in the small intestine of multiple neurotransmitters, e.g. acetylcholine (ACh), noradrenalin and serotonin, that play significant roles in electrochemical and chemo-electrical coupling processes. Formulate a phenomenological mathematical model of cholinergic neurotransmission.
10. The causative mechanism of dumping syndrome is distension of the duodenum with intraluminal content. It affects the gradual reflex – contractions in the longitudinal smooth muscle layer – which is the preliminary component of organized peristaltic movements. Study the effect of overstretching on the propagation of the electrical wave of depolarization in the longitudinal smooth muscle syncytium.
11. Use ABS Technologies© software to study the effect of stiffening of the connective tissue stroma of the wall on the motility of the functional unit.

9

Biomechanics of the large intestine

9.1 Anatomical and physiological background

The human large intestine (colon) is a visceral organ that lies with loops and flexures in varying configurations around the abdomen. The length of the organ is 125–154 cm and its diameter is approximately 4.5 cm. The colon is functionally divided into two parts, the right and left colon. The right colon extends from the caecum and the ascending colon to the mid transverse colon, and the left colon from the mid transverse colon through the descending colon and sigmoid to the rectum.

The wall of the organ consists of four layers – the mucosa, submucosa, circular and longitudinal muscle layers and serosa. The thickness of the wall of the large intestine is relatively constant, $h \approx 0.4$–0.5 mm. Cells lining the mucosa and submucosa resemble those found in the small intestine. However, they contain significantly greater numbers of goblet cells. They secrete viscous mucus into the lumen and thus moisturize and lubricate the passage of the waste. The layers play a major role in digestion and absorption of food, water and electrolytes. It is the absorption of fluids and bacterial processing that transform the intraluminal effluent into solid stool.

The longitudinal muscle is organized in three bands – teniae coli. They run from the caecum to the rectum, where they fuse together to form a uniform outer muscular layer. The circular muscle layer is homogeneous and uniformly covers the entire colon.

The serosa is composed of a thin sheet of epithelial cells and connective tissue.

The innervation of the colon is a complex interaction between the enteric nervous and autonomic nervous systems. The cell bodies of neurons in the enteric nervous system are organized into spatially distributed ganglia with interconnecting fibre tracts. They form the submucosal and myenteric plexi that contain local neural reflex circuits, which modulate multiple functions of the organ. The autonomic nervous system comprises sensory, motor, sympathetic and parasympathetic nerves. The autonomic nerves modulate the intramural enteric neural circuits and provide neural reflexes at the higher organizational levels, including the autonomic ganglia, spinal chord and brain.

Although it is generally accepted that colon movements are regulated by the ICC, the exact mechanisms of coordinated motility are not known. A subpopulation of cells distributed along the submucosal border of the circular smooth muscle layer, ICC_{SM}, is responsible for the myoelectrical activity of the large intestine and plays the key role in its pacemaker activity.

Five types of motor patterns are observed in the organ: (1) haustral churning, (2) peristalsis, (3) propulsive movements, (4) defecation reflex and (5) cooperative abdominal effort. Haustral churning is a combined simultaneous shallow contraction of the longitudinal and circular smooth muscles. The significance of these motor patterns is that they serve to stir up the liquid intraluminal content as fluids are extracted until the stool is formed. The contractions propagate over short distances upstream and downstream from their point of origin. The coordinated reciprocal electromechanical activity of the muscle layers is known as peristalsis. It gently moves the soft content along through the right colon to the rectum, from where it is evacuated. A modified type of peristalsis is propulsive movements. They are characterized by extensive clustered contractions of the colon separated by short dilated regions and propulsion *en masse* of the faecal material. The defecation reflex and cooperative abdominal effort represent the final stage of expulsion of stools from the rectum.

The molecular and electrophysiological processes underlying colonic functions are similar to those described for the stomach and small intestine. The migrating motor complex is a fundamental motility phenomenon and is tightly controlled by the enteric reflex pathways. It is known that they are evoked by chemical or mechanical stimulation of intrinsic primary afferent neurons since their blockade with tetrodotoxin abolishes all types of myoelectrical activity. In contrast, stimulation of nicotinic ganglions and muscarinic receptors with ACh together with substance P and serotonin induces rapid excitatory responses. The major inhibitory neurotransmitter is assumed to be nitric oxide. In the case of a congenital absence of nitric oxide-containing neurons, the colon fails to relax and remains constricted at all times.

The migrating motor complexes have patterns typical of periodic oscillatory activity. Phases of rapid contractions and action-potential production are separated by periods of quiescence. The patterns are reproduced in all parts of the colon at a constant frequency. However, their duration and amplitude vary significantly along the length of the organ. The longest complexes of the highest amplitude are recorded in the right colon, $\approx$3–4 min, and the shortest of small amplitude in the left colon, $\approx$0.7–1.2 min. The high-amplitude action potentials are produced in the caecum and the ascending part of the colon and the lowest in the recto-sigmoid region. It is noteworthy that migrating motor complexes are intimately related to the dynamics of pressure waves produced in the intraluminal content.

A common condition associated with altered motility of the colon is constipation. The underlying mechanisms are poorly understood and may vary among different groups of patients. Some cases result from systemic diseases, others from abnormalities within the colonic wall itself. Transit through particular regions of the large bowel can be measured using a variety of techniques, including colonic scintigraphy, magnetic resonance imaging, use of radio-labelled markers (pellets) and video mapping. The plastic or metallic pellets used usually have the shape of a sphere or ellipsoid. Reported physiological transit times are 7–24 hours for the right colon, 9–30 hours for the left colon and 12–44 hours for the recto-sigmoid. Although the methods allow scientists to appreciate gross variations in motility patterns, they do not offer (i) the desired resolution to establish the relationship between the spatiotemporal distributions of migrating motor complexes and (ii) sufficient depth of accessibility to allow a combined analysis of intricate mechanisms of function of the system.

Over the last decade, considerable effort has been directed towards investigating peristaltic propulsion of, mainly, Newtonian and non-Newtonian fluids in the intestine. There are large numbers of original publications and excellent reviews on the subject, to which the interested reader should refer for details. In contrast, the research into propulsion of solids and biologically active chyme is very limited. In their pioneering work, Bertuzzi *et al.* (1983) formulated a mathematical model and studied the dynamics of propulsion of a non-deformable bolus in a viscoelastic tube – a segment of the small bowel. Even despite severe biological naïveté and the limiting assumption of axially symmetric deformation, they reproduced the propagation of a ring-like electromechanical wave, to simulate *en masse* movement of a solid sphere and to calculate the average velocity of transit. Miftakhov and Abdusheva (1993) studied the general physiological principles of peristaltic transport of a solid bolus in a segment of the gut. The authors analysed the dynamics of propulsion as a result of the electromechanical wave dynamics in the smooth muscle syncytia. Miftahof and Fedotov (2005) and Miftahof and Akhmadeev (2007) recently investigated the motion of the bolus in a segment of the small intestine under normal physiological conditions and after application of drugs that affect peristaltic activity of the organ.

9.2 The colon as a soft shell

A segment of the large intestine can be viewed as a hollow muscular tube – a bioshell of length L and radius (α_1, α_2), where α_1 and α_2 are the orthogonal curvilinear coordinates on the undeformed surface of the bioshell (Fig. 9.1).

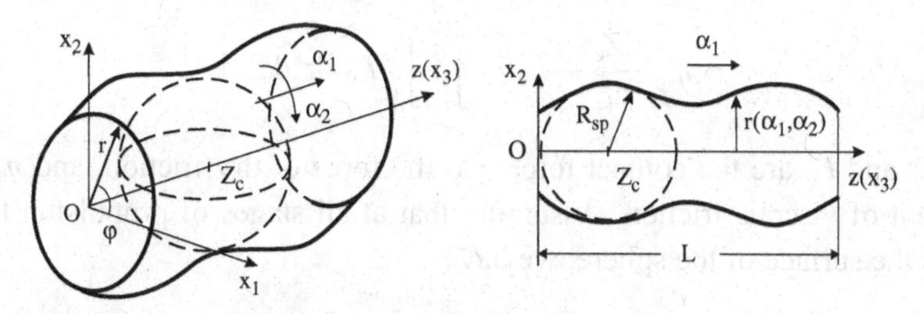

Fig. 9.1 A segment of the large intestine as a soft bioshell.

Let the bioshell contain a solid non-deformable sphere of radius R_{sp} = constant. Given the anatomical and physiological characteristics of the large intestine, we make the following assumptions.

(i) The tissue possesses the property of nonlinear viscoelastic orthotropy (there are no reliable experimental data on uniaxial and biaxial mechanical characteristics of either animal or human colon currently; therefore, our modelling assumptions are based on comparative histomorphological and biomechanical analysis of the small and large intestine).
(ii) The teniae coli and circular smooth muscle layer are excitable syncytia; the teniae have anisotropic and the circular layer has isotropic electrical properties.
(iii) The role of pacemaker cells belongs to the subpopulation of interstitial cells of Cajal – ICC_{SM}; they discharge action potentials V_i of known frequency, amplitude and duration.
(iv) The generation and propagation of the waves of depolarization, V_l^s and V_c^s, along the syncytia are a result of integrated function of voltage-dependent L- and T-type Ca^{2+}, potential-sensitive K^+, Ca^{2+}-activated K^+ and leak Cl^- ion channels.
(v) The waves V_l^s and V_c^s modulate the permeability of L-type Ca^{2+} channels on the smooth muscle membrane.
(vi) Multicascade processes of the activation of intracellular contractile proteins lead to production of active forces, $T_{l,c}^a$, whereas passive forces, $T_{l,c}^p$, are the result of deformation of the viscoelastic connective tissue stroma; the stroma is formed of the collagen and elastin fibres arranged in a regular orthogonal network.
(vii) The bioshell contains a solid spherical pellet and is supported by intraluminal pressure p. The pellet is subjected to dry and viscous friction; the contact forces act perpendicular to the surface of the pellet.

The mathematical formulation of the problem of peristaltic propulsion of a solid non-deformable sphere by a segment of the colon includes the equations of motion of the soft bioshell Eqs. (6.99); Eqs. (7.6)–(7.10) for the myoelectrical activity; Eqs. (7.11)–(7.15) for the dynamics of ICC_{SM}; Eqs. (7.20)–(7.24) for propagation of the wave of excitation within the electrically anisotropic and isotropic smooth muscle syncytia; constitutive relations Eq. (7.1); Eq. (7.36) for the change of intraluminal pressure; and the equation of motion of the pellet

$$\eta_{sp} \frac{dZ_c}{dt} + F_d = \int_{z_1}^{z_2} \int_{r_0}^{r} F_c \, dz \, d\zeta, \tag{9.1}$$

where F_c and F_d are the contact force and the force of dry friction, and η_{sp} is the coefficient of viscous friction. Assuming that at all stages of propulsion the wall contacts the surface of the sphere, we have

$$K_{sp} = (Z_c - u_1)^2 + (r_0 - u_2)^2 + (r_0 - \omega)^2 - R_{sp}^2 < 0, \qquad z \in [z_1, z_2]. \tag{9.2}$$

Here u_1, u_2 and ω are components of the displacement vector, Z_c is the position of the centre of the pellet at time t, and z_1 and z_2 are the boundary points of contact of the pellet and the bioshell. In all simulations the initial position of the pellet is assumed to be known a priori.

The initial conditions specified by Eqs. (7.25), (7.38) and (7.39) state that the bioshell is in the resting state. It is excited by a series of electrical discharges of pacemaker cells located at the left boundary. The impulses have a constant amplitude of $V_i = 100$ mV, duration $t_d = 1.5$ s and frequency $\nu = 0.016$ Hz. Depending on the type of movement, different smooth muscle layers composing the wall of the shell are excited. In peristalsis, a reciprocal relation between the teniae coli and circular muscle syncytium has been ascertained in contraction–relaxation. The first contractions start in the longitudinal layer, being followed by activation of the circular layer.

The left boundary of the bioshell is clamped throughout,

$$\begin{aligned} r(\alpha_1, 0) = r_0, \qquad \varphi(\alpha_1, 0) = z(\alpha_1, 0) = 0, \\ v_r(\alpha_1, 0) = v_\varphi(\alpha_1, 0) = v_z(\alpha_1, 0) = 0, \end{aligned} \tag{9.3}$$

where $v_r = dr/dt$, $v_\varphi = d\varphi/dt$ and $v_z = dz/dt$ are the components of the velocity vector. On the right boundary the following types of conditions are considered:

(i) clamped end,

$$\begin{aligned} r(\alpha_1, L) = \varphi(\alpha_1, L) = 0, \qquad z(\alpha_1, L) = L, \\ v_r(\alpha_1, L) = v_\varphi(\alpha_1, L) = v_z(\alpha_1, L) = 0; \end{aligned} \tag{9.4}$$

(ii) expanding end caused by the propagating pellet,

$$\begin{aligned} r(\alpha_1, L) = K_{sp}(t), \qquad \varphi(\alpha_1, L) = K_{sp}(t), \qquad z(\alpha_1, L) = L, \\ v_r(\alpha_1, L) = v_\varphi(\alpha_1, L) = dK_{sp}(t)/dt, \qquad v_z(\alpha_1, L) = 0; \end{aligned} \tag{9.5}$$

(iii) dilated end,

$$\begin{aligned} r(\alpha_1, L) = R_{sp}, \qquad \varphi(\alpha_1, L) = 0, \qquad z(\alpha_1, L) = L, \\ v_r(\alpha_1, L) = v_\varphi(\alpha_1, L) = v_z(\alpha_1, L) = 0. \end{aligned} \tag{9.6}$$

9.2.1 Haustral churning

In the event of haustral churning the pacemaker cells located in the teniae coli and the circular smooth muscle layer fire simultaneously. Action potentials of amplitude $V_i = 100$ mV and frequency $\nu = 0.33$ Hz are produced. They induce excitation and propagation of the waves of depolarization V_t^s and V_c^s (subscript t refers to the teniae coli) within the wall of the bioshell.

As a result of electromechanical coupling active forces of contraction develop. An initial wave T_c^a of intensity 10 mN/cm and wavelength 0.5 cm is produced (Fig. 9.2). It encases the entire segment with average $T_c^a = 12$ mN/cm. The bioshell experiences a uniform biaxial stress state throughout.

A total force $T_c = 35$ mN/cm is registered in the bioshell upon excitation of the circular smooth muscle layer (Fig. 9.3). In the area of contact with the pellet a maxium $T_c = 41$ mN/cm is produced. The wave T_c does not propagate. As a result the pellet does not move, but rather undergoes small librations about the initial point $Z_c = 0.35$ cm. One can speculate that, if the pellet were deformable, the strong occluding contractions, which are similar to those produced during haustral churning, could break the content into parts, with the subsequent displacement of its fragments along the colonic segment. This event is observed experimentally. However, from the point of view of the mechanics of solids, this problem poses a great mathematical challenge and is not considered here.

9.2.2 Contractions of the teniae coli

Assume that only the teniae coli are myoelectrically active. As a result of excitation a wave T_t^a of average intensity 7.0 mN/cm is produced (Fig. 9.4). It has a length of 0.6 cm and propagates at a velocity of 0.35 cm/s in the aboral direction along the surface of the bioshell. It increases in strength and the maximum force, $T_t^a = 10.5$ mN/cm, is generated in the zone of contact of the wall of the colon with the pellet.

The dynamics of the total force T_t corresponds to the dynamics of development of the wave T_t^a. At the beginning of the process its magnitude is influenced mainly by the intensity of the active force (Fig. 9.5). With the development of forces in the network of connective tissue a uniform stress distribution in the colon is achieved. A maximum total force of 12.5 mN/cm develops in the contact zone and is associated with the intensive propulsion of the pellet. First, it moves backwards by 0.06 cm, followed by incessant propulsion. The pellet is pushed forwards by 0.25 cm at an average velocity of 0.01 cm/s. For $t \geq 25.6$ s the wave T_t pushes the pellet backwards and then, starting from $t > 38$ s, the pellet experiences small displacements ≈ 0.03 cm

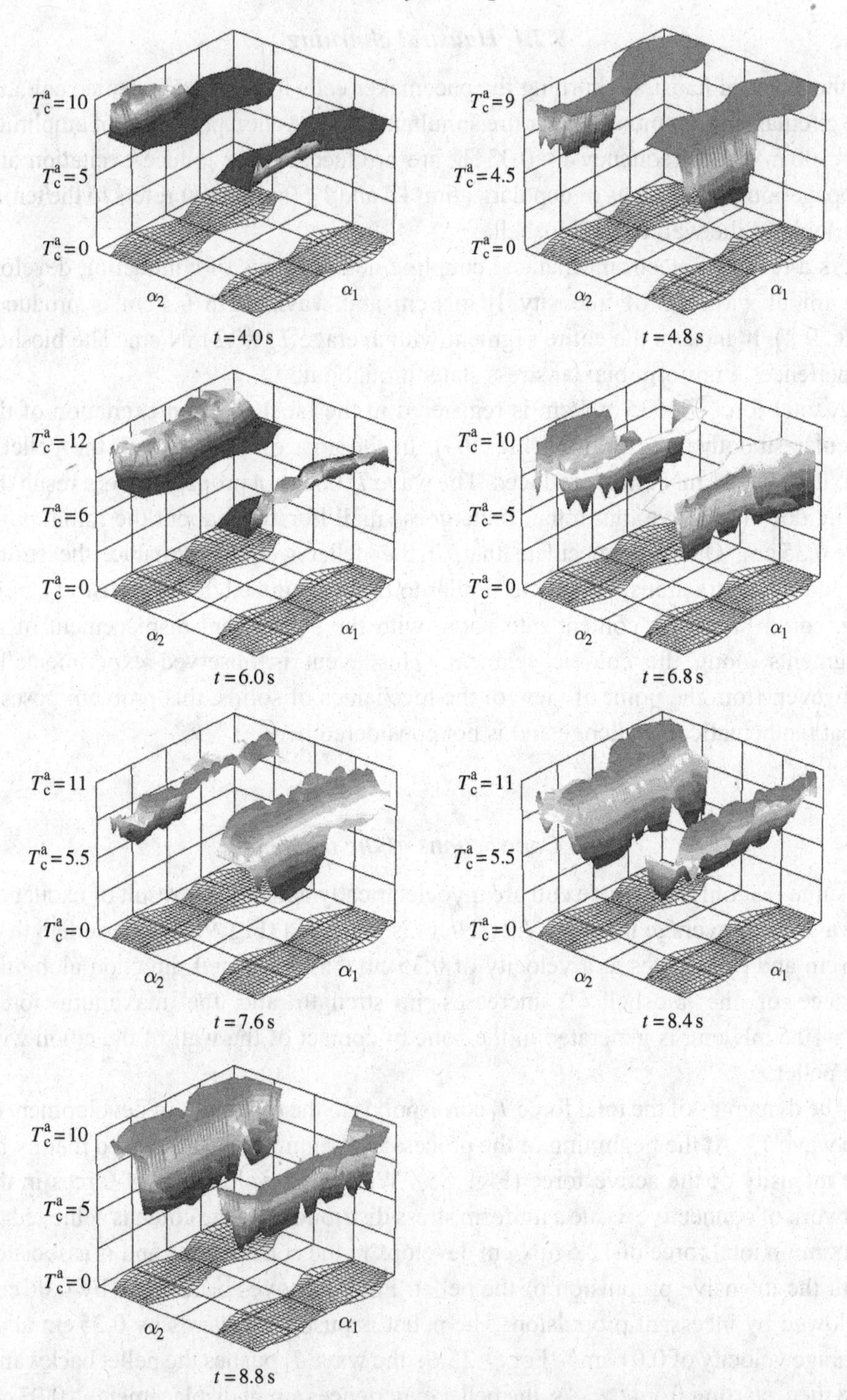

Fig. 9.2 Active-force T_c^a development in a colonic segment during haustral churning.

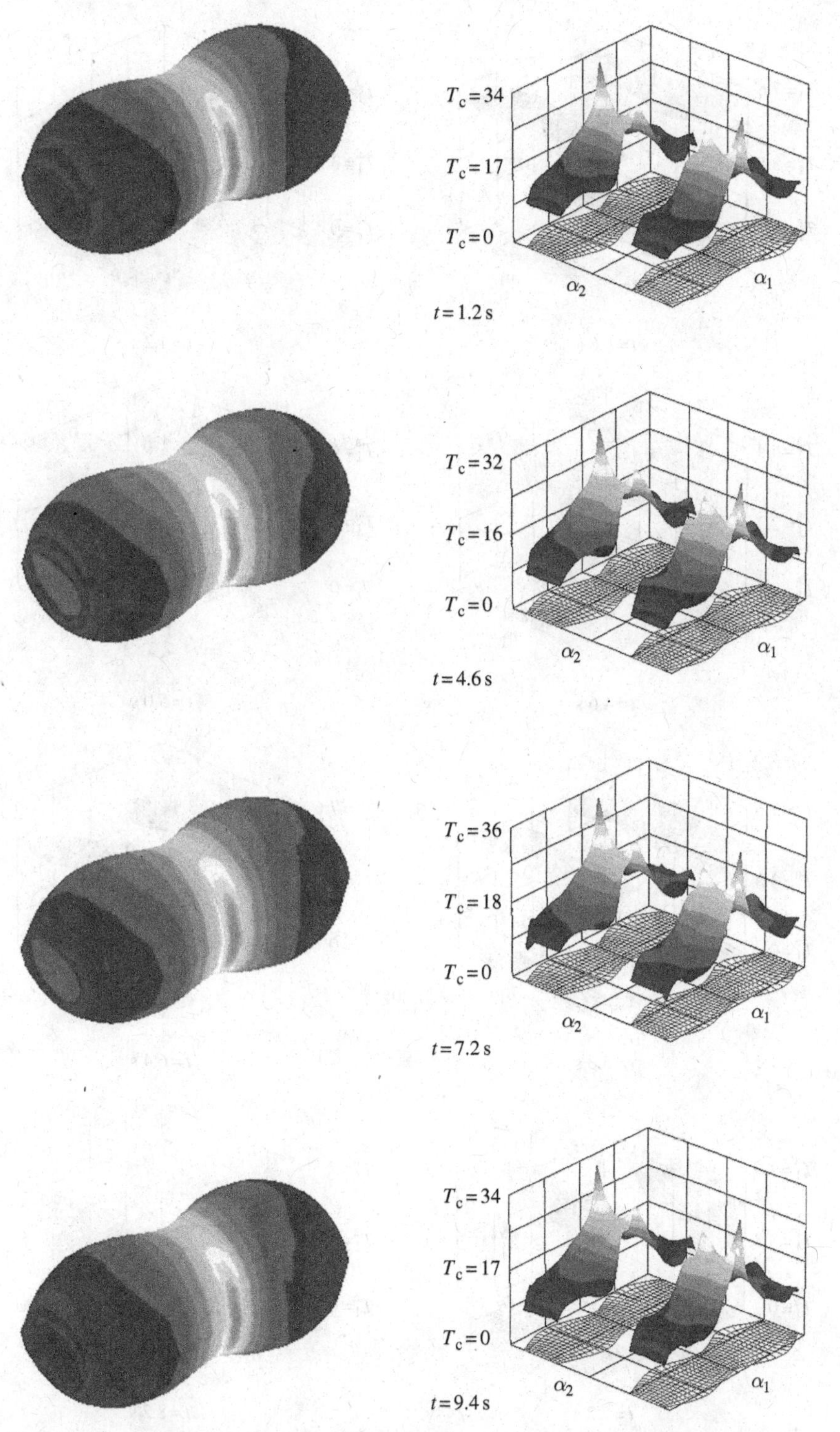

Fig. 9.3 Total-force T_c dynamics during haustral churning. Results are presented for a segment of the colon and its surface envelope.

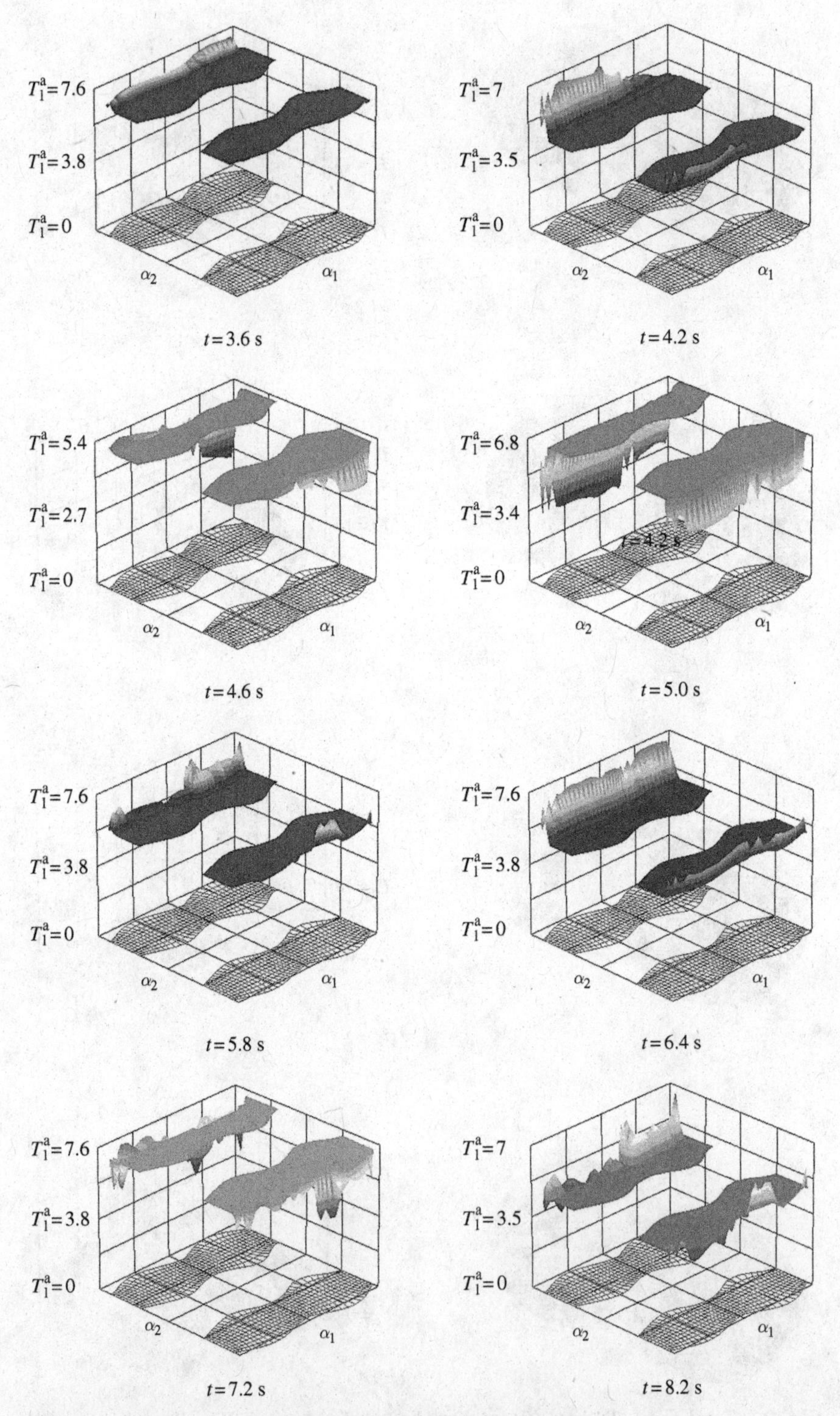

Fig. 9.4 Active-force T_1^a development in a colonic segment during contractions of the teniae coli.

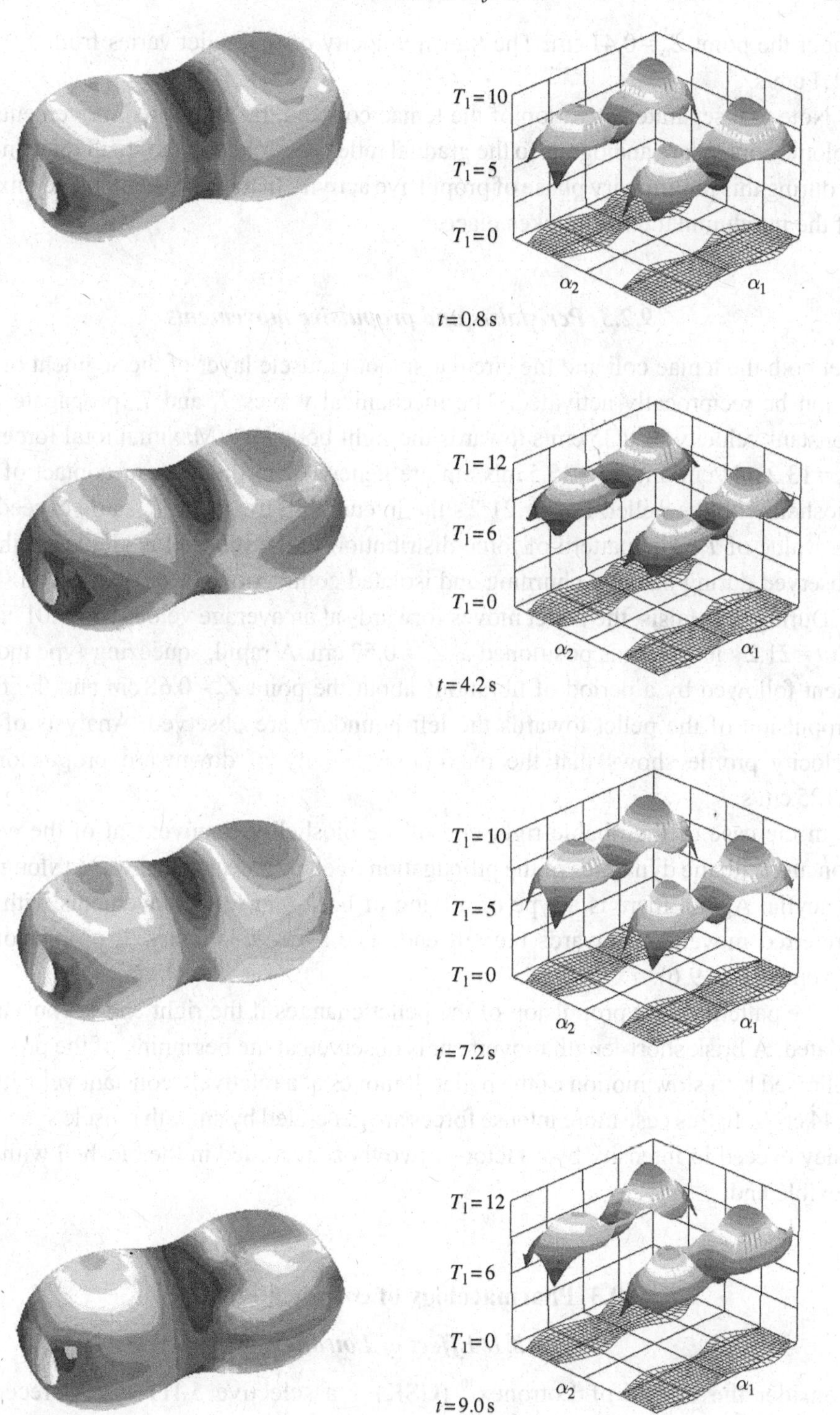

Fig. 9.5 Total-force T_1 dynamics during contractions of the teniae coli. Results are presented for a segment of the colon and its surface envelope.

about the point $Z_c = 0.41$ cm. The transit velocity of the pellet varies from 1.05 to −1.1 cm/s.

Note that separate activation of the teniae coli and the associated movements of colonic content are analogous to the gradual reflex described in the small intestine. It is during this preliminary phase of propulsive activity that the most intensive mixing of the intraluminal content takes place.

9.2.3 Peristalsis and propulsive movements

Let both the teniae coli and the circular smooth muscle layer of the segment of the colon be reciprocally activated. The mechanical waves T_t and T_c propagate at a constant velocity of 0.35 cm/s towards the right boundary. Maximal total forces of $T_t = 13.2$ mN/cm and $T_c = 35.5$ mN/cm are generated in the zone of contact of the bioshell with the pellet. For $t > 21.2$ s the intensity of the wave T_c starts exceeding the value of T_t. The pattern of force distribution in the bioshell is similar to those observed during haustral churning and isolated contractions of the teniae coli.

During peristalsis, the pellet moves forwards at an average velocity of 0.01 cm/s. At $t = 21.2$ s its centre is positioned at $Z_c = 0.57$ cm. A rapid, squeezing-type movement followed by a period of librations about the point $Z_c = 0.68$ cm and the final propulsion of the pellet towards the left boundary are observed. Analysis of the velocity profile shows that the maximum velocity of downward propulsion is 0.125 cm/s.

In the case of the pliable right end of the bioshell the movement of the pellet concurs with the dynamics of the propagation of electromechanical waves along the syncytia. Again, there is a type of mixing of back-and-forth movements with the preferred movement towards the left end. The average velocity of propulsion is 0.8 cm/s (Fig. 9.6).

The pattern of the propulsion of the pellet changes if the right end is constantly dilated. A brisk short-length movement is observed at the beginning of the process, followed by a slow motion of the pellet. It moves at a relatively constant velocity of 0.44 cm/s. In this case more intense forces are generated by smooth muscle syncytia. They exceed in intensity by a factor of two those recorded in the bioshell with the flexible end.

9.3 Pharmacology of colonic motility

9.3.1 Effect of Lotronex

Consider the effects of Lotronex® (GSK) – a selective 5-HT type-3 receptor antagonist – on the biomechanics of pellet propulsion. The mechanism of action

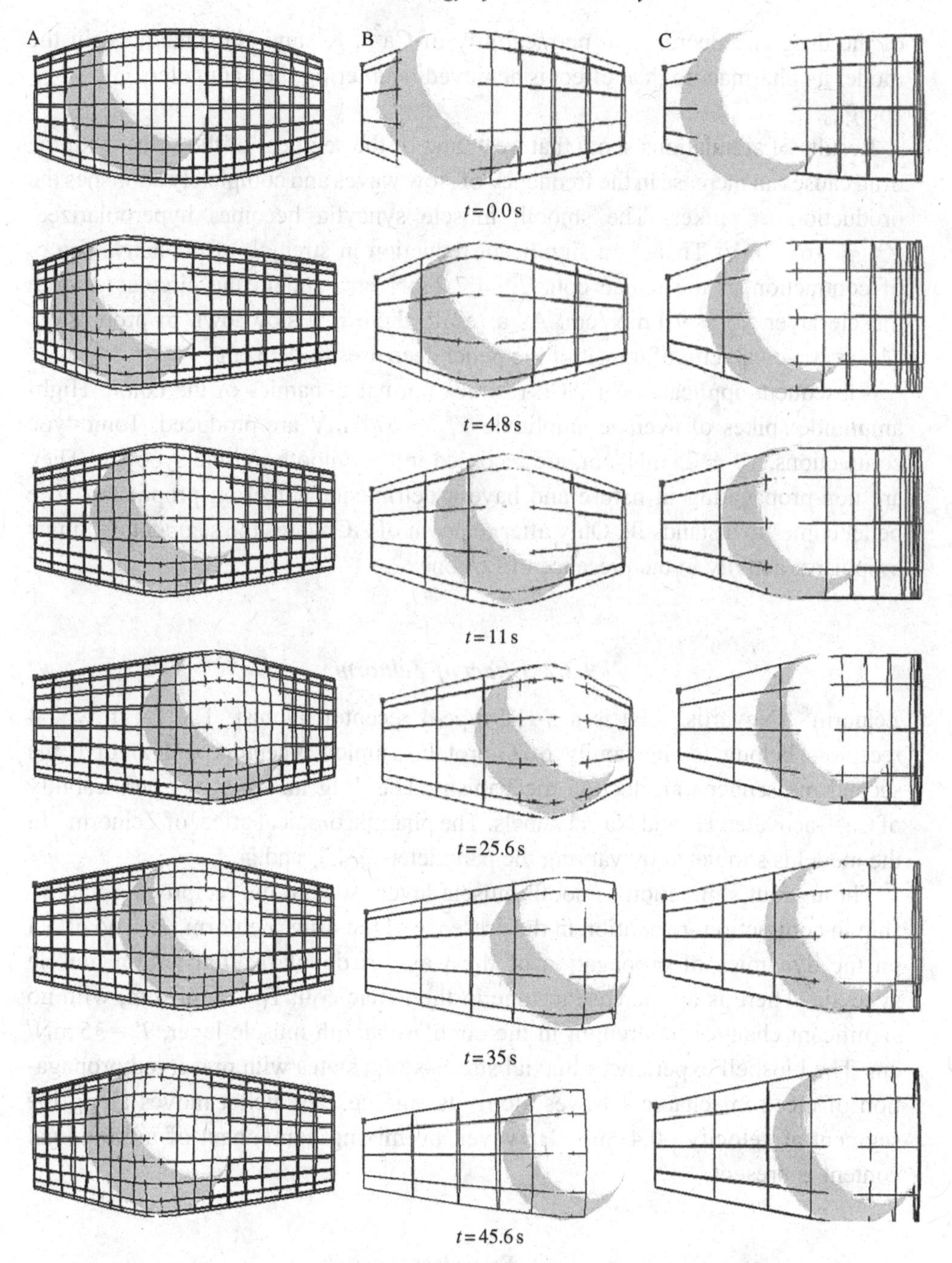

Fig. 9.6 Pellet propulsion in a colonic segment in cases of (A) a clamped end, (B) a pliable end and (C) a dilated end. Adapted from Miftahof *et al.* (2009). With permission from World Scientific Publishing Company.

of the drug is a decrease in permeability of Ca^{2+}, K^+ and Na^+ channels. In the model its pharmacological effect is achieved by altering the parameters $\tilde{g}_{Ca-K}$, $\tilde{g}_K$ and $\tilde{g}_{Na}$.

Results of simulations show that treatment of the segment of the colon with the drug causes an increase in the frequency of slow waves and completely abolishes the production of spikes. The smooth muscle syncytia becomes hyperpolarized, $V^s_{t,c} = -68.2\,\text{mV}$. There is a significant reduction in strength of the active forces of contraction in the teniae coli, $T^a_t = 7.2\,\text{mN/cm}$, and in the circular smooth muscle layer, $T^a_c = 9.1\,\text{mN/cm}$. As a result, there is a slowdown in propulsion. The average velocity of transit of the pellet decreases to 0.23 cm/s.

Subsequent application of ACh restores normal dynamics of the colon. High-amplitude spikes of average amplitude $V^s_{t,c} = 67.4\,\text{mV}$ are produced. Tonic-type contractions, $T^a_c = 25\,\text{mN/cm}$, are recorded in the smooth muscle syncytia. They are non-propagating in nature and have a detrimental effect on propulsion. The pellet comes to a standstill. Only after removal of ACh does the segment regain its propulsive activity in the presence of Lotronex®.

9.3.2 Effect of Zelnorm

Zelnorm® (Novartis, AB) is a 5-HT type-4 receptor agonist. The 5-HT type-4 receptors belong to the family of G-protein-coupled receptors and involve the second messenger transduction mechanism. The drug increases the permeability of Ca^{2+}-activated K^+ and Na^+ channels. The pharmacological effect of Zelnorm® in the model is simulated by varying the parameters $\tilde{g}_{Ca-K}$ and $\tilde{g}_{Na}$.

Throughout simulations smooth muscle layers sustain the reciprocal relationship in contraction–relaxation in the presence of the drug. Zelnorm® has no effect on the dynamics of propagation of the wave of depolarization within muscle syncytia. There is an increase in tone in the teniae coli, $T_t = 18$ mN/cm, with no significant changes in strength in the circular smooth muscle layer, $T_c = 35$ mN/cm. The bioshell experiences biaxial stress–strain states with preserved propagation of electromechanical waves along its surface. The pellet moves along the segment at velocity ~0.4 cm/s. However, no mixing component of intraluminal content is present.

Exercises

1. Compare the patterns of stress distribution in the large and small intestine. Explain the differences.
2. Define the biological and mechanical factors that affect the total transit time of the pellet.

3. The puborectalis muscle surrounding the anorectal junction relaxes to allow the straightening of the anorectal angle. The puborectal sling usually produces an angle of about 90 degrees between the rectal ampulla and the anal canal, so it is closed off. However, as the puborectal sling straightens, the angle increases to about 130–140 degrees, so the canal straightens and the bolus (contents) can be evacuated. The contractility of the puborectalis muscle can be affected by surgery, pelvic trauma, injury to the spinal cord, etc. Study the dynamics of bolus propulsion assuming that the puborectalis muscle is inactive.
4. Hirschsprung's disease, or congenital aganglionic megacolon, is a congenital disorder of the colon in which certain ganglion cells are absent. This results in a persistent overstimulation (cholinergic!) of a segment of the colon within the affected region. Use ABS Technologies© software to simulate the condition. Study the biomechanics of the affected colon.
5. Use ABS Technologies© software to study the effect of the parasympathetic nervous system on bolus propulsion. (*Hint*: cholinergic vagus nerve fibres carry parasympathetic signals to the colon.)
6. Simulate the effect of a drug – Colace – on bolus propulsion. (*Hint*: Colace emulsifies fat in the gastrointestinal tract and decreases reabsorption of water.)
7. Lotronex® (GSK) was a popular drug to treat diarrhoea-predominant irritable bowel syndrome in young women. However, the drug has serious adverse effects including ischaemic colitis and severe constipation, leading to obstruction, ileus and toxic megacolon. Study the effect of co-transmission by substance P, serotonin and Lotronex® on myoelectrical processes in the colonic muscle syncytia (*Hint*: give emphasis to the role of neurokinin type-1 and -3 and 5-HT type-3 receptors expressed on longitudinal and circular smooth muscle cells.)
8. Zelnorm® (Novartis, AB) was a popular drug to treat constipation-predominant irritable-bowel syndrome. Its clinical effectiveness was 10%–12% above the effect of placebo. Suggest 'improvements' – select additional pharmacological targets the drug should reach – that would increase its clinical effectiveness.

10

Biological applications of mathematical modelling

10.1 Biomechanics of hollow abdominal viscera

Mathematical modelling has greatly increased our ability to gain an understanding of many complex biological phenomena. A model can be treated as a hypothesis that can be accepted or rejected on the basis of its ability to predict the experimentally observed results. Numerical simulation techniques are most powerful when a mathematical model is based on understood individual elements of the biological systems, but where their aggregate behaviour cannot be depicted by current theory. In the absence of unexpected interactions, the input–output relationship can be quite accurately calculated. Experimentally inaccessible and sometimes unexpected interactions can be recovered and evaluated by simple comparison of computed versus experimental results. Such use has made mathematical simulation an indispensable tool in the biosciences.

After decades of experimental optimism, there is increasing recognition of the limitations of the *in vivo* and *in vitro* approaches to the study of gastrointestinal function. Possible explanations of these limitations are the size, variable contour and inaccessibility of abdominal viscera and most importantly the fact that existing techniques do not allow us to unravel the multilevel, nonlinear interactions that occur in complex physiological reactions. Today, without employment of the methods of mathematical modelling based on the general principles of computational biology our potential to learn about the complex relationships within the gastrointestinal tract would be totally thwarted. Modern computer-based technology allows us to span many fields of biology (electrophysiology, cell and molecular biology, pharmacology, neurobiology), mechanics (solid and fluid mechanics, the mechanics of shells), medicine (clinical gastroenterology, diagnostic radiology and motility) at various levels of detail and thus provides a thorough analysis of biological phenomena.

The models of the stomach and of the small and large intestine presented in this book are examples of the application of fundamental concepts of the mechanics of solids and the theory of thin shells to study the biomechanics (motility) of the gastrointestinal system. They integrate biomechanical properties and electrophysiological processes in visceral organs and accurately reproduce the combined events of (i) the dynamics of the interstitial cells of Cajal, (ii) propagation of the wave of depolarization along electrically isotropic and anisotropic smooth muscle syncytia, (iii) electromechanical conjugation and the development of forces of contraction–relaxation, (iv) shape and stress–strain changes; and (v) propulsion of the pellet (intraluminal content) in the colon.

The particular classifications of movements observed in abdominal viscera reflect the large repertoire of functions performed by the gastrointestinal system. The movements are broadly divided into (i) local – segmental contractions and pendular movements – and (ii) peristaltic waves. Segmental contractions are produced by the circular smooth muscle layer, do not propagate, occur simultaneously or sequentially in different parts of the organ and promote expulsion of the intraluminal content. In contrast, pendular movements are associated with contraction–relaxation of the longitudinal smooth muscle layer, propagate over short distances aborally and are assumed to mix and grind the content.

Peristalsis is the fundamental phenomenon of gastrointestinal motility. It refers to a coordinated migrating myoelectrical activity of both muscle layers. Peristalsis begins with a preliminary gradual reflex represented by longitudinal contractions, followed by a phase of broadly spread segmentations. Peristaltic waves are recorded in two forms: (i) slowly advancing contractions and (ii) the peristaltic rush. The latter is manifested as rapid contractions that propagate a long distance, which are usually seen in pathological states.

Much of the theoretical work on gastrointestinal motility has concentrated on separate analysis of the mechanical aspects of the phenomenon (Mayo *et al.*, 1992; Gao *et al.*, 2002; Amaris *et al.*, 2002; Rachev *et al.*, 2002). Only a few studies have been dedicated to the modelling of electromechanical processes (Ramon *et al.*, 1976; Miftakhov *et al.*, 1996; Aliev *et al.*, 2000; Miftakhov and Vannier, 2002; Miftakhov and Fedotov, 2004). These models have shed light on some of the general mechanisms of the propagation of action potentials, slow-wave and bursting activity and the migrating motor complex. However, they do not reveal the fundamental biophysical principles of the origination and propagation of the electromechanical waves in abdominal viscera. Thus, it has been postulated that the migrating myoelectrical complex, which is a merely *in vivo* event, is a result of interrelated electrical (slow-wave and spiking) and mechanical (contractions of smooth muscle) processes. Combined at the level of smooth muscle syncytia they form bands of regular contractions, which periodically migrate along the gastrointestinal tract. Starting from purely visual perception and *a priori* assumptions regarding the propagation rather than from a scientific analysis of the event, the terminology and quantitative measures from the physics of waves have been adopted to describe the dynamics of migration. But whether slow waves and the migrating complex are indeed propagating phenomena has never been questioned.

If one assumes that slow waves really do propagate, then a simple analysis of the experimental data indicates that the wavelength of the slow wave varies within the range 2–110 cm! That result is based on the fact that the frequency of the slow waves varies in the range 0.01–0.3 Hz and their velocity in the range 0.6–1.1 cm/s (Szurszewski, 1969). Even if we disregard wide-range variability, which is not unusual in biological observations, the results imply that smooth muscle syncytia could be viewed as an infinite number of dispersed areas of myoelectrical activity, which are either connected over a long distance (110 cm) of the gut or represent spatially distributed units 2 cm in length.

Convincing support for that consideration comes from the results of experiments on isolated preparations of smooth muscle syncytia (Lammers *et al.*, 1993; Lammers *et al.*, 1997; Lammers *et al.*, 2000; Lammers, 2000; Lammers and Slack, 2001). Using a brush of 24 × 10 electrodes arranged in a rectangular array, the authors studied patterns of electrical activity over a large area, and the effect of a single pacemaker and multiple dispersed pacemakers on the spatial conduction of the wave of excitation. However, within the framework of old concepts of understanding of gastrointestinal motility, the authors failed to give any reasonable explanation for their remarkable observations. Thus, a superposition of slow-wave traces obtained simultaneously from 224 points (Lammers *et al.*, 2001) shows that, in the resting state, when no pacemaker activity is present, there are no significant phase and amplitude differences among all of the slow waves recorded. This analysis suggests that (i) smooth muscle syncytium may represent a continuum of spatially distributed autonomous oscillators and (ii) slow waves do not propagate in the silent state!

Further re-evaluation of the above-mentioned findings shows that slow waves recorded at different points of the syncytium have equal frequencies. This condition persists with every excitatory input provided by the pacemaker. This experimental fact confirms our thought that the external excitatory input plays the role of the connector among spatially distributed oscillators. It has also been suggested that a transient influx of calcium ions through the T-type Ca^{2+} channels is responsible for the shift of the membrane potential and the activation of the intracellular contractile protein system. The propagating calcium waves have been proposed as a possible mechanism that sustains the spread of the electromechanical wave. However, the electrical wave propagates at a velocity of 2.3–10.8 cm/s depending on the species and tissue (Lammers, 2000), compared with the mechanical wave of contraction–relaxation that moves at a much lower velocity, 0.2–0.41 cm/s (D'Antona *et al.*, 2001). Also the intracellular calcium waves cannot provide the extensive, tens-of-centimetres conduction within morphologically inhomogeneous smooth muscle syncytia and action potentials in smooth muscle propagate over short distances, 1.3–12.8 cm (Lammers, 2000). Therefore, another mechanism must be involved in the efficient spread of the excitation along smooth muscle syncytia. There is compelling evidence for the crucial role of a planar neural network of the interstitial cells of Cajal, which topographically lies between the elements of the enteric nervous system and smooth muscle. Those who are interested in the detailed analysis of the exact mechanisms of the coordination of slow-wave electrical activity should consult specialized texts (Suzuki, 2000; Akbarali, 2005; Hashitani *et al.*, 2005; Won *et al.*, 2005; Bayguinov *et al.*, 2007; Yin and Chen, 2008).

The attempts to explain the slow-wave dynamics by invoking relaxation oscillators coupled in a network (Sarna *et al.*, 1971; Daniel *et al.*, 1994) and core-conductor models (Publicover and Sanders, 1989) failed because of the drastic simplifications made in construction of the models, which made them biologically irrelevant. With the model of a functional unit of the small intestine, though, we discovered numerically a phenomenon of self-sustained spiral-wave activity in the longitudinal smooth muscle syncytium. The spiral electrical activity abolished coordinated waves of contraction in the longitudinal layer. This situation could occur in diabetes mellitus (Koch *et al.*, 1989), in anorexia nervosa (Abell *et al.*, 1987), with an overabundance of prostaglandins (Sanders, 1984) and with drug overdoses, e.g. erythromycin or atropine (Holle *et al.*, 1992), and is associated with multiple uncoordinated foci of self-sustained myoelectrical activity. Clinically it is manifested by the failure of propulsive activity of the small intestine. As a possible physiological mechanism of initiation, wandering pacemaker activity and abnormal conduction – events analogous to those which are responsible for trigger and re-entry phenomena in the heart – have been proposed. However, the primary pathological feature of cardiac arhythmia is either myocardial ischaemia or an ectopic beat. The first condition is incompatible with the viability of the gut, while the exact role of the second mechanism in a not highly excitable smooth muscle syncytium is uncertain.

None of the existing biological protocols designed to study gastrointestinal propulsion and motility offer the desired depth of accessibility to a combined analysis of the intricate mechanisms of function of the biological system. As a result, most of the conclusions are drawn from 'technically deficient' experiments, since they offer only an implicit partial insight into how the system works under real physiological conditions and in a diseased state. The lack of true understanding of the physiology of the processes involved affects our approach to treating various gastrointestinal disorders, e.g. functional dyspepsia, gastroparesis, the irritable-bowel syndrome, constipation – that remains unsatisfactory.

Only with the accurate biomechanical model of the stomach based on real anatomical and physiological data regarding its structure and function has it become possible to analyse qualitatively and quantitatively the dynamics of the stress–strain distribution in different regions of the organ. The evolution of uniaxially stressed zones in the cardia–fundus area of the stomach serves as a biomechanical explanation for the Mallory–Weiss syndrome – a condition manifested by linear ruptures extending from the mucosa to the submucosal layer and life-threatening internal bleeding. Wrinkling in the pylorus region during peristalsis facilitates longitudinal movements and expulsion of gastric content into the duodenum.

The effective propulsion and mixing of the intestinal content is accomplished by a coordinated mechanical contraction–relaxation reaction of the teniae coli and circular muscle layers. Their anatomically distinct physiological significance still remains a subject of scientific debate. For example, simultaneous intracellular recordings from the layers on an isolated segment of guinea-pig colon have revealed sequential spontaneous rhythmical depolarizations of the two layers and synchronous neuromuscular inputs during ascending excitation (Spencer and Smith, 2001). To test this hypothesis we imposed the reciprocal relationship between mechanical activity of the teniae coli and circular smooth muscle syncytia in the dynamics of peristalsis. The results of numerical simulations reproduced with great accuracy a complex sequence of movements of the pellet: pendular-like movements, followed by a brisk squeeze and finally the intensive mixing. In contrast, pendular contractions provide only back-and-forth movements of the pellet, not mixing, while segmental contractions alone fail to sustain the transit. In a case of synchronous activation of the two smooth muscle layers (results are not shown here) the transit of the pellet resembled movements recorded during the gradual reflex only.

The extensive quantitative analysis of a sequence of physiological events is difficult and sometimes impossible because of the diverse sources of data from different laboratories. Even with this uncertainty in mind, however, the comparison demonstrates good qualitative and quantitative agreement with the model predictions and experiments (Krevsky *et al.*, 1986; Bampton *et al.*, 2000; D'Antona *et al.*, 2001; Gunput, 2001; Marciani *et al.*, 2001).

The results of simulations of the propulsion of the pellet in a segment with a pliable aboral end show that the velocity of the propagation varies between 0.4 and 1.13 cm/s. These values are within the range of the experimentally measured velocity of the movement of content (1–3 cm/s) which was observed in the studies of colonic motility in humans (Bampton *et al.*, 2000; Marciani *et al.*, 2001).

The study of the effect of the constantly dilated aboral end of the segment provided an insight into the pellet propulsion. With the above boundary condition we attempted to model the final act of normal defecation. The total force generated by the segment exceeds by 75% the total force recorded in an isolated segment. Interestingly, such an increase did not speed the movement of the pellet towards the distal end but rather eliminated the 'mixing' component. This dynamics corresponds to a physiological expulsion of the formed faecal mass.

It is noteworthy that in all our simulations we assumed the intraluminal content to be a rigid sphere. Therefore, the results could satisfy only certain *in vivo* or clinical investigations when metallic beads are administered into the gastrointestinal tract to study the transit time. No mathematical models of chyme as a deformable biologically active medium have been proposed so far. Also, in the model we did not consider processes of secretion and absorption. These mechanisms undoubtedly have a significant impact on bolus propulsion. It remains a challenging task to formulate mathematically the above problems and to solve them.

The mathematical modelling allowed us to reproduce the actions of the once-popular bowel drugs Lotronex® and Zelnorm® on colonic motility. Administration of Lotronex® causes a reduction in the total force of contraction and a dilatation of a segment of the gut. This significantly slows down the propulsion of the pellet. The *in vivo* and *in vitro* studies of the effects of the drug on contractility and compliance of the colonic muscle demonstrated (i) the decrease in contractions in the small and large intestines, (ii) the increase in wall compliance of the colon and (iii) the delay of gut transit time (Gunput, 2001; Humphrey *et al.*, 2001). Interestingly, co-activation of the nicotinic and muscarinic types of acetylcholine receptors results in the development of long-lasting tonic-type contractions that bring the pellet to a standstill. The loss of the reciprocal contraction–relaxation relationship between the two muscle layers has been proposed as a possible mechanism.

Treatment of the colon with Zelnorm® does not affect the reciprocal relationships in mechanical activity between the two smooth muscle layers. The drug slightly decreases the transit time of the pellet, increases the total force of contraction of the circular smooth muscle layer, and alters the pattern of propulsion – there is a loss of 'mixing' (Lacy and Yu, 2002).

With the model it was possible to simulate effects of various drugs that are currently being used to treat gastric dysmotility. Although the results of numerical experiments reproduce the 'overall' pharmacological effects of compounds, no affirmative quantitative comparison could be made at this stage. The confirmation of theoretical findings requires accurate verification in biological experimentation.

10.2 Future developments and applications

Despite the fact that the models answer many questions related to the normal and pathological physiology of gut motility, further improvements of basic concepts and mathematical formulation are needed. For example, to enable study of the mechanisms of neuro-neuronal and neuro-muscular signal transduction, the phenomena of receptor polymodality and neurotransmitter co-localization and co-transmission should be included in the model. Another interesting opportunity is in elaboration of sensory pathways of the gut and the addition of chemoreceptors and nociceptors into the system together with high-level regulatory mechanisms, namely the prevertebral ganglia and the central nervous system. Such modifications would increase the biological accuracy of the model and would allow a quantitative study of the intrinsic neurobiological processes of motility, which are not accessible by any currently existing experimental techniques.

Gastrointestinal function is under the control of different hierarchical levels – the enteric nervous plexus, the spinal cord and the central nervous system. The enteric nervous plexus – the 'little brain' of the gut – has been implicated in several prominent aspects of physiological information processing, such as the formation of a space map for coordinated motility patterns and visceral perception. One of the prominent features of the plexus is local ganglion organization, which is reflected in the tremendous complexity of horizontal and vertical connectivity arrangements. Although much is known today about the macroscopic and microscopic characteristics of the structural elements – neurons (Furness, 2006) – it is still hard to derive basic principles underlying their functional mechanisms and to establish functional, rather than neuroanatomical, relationships with higher hierarchical levels, e.g. the brain.

The enteric nervous plexus provides regulatory programmes that sustain the spatio-temporal stability of electrical patterns (the amplitude, duration and frequency of slow waves and spike activity, the direction and velocity of propagation of the myoelectric complex, the activity of the gating mechanism) within the stomach and the small and large intestine itself, and exchanges the intrinsic information with the central nervous system through external sympathetic and parasympathetic innervations.

Electrophysiological combined with histomorphological analyses of neurons have revealed that the pathways of intrinsic reflexes include primary sensory and intestinofugal neurons, a number of secondary neurons and final motor neurons. The primary afferent neurons belong to a class of electrophysiologically defined AH neurons with a distinct Dogiel type-II morphology (Wood, 1989). They have smooth cell bodies, are principally adendritic, pseudo-uniaxonal or multiaxonal with a tendency towards primary and secondary branching of the neurites close to the soma, and exhibit a few or no synaptic inputs. Their receptive fields are free nerve endings and are located in the mucosa and the submucous layer. The fibres are polymodal in nature and respond to more than one type of stimulus: mechanical, chemical or thermal.

Intracellular recordings made from the somas of primary sensory neurons show them to be slowly adapting units. They discharge continuously without adaptation during mechanical distortion, and the frequency of discharge increases as a direct function of the intensity of stimulation. The resting potential of the neurons ranges from −55 to −75 mV and their mean input resistance varies from 20 to 190 MΩ (Nishi and North, 1973; Wood and Mayer, 1978). Action potentials have amplitudes of 60–80 mV and durations of 2.5–3 ms and are followed by characteristic long-lasting after-hyperpolarizations (AHs) of 1–2 s. The waveform of the action potentials is usually complex multiphasic with a prominent shoulder on the repolarizing slope. Simultaneous recordings from a cluster of neurons demonstrated that many primary sensory neurons provide identifiable synaptic inputs to neighbouring cells, which would be predicted for the functional significance of cells.

The identity of the secondary sensory neurons remains uncertain. There are different opinions about their morphological and electrophysiological characteristics: they were described as Dogiel type-I cells with S-type electrical activity by Kuramoto and Furness (1989), while others refer to them, morphologically, as Dogiel type-III neurons and, electrically, as tonic-type mechanosensitive units (Wood, 1989). The neurons are uniaxonal and multidendritic. The dendrites are of intermediate length, relatively little branched and project in the connectives orally and aborally to adjacent cell clusters.

The distinguishing characteristic of their electrical behaviour is that they discharge long trains of spikes ($\simeq$21 s) in a set pattern even after withdrawal of the mechanical stimulus. These patterns resemble all-or-nothing events and are independent of the initial stimulus. Action potentials of amplitude 70–80 mV are generated at a high frequency of 10–40 Hz (Wood and Mayer, 1978). The frequency is relatively constant during the first 3–5-s bursts and then declines linearly. The waveforms of the spikes are biphasic or triphasic and the duration of each spike varies within the range 2.5–4.5 ms. In some cases neurons fire in beating mode, which can last for ~40 min. Multiunit extracellular recordings from the enteric plexus of the small bowel revealed that the discharge of the slowly adapting primary sensory neuron always preceded the discharge of the secondary sensory neuron. This suggests that the secondary neurons may be triggered by synaptic input from a primary sensory neuron (Wood, 1973).

Vertical connections of the enteric nervous plexus with prevertebral ganglia (PVG) are believed to occur via intestinofugal afferent neurons. Their cell bodies lie within the myenteric plexus with axons projecting without synaptic interruption to the PVG neurons. The intestinofugal neurons are classified as having Dogiel type-I and type-II morphology with electrical properties consistent with their being myenteric AH and/or tonic-type neurons (Lomax *et al.*, 2000; Sharkey *et al.*, 1998; Szurszewski *et al.*, 2002). The neurons are uniaxonal with multiple dendrites that possess lamellar expansions. Electrophysiological findings showed that neurons fire (i) action potentials spontaneously and continuously; (ii) a brief burst of spikes (1–10) followed by a short, of duration 520 ± 32 ms, or prolonged after-hyperpolarization, of duration 2.8 ± 0.3 s; and (iii) large-amplitude excitatory post-synaptic potentials of amplitude 85–90 mV and duration <2 ms (Sharkey *et al.*, 1998; Terence *et al.*, 1998; Lomax *et al.*, 1999). The current level of controversy and inconsistency is explained by the significant limitations of experimental methods in studying these cells *in vivo*.

Experimental data on the electrophysiology and neuropharmacology of prevertebral ganglion neurons are sparse and controversial (Porter *et al.*, 2002; Lomax *et al.*, 2000; Furness, 2000; Szurszewski *et al.*, 2002; Gibbins *et al.*, 2003), owing to the highly variable morphology of the structures and their inaccessibility to direct *in vivo* recordings. The ganglia contain a unique population of motor neurons receiving multiple converging synaptic inputs from intrinsic afferents. Neurons have cell bodies that are small in diameter (<20 μm). Their electrical repertoire is diverse and depends entirely on synaptic inputs. The resting potential of the neurons ranges from −50 to −85 mV and their input resistance amounts to 105 MΩ (Jobling and Gibbins, 1999; Gibbins *et al.*, 2003). Action potentials evoked during depolarizing current injection have amplitudes of 60–100 mV and durations of 2.5–3 ms and are followed by characteristic long-lasting after-hyperpolarizations of 1–2 s. The waveform of the action potentials is variable. The principal ganglionic neuron sends its axon back to the gut to synapse with the motor neuron of the enteric nervous plexus.

Interstitial cells of Cajal are distributed throughout the smooth muscle layer of the gastrointestinal tract. Being highly branched, they form networks intimately associated with muscular layers and enteric nervous plexuses. Several subtypes have been identified histomorphologically: ICC_{MY} are located within the ganglia of plexuses; ICC_{IM} are located intramuscularly; ICC_{SEP} are found predominantly in the septa dividing muscle bundles; ICC_{SM} are distributed along the submucosal border of the circular muscle layer; and ICC_{DMP} are present between the inner and the outer portions of the circular muscle layer at the level of the deep muscular plexus. All cell subtypes generate action potentials and function as intermediaries in the signal transduction between the enteric plexus and smooth muscle layers (Camborova *et al.*, 2003; Hirst and Edwards, 2004). They sustain the electrotonic spread of excitation to syncytia through gap junctions. In addition, it has been suggested that they contribute to afferent neural transduction (Fox *et al.*, 2001). However, currently there is no strong experimental evidence to reject or to support this notion.

Morphologically and functionally distinct classes of neurons also have specific chemical coding, which is expressed in the form of combinations of neurotransmitters stored in them. Using double-label immunohistochemical techniques, it has been demonstrated that primary sensory neurons may co-localize ACh, calbindin, tachykinins, glutamate and substance P, motor neurons may co-localize ACh, tachykinins, enkephalin, nitric oxide and vasointestinal peptide, and intestinofugal neurons may co-localize ACh, nitric oxide, vasointestinal peptide, calretinin, glutamate and substance P (Alan *et al.*, 2000; Lomax *et al.*, 1999; Liu *et al.*, 1997;

Brookes, 2001; Hens *et al.*, 2001). The neurotransmitter co-localization correlates with receptor polymodality, the expression of different types of receptors on adjacent neurons. For example, muscarinic- and nicotinic-type ACh receptors, α/β-type adrenoceptors, 5-HT_3 and 5-HT_4 serotonergic receptors, NMDA and AMPA receptors, NK_1, NK_2 and NK_3 neurokinin receptor subtypes and others could be co-present on a cell. Neurotransmitters and their corresponding receptors contribute differently to the processes of signal transduction mechanisms. These intrinsic dynamic differences and interactions become critical in explaining the multifaceted symptoms of gastrointestinal disorders (Furness and Sanger, 2002; Blackshaw and Gebhart, 2002; Thielecke *et al.*, 2004; Callahan, 2002; Sanger and Hicks, 2002). However, the exact control mechanisms of these processes are unknown as well, since it is impossible to predict from the results of classical *in vivo* and *in vitro* experiments the whole repertoire of electrical and pharmacological responses.

Physical models have been proposed to reproduce the physiological cognitive and coherent phenomena of the myenteric plexus: (a) a self-oscillatory neural-network and (b) an abstract energy-landscape model. The standard approach to the modelling is based on synthesis of firing neurons with the introduction of a weight matrix representing the connections between these neurons. Such an approach has been proven extremely useful for the modelling of limited sets of neurons. However, the approach is not easily applied to multilevel hierarchical structures of large-scale networks. Existing models and their cognitive capacities only emphasize space-size and comparative neuroanatomical approaches and do not incorporate neurochemical aspects of signal transmission and processing at synaptic levels. This adds enormous plasticity, allowing individual neurons to serve several functions, to adapt to constantly changing physiological demands and to process information independently of their connectivity. Therefore, it is an important task to analyse how the multilevel neuron assembly can be applied to the reality of the regulatory system of the gastrointestinal tract and to elucidate the roles of various elements – neurons, neurotransmitters and receptors – in the assembly formation.

The intramural nervous plexuses function independently and in concert with the central nervous system. They communicate with the brain through the sympathetic and parasympathetic nervous divisions of the autonomic nervous system. In the sympathetic nervous system, there are ganglia situated close to the abdominal organ, which participate in the control of contractions. These ganglia are connected to the spinal cord and to higher centres in the brain. The parasympathetic innervation is provided by the vagus nerve. Painful signals from the abdominal viscera are assessed by the thalami-cortical system, central amygdala and hypothalamus. Although new physiological methodologies, including cerebrally evoked responses, positron-emission-tomography scanning and functional magnetic resonance imaging, are being used to study the central mechanisms of pain perception, our understanding of the functionality of gastrointestinal-tract–brain and brain–gastrointestinal-tract communications is limited (Casey *et al.*, 2001; Bernstein *et al.*, 2002). Despite there having been a tremendous amount of research investigations in this area, there has been no effort made to simulate the brain–gut interaction.

The fact that gastrointestinal motility changes in response to mechanical and chemical stimulation of the organ indicates the existence of feedback regulation in the system. These effects, which are commonly attributed to reflex pathways and hormonal actions, are considered to be critical to the normal operation of the system. The integration of such feedback control mechanisms into a model of the neuromuscular apparatus of the abdominal viscera constitutes a major problem in this area.

An interesting area of research and a possible expansion is the role of cytokines, hormones and inflammatory agents in the pathogenesis of visceral allodynia and hyperalgesia. It is a great challenge to model their co-joint integrative effects.

When the above 'improvements and adjustments' are incorporated into the models of the abdominal viscera, they will significantly improve their biological and clinical feasibility. It will become a challenge then to try to solve problems related to a broad spectrum of functional gastrointestinal disorders, including functional chest pain, dyspepsia, biliary dyskinesia, irritable-bowel syndrome and proctalgia fugax. The biological processes involved in the pathogenesis of functional gastrointestinal disorders are complex and dynamically variable (Spiller, 2004; Hunt, 2002). Visceral hypersensitivity is the ubiquitous finding in these diseases and presents in the form of either (i) allodynia, in which a previously non-painful stimulus induces pain, or (ii) hyperalgesia, when a previously painful stimulus causes an exaggerated pain. It is generally agreed that the location and the functionality of receptive fibres of the primary sensory and intestinofugal neurons determine sensitivity, although the pathways responsible for visceral nociception are not well established. Mucosal receptors and muscular afferents with mechanonociceptive and chemonociceptive properties have been described anatomically and electrophysiologically (Blackshaw and Gebhart, 2002).

Various 'blueprints' have been suggested to explain the possible nociceptive signal transduction from the abdominal viscera to the brain (Furness and Sanger, 2002; Holzer *et al.*, 2001; Hens *et al.*, 2001; Szurszewski *et al.*, 2002). The current concepts suggest that signals originating in intrinsic primary sensory neurons of the enteric nervous plexus and extrinsic sensory afferent pathways, with cell bodies located in the dorsal root ganglia, pass to neurons located in the dorsal horn – lamina II and IV – of the spinal cord. They are conveyed further along the spinothalamic and spinoreticular tracts to neurons of the reticular system and vagal nuclei. The latter interact with the thalamus, limbic pain centres and the pre-frontal cortex of the brain where visceral signals are decoded as pleasurable or painful. The levels are reciprocally connected with forward-feedback regulatory mechanisms through ascending pathways and descending tracts (Holzer *et al.*, 2001; Grundy and Schemann, 2005; Hens *et al.*, 2001; Ishiguchi *et al.*, 2003).

The associated dysfunction of the abdominal viscera is recorded primarily as altered motility patterns, which are neither constant nor specific.

Approaches to treat functional gastrointestinal disorders demand that a biological target(s) that has a causative role in either the onset or the progression of a disease must be identified, and pharmacological agents that appropriately modulate this target(s) with limited or no adverse effects must be designed. Modern technologies, e.g. genome-wide forward and reverse genetics screens, small-interfering RNAs (siRNAs), gene-expression profiling (Butcher, 2005; Van der Greef and McBurney, 2005; Hardy and Peet, 2004; Betz *et al.*, 2005; Zambrowicz and Sands, 2003), combinatorial chemistry, ultra-high-throughput screening, virtual-ligand screening and structure-based drug design (Macarron, 2006; Silverman *et al.*, 1998; Oprea and Matter, 2004; Shoichet, 2004; Scapin, 2006; Cheng *et al.*, 2007) have markedly improved the quality and efficiency of the drug-discovery processes. On the basis of the pharmacological profiles and clinical effectiveness of acetylcholine, serotonin, glutamate, substance P, purine amines and other neurotransmitters that have been implicated in the induction and enhancement of allodynia and hyperalgesia (Gaudreu and Plourde, 2004; Willert *et al.*, 2006), attempts have been made to develop drugs to modulate and to control pathological processes that are manifested in various forms of functional gastrointestinal disorders. There is a remarkably wide range of therapeutic modalities that act primarily at single targets. For example, tegaserod – a partial 5-HT_4 agonist – reduces the symptoms of abdominal pain, bloating and constipation; Lotronex® – a 5-HT_3 antagonist – improves symptoms in patients with diarrhoea-predominant irritable-bowel syndrome; norcisapride – a 5-HT_3 antagonist/5-HT_4 agonist – and dicyclomine bromide – a non-selective cholinergic antagonist – decrease pain and bowel motility; loperamide – a μ-opioid agonist – effectively decreases intestinal transit and increases absorption of water (Parsons, 2001; Callahan, 2002; Blackshaw and Gebhart, 2002; Thielecke *et al.*, 2004; Galligan, 2004); memantine – an NMDA antagonist – exhibits antinociceptive properties; enteric P2X-receptor agonists have demonstrated their potential usefulness for treating constipation-predominant irritable-bowel syndrome; and P2X-receptor antagonists have shown their effectiveness in the diarrhoea-predominant state (Callahan, 2002). However, all existing therapeutic modalities have a limited clinical effectiveness. The main reason for that is the erroneous methodology of selection of therapeutics. It is based on the classical reductionist approach to biological phenomena and neglects the multicomponent and multi-hierarchical complexity of the pathophysiological mechanisms involved.

The method of mathematical modelling, as a new tool, will enable a researcher to gain the insight into intricate physiological effects of various drugs which cannot be achieved in either *in vivo* or *in vitro* preparations, and help him/her to develop an integrative strategy for the drug-discovery process.

Care should be taken, though, in transferring the results of simulations to explain the real biomechanics of the organ. One has to bear in mind that the biological plausibility of the model is constrained by the model assumptions, despite the fact that the theoretical results resemble qualitatively patterns of electrical and mechanical activity that are observed in mainly animal studies *in vivo* and *in vitro*. At the moment there is no direct affirmative experimental evidence obtained on human subjects with which to carry out a detailed quantitative evaluation and comparison of the computational data.

The models of the abdominal viscera studied in this book constitute a first step in an integrative approach to build a theoretical framework, which is still missing, of the gastrointestinal system. This approach is based on real anatomical, electrophysiological and mechanical data about the arrangements and functions of the stomach and of the small and large intestine. Concepts of the theory of soft thin shells, electromechanical conjugation and electromechanical wave activity have been implemented in biomechanical models of various visceral organs and studied numerically. Further developments and improvements of the models are needed. When they are 'complete', they will offer unique insights, provide otherwise inaccessible information on intrinsic physiological processes and serve as a tool to facilitate the development of novel diagnostic tests by experimentalists and bioengineers. Such an approach will have enormous implications for our understanding of the mechanisms of certain diseases, for improving diagnostic accuracy and for the planning of therapeutic interventions.

Exercises

1. Gastrointestinal motor function is complex and requires constant coordination of the central nervous system, extrinsic parasympathetic and sympathetic nerves, the enteric nervous system and the longitudinal and circular smooth muscles. Identify the anatomical functional centres in the brain that are components in the brain–gut axis.
2. The difficulty in collecting high-resolution temporal and spatial data from patients with gastrointestinal disorders has limited the success of mathematical modelling approaches used to study these conditions. To investigate clinically questions regarding visceral nociception requires simultaneous recording of physiological responses from the gut and the brain. How can brain activity be assessed in a clinical setting?
3. The enteric nervous system – the little brain of the gut – sustains a space map for gastrointestinal function. Its morphostructural element is a ganglion composed of the primary neurons, motor neurons and inter-neurons. Construct an anatomically realistic functional neuronal circuit of the ganglion. Use the Hodgkin–Huxley formalism to formulate a mathematical model of the ganglion.

4. Double-label immunohistochemical techniques show that various classes of neurons in ganglia contain multiple neurotransmitters. Formulate a mathematical model of serotonergic neurotransmission. Derive a model of co-transmission by acetylcholine and serotonin.
5. Substance P is a neurotransmitter that is implicated in visceral nociception. Derive a mathematical model of co-localization and co-transmission by acetylcholine and substance P at the neuro-neuronal synapse. (*Hint*: consider muscarinic and nicotinic type-1 and -2 receptors on the postsynaptic membrane.)
6. Intestinofugal neurons are believed to provide vertical connections between the enteric nervous plexus and prevertebral ganglia. On the basis of the neuroanatomy of connections and types of neurotransmitters involved, formulate a mathematical model of the enteric–prevertebral ganglia.
7. Construct a 'blueprint' of the possible brain–gut axis. Specify the cell and neurotransmitter types involved.
8. Despite the enormous advances of modern experimental and computer-assisted technologies in drug discovery and drug development, there has been limited success at finding effective remedies to treat gastrointestinal motor dysfunctions. What is going wrong?
9. Mathematics does not have a long-standing relation to the life sciences. What is the future of the applied mathematical sciences in biology and medicine?
10. R. M. May (2007) wrote that '… most common abuses [of mathematics in biology] are situations when mathematical models are constructed with an excruciating abundance of detail in some aspects, whilst other important facets of the problem are misty or a vital parameter is uncertain to within, at best, an order of magnitude'. What are other abuses one should be aware of?

References

Abell, T. L., Malagelada, J. R., Lucas, A. R. *et al.* (1987). Gastric electromechanical and neurohormonal function in anorexia nervosa. *Gastroenterology*, **93**, 958–965.

Akbarali, H. I. (2005). Signal-transduction pathways that regulate smooth muscle function II. Receptor–ion channel coupling mechanisms in gastrointestinal smooth muscle. *American Journal of Physiology, Cell Physiology*, **288**, C598–C602.

Alan, E., Lomax, A. E. and Furness, J. B. (2000). Neurochemical classification of enteric neurons in the guinea-pig distal colon. *Cell Tissue Research*, **302**, 59–72.

Aliev, R. R., Richards, W. and Wikswo, J. P. (2000). A simple nonlinear model of electrical activity in the intestine. *Journal of Theoretical Biology*, **204**, 21–28.

Alvarez, W. C. and Zimmermann, A. (1928). Movements of the stomach. *American Journal of Physiology*, **84**, 261–270.

Amaris, M. A., Rashev, P. Z., Mintchev, M. P. and Bowes, K. L. (2002). Micro-processor controlled movement of solid colonic content using sequential neural electrical stimulation. *Gut*, **50**, 475–479.

Bampton, P. A., Dinning, P. G., Kennedy, M. L. *et al.* (2000). Spatial and temporal organization of pressure patterns throughout the unprepared colon during spontaneous defecation. *American Jouranl of Gastroenterology*, **95**, 1027–1035.

Bayguinov, O., Ward, S. M., Kenyon, J. L. and Sanders, K. M. (2007). Voltage-gated Ca^{2+} currents are necessary for slow-wave propagation in the canine gastric antrum. *American Journal of Physiology, Cell Physiology*, **293**, C1645–C1659.

Bernstein, C. N., Frankenstein, U. N., Rawsthorne, P. *et al.* (2002). Cortical mapping of visceral pain in patients with GI disorders using functional magnetic resonance imaging. *American Journal of Gastroenterology*, **97**(2), 319–327.

Bertuzzi, A., Marcinelli, R., Ronzoni, G. and Salinari, S. (1983). Peristaltic transport of a solid bolus. *Journal of Biomechanics*, **16**, 459–464.

Betz, U. A., Farquhar, R. and Ziegelbauer, K. (2005). Genomics: success or failure to deliver drug targets? *Current Opinion in Chemical Biology*, **9**, 387–391.

Blackshaw, L. A. and Gebhart, G. F. (2002). The pharmacology of gastrointestinal nociceptive pathways. *Current Opinion in Pharmacology*, **2**, 642–649.

Brookes, S. J. (2001). Classes of enteric nerve cells in the guinea-pig small intestine. *Anatomy Research*, **262**, 58–70.

Butcher, E. C. (2005). Can cell systems biology rescue drug discovery? *Nature Review: Drug Discovery*, **4**, 461–467.

Callahan, M. J. (2002). Irritable bowel syndrome neuropharmacology. *Journal of Clinical Gastroenterology*, **35**, S58–S67.

Camborova, P., Hubka, P., Sulkova, I. and Hulin, I. (2003). The pacemaker activity of interstitial cells of Cajal and gastric electrical activity. *Physiological Research*, **52**, 275–284.

Carniero, A. A., Baffa, O. and Oliveira, R. B. (1999). Study of stomach motility using relaxation of magnetic tracers. *Physics and Medical Biology*, **44**, 1691–1697.

Casey, K. L., Morrow, T. J., Lorenz, J. and Minoshima, S. (2001). Temporal and spatial dynamics of human forebrain activity during heat pain: analysis of positron emission tomography. *Journal of Neurophysioogy*, **85**, 951–959.

Cheng, L., Komuro, R., Austin, T. M., Buist, M. L. and Pullan, A. J. (2007). Anatomically realistic multiscale models of normal and abnormal gastrointestinal electrical activity. *World Journal of Gastroenterology*, **7** (March), 1378–1383.

Corrias, A. and Buist, M. L. (2007). A quantitative model of gastric smooth muscle cellular activation. *Annals of Biomedical Engineering*, **35**, 1595–1607.

Cowin, S. C. (2000). How is a tissue built? *Transactions of the ASME, Journal of Biomechanical Engineering*, **122**, 553–559.

Daniel, E. E., Bardakjian, B. L., Huizinga, J. D. and Diamant, N. E. (1994). Relaxation oscillator and core conduction models are needed for understanding of GI electrical activities. *American Journal of Physiology*, **266**, G339–G349.

D'Antona, G., Hennig, G. W., Costa, M., Humphreys, C. M. and Brookes, S. J. H. (2001). Analysis of motor patterns in the isolated guinea-pig large intestine by spatiotemporal maps. *Neurogastroenterology and Motility*, **13**, 483–492.

Dickens, E. J., Edwards, F. R. and Hirst, G. D. S. (2001). Selective knockout of intramuscular interstitial cells reveals their role in the generation of slow waves in mouse stomach. *Journal of Physiology*, **531**, 827–833.

Fox, E. A., Phillips, R. J. and Baronowsky, E. A. (2001). Neurotrophin-4 deficient mice have a loss of vagal intraganglionic mechanoreceptors from the small intestine and a disruption of short-term satiety. *Journal of Neuroscience*, **21**, 8602–8615.

Fung, Y. C. (1993). *Biomechanics: Material Properties of Living Tissues*. Berlin: Springer.

Furness, J. and Sanger, G. J. (2002). Intrinsic nerve circuits of the gastrointestinal tract: identification of drug targets. *Current Opinion in Pharmacology*, **2**, 612–622.

Furness, J. B. (2000). Novel gut afferents: intrinsic afferent neurons and intestinofugal neurons. *Autonomic Neuroscience: Basic and Clinical*, **125**, 81–85.

(2006). *The Enteric Nervous System*. New York: Wiley–Blackwell.

Galimov, K. Z. (1975). *Foundations of the Nonlinear Theory of Thin Shells*. Kazan: Kazan University Publisher.

Galimov, K. Z., Paimushin V. N. and Teregulov I. G. (1996). *Foundations of the Nonlinear Theory of Shells*. Kazan: ФЭН (FÉN).

Galligan, J. J. (2004). Enteric P2X receptors as potential targets for drug treatment of the irritable bowel syndrome. *British Journal of Pharmacology*, **141**, 1294–1302.

Gao, C., Petersen, P., Liu, W., Arendt-Nielsen, L., Drewes, A. M. and Gregersen, H. (2002). Sensory motor responses to volume controlled duodenal distension. *Neurogastroenterology and Motility*, **14**, 365–374.

Gaudreu, G.-A. and Plourde, V. (2004). Involvement of N-methyl-D-áspartate (NMDA) receptors in a rat model of visceral hypersensitivity. *Behavioral Brain Research*, **150**, 185–189.

Gibbins, I. L., Hiok, E. E., Jobling, P. and Morris, J. L. (2003). Synaptic density, convergence, and dendritic complexity of prevertebral sympathetic neurons. *Journal of Comparative Neurology*, **455**, 285–298.

Grundy, D. and Schemann, M. (2005). Enteric nervous system. *Current Opinion in Gastroenterology*, **21**, 176–182.

Gunput, M. D. (2001). Clinical pharmacology of alosetron. *Alimentary Pharmacology and Therapeutics*, **13** (2), 70–76.

Hardy, L. W. and Peet, N. P. (2004). The multiple orthogonal tools approach to define molecular causation in the validation of druggable targets. *Drug Discovery Today*, **9**, 117–126.

Hashitani, H., Garcia-Londoño, A., Hirst, G. D. S. and Edwards, F. R. (2005). Atypical slow waves generated in gastric corpus dominant pacemaker activity in guinea pig stomach. *Journal of Physiology*, **569**, 459–465.

Hennig, G. W., Hirst, G. D. S., Park, K. J. *et al.* (2004). Propagation of pacemaker activity in the guinea-pig antrum. *Journal of Physiology*, **556**, 585–599.

Hens, J., Vanderwinden, J.-M., De Laet, M.-H., Scheuermann, D. W. and Timmermans, J.-P. (2001). Morphological and neurochemical identification of enteric neurones with mucosal projections in the human small intestine. *Journal of Neurochemistry*, **76**, 464–471.

Hirst, D. G. S. and Suzuki, H. (2006). Involvement of interstitial cells of Cajal in the control of smooth muscle excitability. *Journal of Physiology*, **576**, 651–652.

Hirst, G. D. and Edwards, F. R. (2004). Role of interstitial cells of Cajal in the control of gastric motility. *Journal of Pharmacological Science*, **96**, 1–10.

Hirst, G. D. S., Garcia-Londoño, A. P. and Edwards, F. R. (2006). Propagation of slow waves in the guinea-pig gastric antrum. *Journal of Physiology*, **571**, 165–177.

Hodgkin, A. and Huxley, A. (1952). A quantitative description of membrane current and its application to conduction and excitation in nerve. *Journal of Physiology (London)*, **117**, 500–544.

Holle, G. E., Steinbach, E. and Forth, W. (1992). Effects of erythromycin in the dog upper gastrointestinal tract. *American Journal of Physiology*, **263**, G52–G59.

Holzapfel, G. A., Stadler, M. and Schulze-Bauer, C. A. J. (2002). A structural model for the viscoelastic behavior of arterial walls: continuum formulation and finite element analysis. *European Journal of Mechanics*, **A21**, 441–463.

Holzer, P., Michl, T., Danzer, M. *et al.* (2001). Surveillance of the gastrointestinal mucosa by sensory neurons. *Journal of Physiological Pharmacology*, **52**(4), 505–521.

Humphrey, P. P. A., Bountra, C., Clayton, N. and Kozlowski, K. (2001). The therapeutic potential of 5-HT$_3$ receptor antagonists in the treatment of irritable bowel syndrome. *Alimentary Pharmacology and Therapeutics*, **13** (2), 31–38.

Hunt, R. (2002). Evolving concepts in functional gastrointestinal disorders: promising directions for novel pharmaceutical treatments. *Best Practice & Research in Clinical Gastroenterology*, **16**(6), 869–883.

Ishiguchi, T., Itoh, H. and Ichinose, M. (2003). Gastrointestinal motility and the brain–gut axis. *Digestive Endoscopy*, **15**, 81–86.

Jobling, P. and Gibbins, I. L. (1999). Electrophysiological and morphological diversity of mouse sympathetic neurons. *Journal of Neurophysiology*, **82**, 2747–2764.

Johnson, G. A., Livesay, G. A., Woo, S. L. Y. and Rajagopal, K. R. (1996). A single integral finite-strain viscoelastic model of ligaments and tendons. *ASME Journal of Biomechanical Engineering*, **118**, 221–226.

Koch, K. L., Stern, R. M., Steward, W. R. and Vasey, M. W. (1989). Gastric emptying and gastric myoelectrical activtiy in patients with diabetic gastroparesis: effect of long-term domperidone treatment. *American Journal of Gastroenterology*, **84**, 1069–1075.

Koh, S. D., Ward, S. M., Tamas, O., Sanders, K. M. and Horowitz, B. (2003). Conductances responsible for slow wave generation and propagation in interstitial cells of Cajal. *Current Opinion in Pharmacology*, **3**, 579–582.

Krevsky, B. L., Malmud, L. S., D'Ercole, F., Maurer, A. H. and Fisher, R. S. (1986). Colonic transit scintigraphy: a physiologic approach to the quantitative measurement of colonic transit in humans. *Gastroenterologia*, **91**, 1102–1112.

Kuramoto, H and Furness, J. B. (1989). Distribution of enteric nerve cells that project from the small intestine to the coeliac ganglion in the guinea-pig. *Journal of the Autonomous Nervous System*, **27**, 241–248.

Lacy, B. E. and Yu, S. (2002). Tegaserod. A new 5-HT_4 agonist. *Journal of Clinical Gastroenterology*, **34** (1), 27–33.

Lammers, W. J. E. P. (2000). Propagation of individual spikes as "patches" of activation in the isolated feline duodenum. *American Journal of Physiology, Gastrointestinal and Liver Physiology*, **278**, G297–G307.

Lammers, W. J. E. P. and Slack, J. R. (2001). Of slow waves and spike patches. *News in Physiological Sciences*, **16**, 138–144.

Lammers, W. J. E. P., al-Kais, A., Singh, S., Arafat, K. and el-Sharkawy, T. Y. (1993). Multielectrode mapping of slow-wave activity in the isolated rabbit duodenum. *Journal of Applied Physiology*, **74**, 1454–1461.

Lammers, W. J. E. P., Dhanasekaran, S., Slack, J. R. and Stephen, B. (2001). Two-dimensional high-resolution motility mapping in the isolated feline duodenum: methodology and initial results. *Neurogastroenterology and Motility*, **13**, 309–323.

Lammers, W. J. E. P., el-Kays, A., Manefield, G. W., Arafat, K. and El-Sharkawy, T. Y. (1997). Disturbances in the propagation of the slow wave during acute local ischemia in the feline small intestine. *European Journal of Gastroenterology and Hepatology*, **9**, 381–388.

Lammers, W. J. E. P., Slack, J. R., Stephen, B. and Pozzan, O. (2000). The spatial behaviour of spike patches in the feline gastroduodenal junction *in vitro*. *Neurogastroenterology and Motility*, **12**, 467–473.

Liao, D., Gregersen, H., Hausken, T. *et al.* (2004). Analysis of surface geometry of the human stomach using real-time 3-D ultrasonography *in vivo*. *Neurogastroenterology and Motility*, **16**, 315–324.

Libai, A. and Simmonds, J. G. (2005). *The Nonlinear Theory of Elastic Shells*. Cambridge: Cambridge University Press.

Liu, M.-T., Rothstein, J. D., Gershon, M. D. and Kirchgessner, A. L. (1997). Glutamatergic enteric neurons. *Journal of Neurosciences*, **17**(2), 4764–4784.

Lomax, A. E., Sharkey, K. A., Bertrand, P. P. *et al.* (1999). Correlation of morphology, electrophysiology and chemistry of neurons in the myenteric plexus of the guinea-pig distal colon. *Journal of the Autonomous Nervous System*, **76**, 45–61.

Lomax, A. E., Zhang, J. Y. and Furness, J. B. (2000). Origins of cholinergic inputs to the cell bodies of intestinofugal neurons in the guinea pig distal colon. *Journal of Comparative Neurology*, **416**, 451–460.

Lyford, G. L. and Farrugia, G. (2003). Ion channels in gastrointestinal smooth muscle and interstitial cells of Cajal. *Current Opinion in Pharamcology*, **3**, 583–587.

Lyford, G. L., Strege, P. R., Shepard, A. *et al.* (2002). Alpha 1C (Cav1.2) L-type calcium channel mediates mechanosensitive calcium regulation. *American Journal of Physiology, Cell Physiology*, **283**, C1001–C1008.

Macarron, R. (2006). Critical review of the role of HTS in drug discovery. *Drug Discovery Today*, **11**, 277–279.

Marciani, L., Young, P., Wright, J. *et al.* (2001). Antral motility measurements by magnetic resonance imaging. *Neurogastroenterology and Motility*, **13**, 511–518.

May, R. M. (2007). Uses and abuses of mathematics in biology. *Science*, **303**, 790–793.

Mayo, C. R., Brookes, S. J. H. and Costa, M. (1992). A computer simulation of intestinal motor activity, in *Proceedings of the Third Australian Conference on Neural Networks '92 (ACNN '92)*, ed. P. Leong and M. Jabri, pp. 48–51.

Miftahof, R. and Akhmadeev, N. (2007). Dynamics of intestinal propulsion. *Journal of Theoretical Biology*, **246**, 377–393.

Miftahof, R. and Fedotov, E. M. (2005). Intestinal propulsion of a solid non-deformable bolus. *Journal of Theoretical Biology*, **235**, 57–70.

Miftahof, R. N., Nam, H. G. and Wingate, D. L. (2009). *Mathematical Modeling and Simulation in Enteric Neurobiology*. Singapore: World Scientific Publishing Company.

Miftakhov, R. and Abdusheva, G. (1993). Small bowel propulsion: transit of a solid bolus, in *Biophysics and Life Sciences*, ed. D. Ghista. Munich: Springer, pp. 253–260.

Miftakhov, R. and Fedotov, E. (2004). The concept of the functional unit of the gut, in *Advances in Fluid Mechanics IV*, ed. A. Mendes, M. Rahman and C. A. Brebbia. Southampton: WIT Press, pp. 353–361.

Miftakhov, R. and Vannier, M. V. (2002). Nonlinear dynamic waves in electromechanical excitable biological media, in *Advances in Fluid Mechanics V*, ed. M. Rahman, R. Verhoeven and C. A. Brebbia. Southampton: WIT Press, pp. 725–735.

Miftakhov, R. N., Abdusheva, G. R. and Wingate, D. L. (1996). Model predictions of myoelectrical activity of the small bowel. *Biological Cybernetics*, **74**, 167–179.

Miftakhov, R. N. (1981). Micromechanics of tissue fracture in uniaxial elongation, in *Shell Interactions with Fluids*. Moscow: Academy of Sciences of the USSR, pp. 205–214.

(1983a) Investigation of the human stomach tissue in uniaxial loading, in *Hydroelasticity of Shells*. Moscow: Academy of Sciences of the USSR, pp. 163–171.

(1983b) Experimental investigations of the stomach under complex loading, in *Hydroelasticvity of shells*. Moscow: Academy of Sciences of the USSR, pp. 172–181.

(1983c). Experimental and numerical investigations of soft shells. Unpublished Ph.D. thesis, Kazan State University.

(1985). Experimental investigation of the stomach tissue in biaxial loading, in *Investigations in the Theory of Plates and Shells*. Kazan: Kazan University Press, pp. 35–46.

(1988). Applications of the theory of soft thin shells in problems of biomechanics, in *Biomechanics: Problems and Investigations*, vol. VI. Riga: Zinatne, pp. 51–56.

Muraki, K., Imaizumi, Y. and Watanabe, M. (1991). Sodium currents in smooth muscle cells freshly isolated from stomach fundus of the rat and ureter of the guinea-pig. *Journal of Physiology*, **442**, 351–375.

Nikitin, N. L. (1980). A model of the muscle tissue with alternating number of contracting fibers. *Mechanics of Composite Materials*, **1**, 113–120.

Nishi, S. and North, R. A. (1973). Intracellular recording from the myenteric plexus of the guinea-pig ileum. *Journal of Physiology (London)*, **231**, 471–491.

Oprea, T. I. and Matter, H. (2004). Integrating virtual screening in lead discovery. *Current Opinion in Chemical Biology*, **8**, 349–358.

Ou, Y., Strege, P., Miller, S. M. *et al.* (2003). Syntrophin gamma 2 regulates SCN5A gating by a PDZ domain-mediated interaction. *Journal of Biological Chemistry*, **278**, 1915–1923.

Pal, A., Brasseur, J. and Abrahamsson, B. (2007). A stomach road or «Magenstrasse» for gastric emptying. *Journal of Biomechanics*, **40**, 1202–1210.

Pal, A., Indireshkumar, K., Schwizer, W. *et al.* (2004). Gastric flow and mixing studied using computer simulation. *Proceedings of the Royal Society of London, Series B*, **271**, 2587–2594.

Parsons, C. G. (2001). NMDA receptors as targets for drug action in neuropathic pain. *European Journal of Pharmacology*, **429**, 71–78.

Plonsey, R. L. and Barr, R. G. (1984). Current flow patterns in two-dimensional anisotropic bisyncytia with normal and extreme conductivities. *Biophysical Journal*, **43**, 557–571.

Porter, A. J., Wattchow, D. A., Brookes, S. J. H. and Costa, M. (2002). Cholinergic and nitrergic interneurones in the myenteric plexus of the human colon. *Gut*, **51**, 70–75.

Provenzano, P. P., Lakes, R. S., Corr, D. T. and Vanderby Jr, R. (2002). Application of nonlinear viscoelastic models to describe ligament behavior. *Biomechanical Modeling Mechanobiology*, **1**, 45–57.

Publicover, N. G. and Sanders, K. M. (1989). Are relaxation oscillators an appropriate model of gastrointestinal electrical activity? *American Journal of Physiology*, **256**, G256–G274.

Pullan, A., Cheng, L., Yassi, R. and Buist, M. (2004). Modelling gastrointestinal bioelectric activity. *Progress in Biophysics & Molecular Biology*, **85**, 523–550.

Rachev, P. Z., Amaris, M., Bowes, K. L. and Mintchev, M. P. (2002). Micro-processor-controlled colonic peristalsis: dynamic parametric modeling in dogs. *Digestive Diseases*, **47**, 1034–1048.

Ramon, F., Anderson, N. C., Joyner, R. W. and Moore, J. W. (1976). A model for propagation of action potentials in smooth muscle. *Journal of Theoretical Biology*, **59**, 381–408.

Ridel, V. V. and Gulin B. V. (1990). *Dynamics of Soft Shells*. Moscow: Nauka.

Sanders, K. M. (1984). Role of prostaglandins in regulating gastric motility. *American Journal of Physiology*, **247**, G117–G126.

Sanger, G. J. and Hicks, G. (2002). Drugs targeting functional bowel disorders: insights from animal studies. *Current Opinion in Pharmacology*, **2**, 678–683.

Sarna, S. K., Daniel, E. E. and Kingma, Y. J. (1971). Simulation of slow-wave electrical activity of small intestine. *American Journal of Physiology*, **221**, 166–175.

Scapin, G. (2006). Structural biology and drug discovery. *Current Pharmacological Design*, **12**, 2087–2097.

Sharkey, K. A., Lomax, A. E. and Bertrand, P. P. (1998). Electrophysiology, shape, and chemistry of neurons that project from guinea pig colon to inferior mesenteric ganglia. *Gastroenterology*, **115**, 909–918.

Shoichet, B. K. (2004). Virtual screening of chemical libraries. *Nature*, **432**, 862–865.

Silverman, L., Campbell, R. and Broach, J. R. (1998). New assay technologies for high-throughput screening. *Current Opinion in Chemical Biology*, **2**, 397–403.

Spencer, N. J. and Smith, T. K. (2001). Simultaneous intracellular recordings from longitudinal and circular muscle during the peristaltic reflex in guinea-pig distal colon. *American Journal of Physiology*, **533**, 787–799.

Spiller, R. (2004). Irritable bowel syndrome. *British Medical Bulletin*, **72**, 15–29.

Suzuki, H. (2000). Cellular mechanisms of myogenic activity in gastric smooth muscle. *Japanese Journal of Physiology*, **50**, 289–301.

Szurszewski, J. H. (1969). A migrating electric complex of the canine small intestine. *American Journal of Physiology*, **217**, 1757–1763.

Szurszewski, J. H., Ermilov, L. G. and Miller, S. M. (2002). Prevertebral ganglia and intestinofugal afferent neurons. *Gut*, **51**, i6–i10.

Taber, L. A. (2004). *Nonlinear Theory of Elasticity: Applications in Biomechanics*. Singapore: World Scientific Publishing Company.

Terence, K., Smith, T. K. and Lunam, C. A. (1998). Electrical characteristics and responses to jejunal distension of neurons in Remak's juxta-jejunal ganglia of the domestic fowl. *Journal of Physiology*, **510**(2), 563–575.

Thielecke, F., Maxion-Bergemann, S., Abel, F. and Gonschior, A. K. (2004). Update in the pharmaceutical therapy of the irritable bowel syndrome. *International Journal of Clinical Practice*, **58**(4), 374–381.

Usik, P. I. (1973). Continual mechanochemical model of muscular tissue. *Journal of Applied Mathematics and Mechanics* (Trans. *Prikladnaya Mathematika i Mehanika*) **37**(3), 428–439.

Van der Greef, J. and McBurney, R. N. (2005). Innovation: rescuing drug discovery: *in vivo* systems pathology and systems pharmacology. *Nature Reviews: Drug Discovery*, **4**, 961–967.

Ventsel, E. and Krauthammer, T. (2001). *Thin Plates and Shells: Theory, Analysis and Applications*. Boca Raton, FL: CRC Press.

Willert, R. P., Woolf, C. J., Hobson, A. R. *et al.* (2006). The development and maintenance of human hypersensitivity is dependent on the N-methyl-D-aspartate receptor. *Gastroenterology*, **126**, 683–692.

Won, K. J., Sanders, K. M. and Ward, S. M. (2005). Interstitial cells of Cajal mediate mechanosensitive responses in the stomach. *The Proceedings of the National Academy of Sciences of the USA*, **102**, 14 913–14 918.

Wood, J. D. (1973). Electrical discharge of single enteric neurons of guinea pig small intestine. *American Journal of Physiology*, **225**, 1107–1113.

(1989). Electrical and synaptic behavior of enteric neurons, in *Handbook of Physiology*, ed. J. Wood. Washington: American Physiological Society, pp. 465–517.

Wood, J. D. and Mayer, C. J. (1978). Intracellular study of electrical activity of Auerbach's plexus in the guinea-pig small intestine. *Pflügers Archiv*, **374**, 225–275.

Yin, J. and Chen, J. D. Z. (2008). Roles of interstitial cells of Cajal in regulating gastrointestinal motility: *in vitro versus in vivo* studies. *Journal of Cellular and Molecular Medicine*, **12**(4), 1118–1129.

Zambrowicz, B. P. and Sands, A. T. (2003). Knockouts model the 100 best-selling drugs – will they model the next 100? *Nature Reviews: Drug Discovery*, **2**, 38–51.

Index

acetylcholine 135, 181, 201, 207, 209
azithromycin 155

base xvi, 7, 18, 26, 53, 55, 79, 85, 86, 95, 110, 112
  contravariant xv
  covariant xv, 7, 10
  local 15
  orthonormal xv, 34, 59, 76
Bay K 8644 135
biofactor xix, 70, 119
biomaterial xviii, 63, 121, 142

channels xix, 62, 73, 117, 118, 125, 126, 130, 133, 194
  Ca^{2+} 117, 118, 124, 126, 134, 135, 142, 145, 155, 158, 159, 160, 164, 166, 194, 198
  Cl^- xix, 126, 158, 180
  K^+ xix, 117, 118, 124, 126, 136, 158, 160, 161, 163, 166, 194
  L-type 117, 118, 124, 126, 134, 135, 136, 142, 155, 158, 160, 164, 166
  Na^+ 117, 118, 126, 194
  T-type 117, 118, 124, 134, 135, 155, 158, 159, 160, 166, 198
Christoffel symbol xvi, 15, 24, 33, 37, 46
  first kind xvi, 33
  second kind xvi, 11, 33
colon (*see* large intestine) xii, xiii, 74, 157, 158, 182, 183, 184, 185, 187, 194, 197, 200, 201
concentration xix, 64, 118, 119, 124, 129, 130, 133, 134, 135, 145, 156
  effective 64, 65
  mass xviii
conductance xix, 118, 125, 126, 128, 155, 160, 177
  extracellular xix, 127
  intracellular xix, 127
connective tissue 61, 62, 130, 157, 166, 182, 185
  fibres 139, 157
  network 62, 63, 121, 137, 147, 165, 192
  matrix 74
conservation 139
  mass 65, 107
  momentum 66, 108
constitutive relations xii, 59, 69, 71, 74, 102, 103, 109, 111, 113, 138, 166, 185
continuum xii, 63, 75, 102, 105, 130, 180, 198
contour curve 76, 77
coordinate xv
  Cartesian 39, 66, 109, 111, 138
  curvilinear xii, xv, 1, 16, 26, 47, 82, 89
  cylindrical 26, 46, 111, 112
  Lagrange xv, 6, 7, 10, 13, 15, 16, 17, 19, 20, 30, 34, 37, 42, 47, 48, 71, 77, 89, 90, 96, 128, 130
  line xv
  line of curvature 10, 15, 16, 17, 37
  orthogonal xii, 12, 14, 17, 42
crack 121, 122
  growth 121
  nucleation 121
current 125, 204
  extracellular xviii, 71, 126
  intracellular xviii, 71
  ion xviii, xix, 73, 74, 125, 126, 128, 131, 159, 165
  transmembrane xviii, 71, 127
curvature 1, 2, 4, 9, 10, 15, 20, 22, 40, 49, 55, 59, 71, 78, 87, 115, 118, 138, 140, 142, 145, 147, 155
  Gaussian xvii, 13, 24, 32
  geodesic 86
  normal xvi, xvii, 76
  principal xvii, 10, 13, 38

deformation xii, xvi, 1, 18, 19, 20, 22, 30, 47, 48, 49, 55, 62, 63, 64, 70, 71, 77, 84, 85, 89, 92, 95, 96, 100, 101, 109, 128, 140, 173, 184, 185
  bending xvi, 31, 84, 87
  fictitious xvi, 30, 31, 35, 46
  shear 19, 48, 49, 84
  tangent 19, 42, 43, 49
  viscous 66
density xviii, 64, 81, 82, 107, 118, 130
  deformed xviii
  partial 64
  undeformed xviii, 121
direction cosine 110, 112
dissipative function xviii, 67

edge 63, 78, 80, 81, 84, 85, 120
clamped 83
free 83
freely supported 83
simply supported 83
element xv, xvi, xvii, 8, 17, 19, 29, 30, 47, 50, 51, 52, 62, 73, 78, 84, 93, 96, 97, 105, 106, 107, 108, 116, 131
histomorphological 116, 122, 157, 199, 202, 205, 208
surface area xvi, xvii, 32, 46, 51
energy density function xx, 121
entropy xviii, 63, 66
balance 67
partial 67
equilibrium 55, 75, 97, 108, 129, 140, 145
equation xii, 56, 57, 59, 60, 70
solid 56
thin-shell 56, 60, 97

fibre xvii, 61, 62, 103, 105, 109, 116, 130, 131, 139, 159, 160, 182, 195, 202, 206
collagen 63, 69, 103, 116, 119, 121, 122, 137, 166, 185
elastin 63, 69, 103, 116, 119, 121, 137, 166, 185
smooth muscle xviii, xix, 61, 62, 63, 120, 130, 131, 132, 133, 135, 136, 137, 138, 142, 158, 159, 160, 161, 162, 163, 164, 165
flux 65, 160
external sources xviii, 64
general heat xviii, 67
thermodynamic 67, 68
force xvii, 51, 59, 63, 75, 80, 81, 82, 93, 97, 98, 100, 101, 102, 103, 104, 105, 106, 107, 109, 112, 119, 120, 121, 124, 130, 137, 139, 142, 147, 155, 160, 161, 166, 173, 187, 192, 197
active xvii, 63, 64, 124, 129, 138, 145, 155, 162, 166, 173, 185, 187, 194
contact xvii, 185, 186
external xvii, 51, 52, 55, 60
in-plane 54, 70, 71, 79, 121, 138, 139
internal xvii, 50, 51
lateral xvii, 54, 59, 70, 79
mass xvii, 51, 52, 66, 107
normal 54
passive xvii, 166, 185
resultant xvii, 50, 51, 55, 79, 80, 107
shear 54, 105, 121
thermodynamic 68
total xvii, 119, 130, 142, 145, 155, 160, 162, 165, 168, 173, 187, 200, 201
forskolin 160, 161, 162
Frenet–Serra formula 9
functional unit xiii, 158, 165, 166, 180, 199
fundamental form
first 8, 47, 84
second xvi, 9, 12, 13, 31, 32, 36, 40

Gauss–Codazzi 13, 32, 87
Gibbs relation 66

haustral churning 183, 187, 192
Hodgkin–Huxley 73, 75, 156, 208
hysteresis 120

interstitial cells of Cajal (ICC) xix, xx, 116, 117, 118, 126, 129, 155, 158, 166, 185, 197, 199, 204
invariant (tensor) xvi, xvii, 101
first 96, 101
second 96, 101

Kirchhoff–Love 22, 55, 60, 70
Kronecker delta 7, 31

Lamé parameters xvi, 7, 16, 26, 38, 40, 47, 84
large intestine xiii, 4, 26, 33, 182, 183, 184, 185, 196, 201, 202, 208
lemakalim 161, 162
librations 187, 192
lidocaine 177, 180
Lotronex 192, 195, 201, 207

Mallory–Weiss 3, 200
methoxyverapamil 165
metoclopramide 155, 160
mibefradil 134
migrating myoelectrical complex 180, 197
moment xvii, 51, 52, 54, 57, 59, 60, 70, 71, 79, 80, 81, 107, 108
bending xviii, 80, 81, 82, 83
external xvii, 52, 55
internal 51
resultant xvii, xviii, 51, 52, 55, 79, 80, 81, 108
twisting xvii, xviii, 54, 80, 81, 82
movements 3, 117, 119, 173, 183, 192, 197, 200
pendular 3, 158, 166, 168, 173, 197, 200
peristalsis 3, 61, 130, 145, 158, 173, 180, 183, 186, 192, 197, 200
segmentation 3, 158, 168, 173, 197
myosin light chain 118
phosphorylation 118
kinase 118

net 62, 63, 105, 106, 166
neuron 61, 62, 116, 118, 182, 183, 202, 203, 204, 205, 206
AH-type 202, 203
Dogiel type I 203
Dogiel type II 202, 203
motor 62, 202, 204
primary sensory 202, 203, 204

Ohm's law 71, 127
Onsager relation 69

pacemaker xiii, 117, 118, 126, 127, 129, 131, 138, 140, 142, 145, 147, 156, 158, 159, 166, 168, 173, 180, 183, 185, 186, 198, 199
parameterization xii, 28, 30, 37, 44, 100, 103, 105
pellet 184, 185, 187, 192, 197, 200, 201

phase xii, xviii, 63, 64, 65, 66, 67, 117, 131, 137, 145, 156, 158, 159, 160, 183, 192, 197, 198
  porosity xviii, 64, 163
phencyclidine 163
plexus 3, 118, 202, 203, 204, 205
  enteric 156, 158, 202, 203, 204, 206
potential xix, 71, 117, 118, 129, 132, 158, 166, 185, 203
  action 75, 117, 131, 132, 134, 136, 142, 156, 158, 160, 162, 173, 177, 183, 185, 187, 197, 198, 202, 203, 204
  chemical xviii, 66
  electrical xix, 71
  membrane xix, 117, 124, 134, 135, 159, 160, 161, 198
  resting xix, 125, 126, 128, 202, 204
principal xvi, xvii, 13, 38, 62, 95, 96, 97
  axis 95, 96, 101, 109
  deformation 100, 103
  direction 10, 95, 96, 100, 103
  membrane forces 100, 101, 103, 106
  section 10
product 17
  scalar triple 7, 9, 12, 16, 20, 38, 82, 98, 110
  vector 7, 32, 56, 90
propulsion xii, xiii, 61, 116, 130, 184, 185, 187, 194, 195, 199, 200, 201
  en masse 183
  pellet 192, 197, 200, 201

reciprocal relation 69, 75, 173, 180, 186, 200, 201
relative elongation 19
resistance xix, 89, 105, 127, 177, 202, 204
  cellular xix, 126
  specific xix, 131
rotation xii, xvi, 20, 43, 83, 86

Saint-Venant 80
shear angle xvi, 19, 94, 97, 103
shell xi, xii, 1, 28, 39, 41, 48, 49, 50, 51, 52, 55, 56, 57, 59, 63, 70, 71, 76, 77, 78, 80, 81, 83, 84, 85, 88, 89, 92, 100, 107, 108, 109, 165
  bioshell 2, 3, 4, 28, 33, 138, 139, 140, 142, 145, 147, 155, 165, 166, 168, 173, 177, 184, 185, 187, 192
  soft xii, xiii, 2, 3, 89, 92, 94, 97, 101, 105, 106, 108, 109, 110, 111, 112, 130, 138, 208
  thin xi, xii, 1, 2, 4, 22, 47, 49, 54, 55, 60, 80, 82, 87, 89, 93, 197
slow wave 117, 118, 123, 129, 131, 132, 133, 134, 135, 136, 155, 158, 159, 160, 163, 164, 166, 194, 197, 198, 199, 202
small intestine xii, xiii, 3, 74, 115, 157, 158, 165, 166, 168, 177, 181, 182, 183, 184, 199
smooth muscle xix, 61, 62, 63, 64, 74, 103, 116, 117, 118, 119, 121, 122, 124, 127, 130, 132, 133, 134, 135, 142, 155, 157, 160, 163, 166, 185, 199
  circular xix, 117, 119, 120, 124, 128, 129, 130, 137, 138, 140, 142, 145, 158, 166, 173, 177, 180, 183, 185, 187, 192, 194, 195, 197, 200, 201
  fibre xviii, 62, 63, 120, 130, 131, 136, 137, 138, 158, 159, 160, 164
  layer 62, 117, 140, 142, 145, 147, 155, 157, 158, 165, 173, 180, 181, 183, 185, 186, 187, 192, 194, 197, 199, 200, 201, 204
  longitudinal xix, 116, 117, 118, 119, 120, 123, 124, 127, 128, 129, 130, 137, 138, 140, 142, 145, 147, 155, 158, 166, 173, 177, 182, 183, 195, 197, 199
  syncytium xviii, xx, 61, 71, 118, 123, 127, 129, 137, 138, 145, 166, 173, 177, 184, 185, 192, 194, 197, 198, 199, 200
spike 117, 129, 131, 132, 135, 136, 158, 160, 194, 202, 203
stoichiometric coefficient xviii, 65
stomach xii, xiii, 1, 2, 3, 4, 28, 33, 102, 115, 116, 117, 118, 119, 120, 121, 123, 124, 129, 137, 138, 140, 142, 145, 147, 155, 157, 158, 183, 196, 199, 200, 202, 208
  antrum 115, 116, 118, 140, 142, 145, 147, 155
  body 115, 120, 140, 142, 145, 147, 155
  cardia 2, 3, 115, 116, 140, 142, 145, 147, 200
  greater curvature 140, 142, 145
  lesser curvature 115, 118, 140, 142, 145, 147, 155
  pylorus 2, 116, 118, 142, 145, 147, 155, 200
stress xvii, xviii, 1, 55, 56, 62, 66, 67, 69, 70, 74, 78, 81, 82, 93, 101, 102, 119, 187
  internal 53
stress–strain xii, xiii, 2, 4, 62, 89, 102, 109, 129, 156, 194, 197, 199
  biaxial 102, 109, 129, 139, 142, 168, 177, 187, 194
  uniaxial 2, 102, 109, 129, 139, 142, 200
stretch ratio xvi, 29, 89, 93, 94, 96, 97, 103, 105, 116, 119, 120, 121, 127, 130, 138, 142, 145, 155
surface
  area xvi, xvii, 8, 50, 51, 52, 78, 90, 93, 97, 106
  cut xv, 89
  deformed 19, 24, 96, 109
  equidistant 31, 34, 38, 49, 77
  free xvii, 51
  middle xv, 1, 2, 20, 22, 28, 47, 49, 51, 52, 54, 55, 71, 76, 77, 82, 84, 89, 138
  undeformed xv, 22, 31, 49, 77, 84, 105, 184
syncytium 127
  anisotropic 73, 127, 177, 185, 197
  isotropic 166, 177, 185, 197

tensor 69, 70
  deformation xvi, 31, 35, 43, 92, 96, 101
  membrane force 98
  metric xvi, 7, 31, 32, 34, 38, 92, 139
tetraethylammonium chloride 136, 137
thapsigargin 164
transformation 28, 29, 30, 32, 34, 39, 43, 44, 73, 89, 92

vector xvii, xviii, 6, 7, 8, 9, 15, 20, 30, 31, 34, 35, 36, 37, 38, 43, 47, 48, 50, 51, 52, 54, 60, 77, 80, 81, 82, 83, 85, 90, 91, 95, 107, 108, 110, 112, 138, 186
  displacement xvi, 18, 29, 33, 47, 66, 83, 84, 86, 186
  normal xv, xvii, 1, 7, 9, 10, 12, 31, 33, 34, 35, 38, 76, 80, 84, 85, 88, 110, 111
  position xv, 6, 10, 15, 31, 47, 89, 109, 111
  stress xvii, 50, 78, 107
  tangent xv, 42, 43, 76, 77

vector (cont.)
  unit xvi, 6, 7, 13, 16, 17, 38, 45, 53, 71, 76, 84, 90, 92, 110, 112
velocity xviii, 117, 130, 132, 134, 138, 142, 158, 168, 173, 184, 186, 187, 194, 198, 200
  flux 65
  slow-wave 158, 202
visceral hypersensitivity 180, 206
volume xvii, xviii, xx, 51, 64, 115, 139
  element xvii
  gastric 140, 155
wave xii, xiii, 3, 4, 71, 116, 117, 124, 132, 138, 142, 145, 155, 158, 159, 166, 168, 173, 177, 184, 185, 187, 194, 197, 198, 199
  depolarization 61, 129, 142, 166, 168, 177, 185, 187, 194, 197
  electrical 127, 129, 131, 138, 155, 168, 198
  mechanical 116, 168, 173, 177, 192, 198
wrinkle 1, 109, 139, 145, 147

Zelnorm 194, 201